# An Assessment of Research-Doctorate Programs in the United States: *Mathematical & Physical Sciences*

Committee on an Assessment of Quality-Related Characteristics
of Research-Doctorate Programs in the United States

Lyle V. Jones, Gardner Lindzey, and
Porter E. Coggeshall, *Editors*

*Sponsored by*

The Conference Board of Associated Research Councils

American Council of Learned Societies
American Council on Education
National Research Council
Social Science Research Council

NATIONAL ACADEMY PRESS
Washington, D.C. 1982

NOTICE: The project that is the subject of this report was approved
by the Conference Board of Associated Research Councils, whose members
are drawn from the American Council of Learned Societies, the American
Council on Education, the National Research Council, and the Social
Science Research Council. The members of the committee responsible
for the report were chosen for their special competences and with
regard for appropriate balance.

This report has been reviewed by a group other than the authors
and editors according to procedures approved by each of the four
member Councils of the Conference Board.

The Conference Board of Associated Research Councils was created
to foster discussion of issues of mutual interest; to determine the
extent to which a common viewpoint on such issues prevails within the
academic community of the United States; to foster specific
investigations when so desired; and, when the Conference Board finds
joint, common, or other action desirable, to make recommendations to
the appropriate Councils.

Blue page insert duplicates page 15 and may be used as a portable
guide to program measures by placing it beside the tables under
examination.

Library of Congress Catalog Card Number 82-61277

International Standard Book Number 0-309-03299-7

Available from

NATIONAL ACADEMY PRESS
2101 Constitution Avenue, N.W.
Washington, D.C. 20418

Printed in the United States of America

# Acknowledgments

In conducting this assessment the committee has benefited from the support and advice of many individuals and organizations. The assessment was conducted under the aegis of the Conference Board of Associated Research Councils, and special thanks go to Roger Heyns, Robert M. Lumiansky, Jack W. Peltason, Frank Press, Kenneth Prewitt, Eleanor Sheldon, John William Ward, and the late Philip Handler for their efforts in overseeing the planning and execution of this project. Financial support was provided by the Andrew W. Mellon Foundation, the Ford Foundation, the Alfred P. Sloan Foundation, the National Institutes of Health (NIH), the National Science Foundation (NSF), and the National Academy of Sciences. Without the combined support from these organizations the project would not have been undertaken. The committee appreciates the excellent cooperation it received from the staff officers at these organizations--including John Sawyer and James Morris at Mellon; Mariam Chamberlain, Gladys Chang Hardy, and Sheila Biddle at Ford; Albert Rees and James Koerner at Sloan; Helen Gee at NIH; and Bernard Stein at NSF. Some supplemental funds to enhance the study were furnished by the Association of American Geographers, the American Psychological Association, and the American Psychological Foundation.

The committee is most appreciative of the cooperation it received from individuals in the 228 universities participating in the assessment. In particular we thank the university presidents and chancellors who agreed to participate and offered the assistance of staff members at their institutions; the graduate deans, department chairmen, and many other university personnel who helped to compile information about the research-doctorate programs at their own institutions; and the nearly 5,000 faculty members who took the time to complete and return reputational survey forms. This assessment would not have been feasible without the participation of these individuals. Nor would it have been complete without the suggestions from many individuals within and outside the academic community who reviewed the study plans and committee reports.

The committee also acknowledges the contributions of Francis Narin and Paul R. McAllister, whose innovative work in the area of publication productivity in science and engineering fields has been a valuable resource. We thank H. Roberts Coward and his colleagues at the Institute for Scientific Information for their help in compiling publications data as well as William Batchelor and John James at NIH and David Staudt at NSF for their help in acquiring data on individual research grant awards.

Within the National Research Council many individuals have assisted in the planning and completion of this project. Robert A. Alberty, Harrison Shull, and W. K. Estes, former chairmen of the Commission on Human Resources, and William C. Kelly, Executive Director of the commission (now the Office of Scientific and Engineering Personnel), offered assistance and helpful counsel during all phases of the study. Lindsey R. Harmon and C. Alan Boneau contributed greatly to the planning of the assessment.

To Porter E. Coggeshall, Study Director, the committee expresses thanks for a job extremely well done. His ability to translate the committee's directions into compiled data and analyses must be given a large share of the credit for the completion of this project. He has been ably assisted by Prudence W. Brown, who supervised the data collection activities; Dorothy G. Cooper, who provided excellent secretarial support; and George A. Boyce, whose programming expertise was invaluable.

Committee on an Assessment of Quality-Related
Characteristics of Research-Doctorate Programs
in the United States

# Preface

The genius of American higher education is often said to be in the close association of training and research—that is, in the nation's research-doctorate programs. Consequently, we are not surprised at the amount of worried talk about the quality of the research doctorate, for deterioration at that level will inevitably spread to wherever research skills are needed—and that indeed is a far-flung network of laboratories, institutes, firms, agencies, bureaus, and departments. What might surprise us, however, is the imbalance between the putative national importance of research-doctorate programs and the amount of sustained evaluative attention they themselves receive.

The present assessment, sponsored by the Conference Board of Associated Research Councils—comprised of the American Council of Learned Societies, the American Council on Education, the National Research Council (NRC), and the Social Science Research Council—seeks to correct the imbalance between worried talk and systematic study. In this effort the Conference Board continues a tradition pioneered by the American Council on Education, which in 1966 published An Assessment of Quality in Graduate Education, the report of a study conducted by Allan M. Cartter, and in 1970 published A Rating of Graduate Programs, by Kenneth D. Roose and Charles J. Andersen. The Cartter and Roose-Andersen reports have been widely used and frequently cited.

Some years after the release of the Roose-Andersen report, it was decided that the effort to assess the quality of research-doctorate programs should be renewed, and the Conference Board of Associated Research Councils agreed to sponsor an assessment. The Board of Directors of the American Council on Education concurred with the notion that the next study should be issued under these broader auspices. The NRC agreed to serve as secretariat for a new study. The responsible staff of the NRC earned the appreciation of the Conference Board for the skill and dedication shown during the course of securing funding and implementing the study. Special mention should also be made of the financial contribution of the National Academy of Sciences which, by supplementing funds available from external sources, made it possible for the study to get under way.

To sponsor a study comparing the quality of programs in 32

disciplines and from more than 200 doctorate-granting universities is to invite critics, friendly and otherwise. Such was the fate of the previous studies; such has been the fate of the present study. Scholarship, fortunately, can put criticism to creative use and has done so in this project. The study committee appointed by the Conference Board reviewed the criticisms of earlier efforts to assess research-doctorate programs, and it actively solicited criticisms and suggestions for improvements of its own design. Although constrained by limited funds, the committee applied state-of-the-art methodology in a design that incorporated the lessons learned from previous studies as well as attending to many critics of the present effort. Not all criticism has thus been stilled; nor could it ever be. Additional criticisms will be voiced by as many persons as begin to use the results of this effort in ways not anticipated by its authors. These criticisms will be welcome. The Conference Board believes that the present study, building on earlier criticisms and adopting a multidimensional approach to the assessment of research-doctorate programs, represents a substantial improvement over past reports. Nevertheless, each of the diverse measures used here has its own limitations, and none provides a precise index of the quality of a program for educating students for careers in research. No doubt a future study, taking into account the weaknesses as well as strengths of this effort, will represent still further improvement. One mark of success for the present study would be for it to take its place in a continuing series, thereby contributing to the indicator base necessary for informed policies that will maintain and perhaps enhance the quality of the nation's research-doctorate programs.

For the more immediate future the purposes of this assessment are to assist students and student advisers seeking the best match possible between individual career goals and the choice of an advanced degree program; to serve scholars whose study site is higher education and the nation's research enterprise; and to inform the practical judgment of the administrators, funders, and policymakers responsible for protecting the quality of scholarly education in the United States.

A remarkably hard-working and competent group, whose names appear on p. vii, oversaw the long process by which this study moved from the planning stage to the completion of these reports. The Conference Board expresses its warmest thanks to the members of its committee and especially to their co-chairmen, Lyle V. Jones and Gardner Lindzey.

Conference Board of Associated<br>Research Councils

# Committee on an Assessment of Quality-Related Characteristics of Research-Doctorate Programs in the United States

LYLE V. JONES (Co-Chairman), Director of the L. L. Thurstone Psychometric Laboratory, University of North Carolina at Chapel Hill

GARDNER LINDZEY (Co-Chairman), Director, Center for Advanced Study in the Behavioral Sciences, Stanford, California

PAUL A. ALBRECHT, Vice-President and Dean, Claremont Graduate School

MARCUS ALEXIS, Department of Economics, Northwestern University

ROBERT M. BOCK, Dean of the Graduate School, University of Wisconsin at Madison

PHILIP E. CONVERSE, Institute for Social Research, University of Michigan

JAMES H. M. HENDERSON, Department of Plant Physiology, Tuskegee Institute of Alabama

ERNEST S. KUH, Department of Electrical Engineering and Computer Sciences, University of California, Berkeley

WINFRED P. LEHMANN, Department of Linguistics, University of Texas at Austin

SAUNDERS MAC LANE, Department of Mathematics, University of Chicago

NANCY S. MILBURN, Dean, College of Liberal Arts and Jackson College for Women, Tufts University

LINCOLN E. MOSES, Department of Statistics, Stanford University

JAMES C. OLSON, President, University of Missouri

C. K. N. PATEL, Director, Physical Research Laboratory, Bell Laboratories

MICHAEL J. PELCZAR, JR., President, The Council of Graduate Schools in the United States

JEROME B. SCHNEEWIND, Department of Philosophy, Johns Hopkins University

DUANE C. SPRIESTERSBACH, Vice-President, Educational Development and Research, University of Iowa

HARRIET A. ZUCKERMAN, Sociology Department, Columbia University


Study Director

PORTER E. COGGESHALL, Office of Scientific and Engineering Personnel, National Research Council

# Contents

## LIST OF FIGURES

LIST OF TABLES

# I
# Origins of Study and Selection of Programs

Each year more than 22,000 candidates are awarded doctorates in engineering, the humanities, and the sciences from approximately 250 U.S. universities. They have spent, on the average, five-and-a-half years in intensive education in preparation for research careers either in universities or in settings outside the academic sector, and many will make significant contributions to research. Yet we are poorly informed concerning the quality of the programs producing these graduates. This study is intended to provide information pertinent to this complex and controversial subject.

The charge to the study committee directed it to build upon the planning that preceded it. The planning stages included a detailed review of the methodologies and the results of past studies that had focused on the assessment of doctoral-level programs. The committee has taken into consideration the reactions of various groups and individuals to those studies. The present assessment draws upon previous experience with program evaluation, with the aim of improving what was useful and avoiding some of the difficulties encountered in past studies. The present study, nevertheless, is not purely reactive: it has its own distinctive features. First, it focuses only on programs awarding research doctorates and their effectiveness in preparing students for careers in research. Although other purposes of graduate education are acknowledged to be important, they are outside the scope of this assessment. Second, the study examines a variety of different indices that may be relevant to the program quality. This multidimensional approach represents an explicit recognition of the limitations of studies that rely entirely on peer ratings of perceived quality--the so-called reputational ratings. Finally, in the compilation of reputational ratings in this study, evaluators were provided the names of faculty members involved with each program to be rated and the number of research doctorates awarded in the last five years. In previous reputational studies evaluators were not supplied such information.

During the past two decades increasing attention has been given to describing and measuring the quality of programs in graduate education. It is evident that the assessment of graduate programs is highly important for university administrators and faculty, for employers in industrial and government laboratories, for graduate

1

students and prospective graduate students, for policymakers in state and national organizations, and for private and public funding agencies. Past experience, however, has demonstrated the difficulties with such assessments and their potentially controversial nature. As one critic has asserted:

> . . . the overall _effect_ of these reports seems quite clear. They tend, first, to make the rich richer and the poor poorer; second, the example of the highly ranked clearly imposes constraints on those institutions lower down the scale (the "Hertz-Avis" effect). And the effect of such constraints is to reduce diversity, to reward conformity or respectability, to penalize genuine experiment or risk. There is, also, I believe, an obvious tendency to promote the prevalence of disciplinary dogma and orthodoxy. All of this might be tolerable if the reports were tolerably accurate and judicious, if they were less _prescriptive_ and more _descriptive_; if they did not pretend to "objectivity" and if the very fact of ranking were not pernicious and invidious; if they genuinely promoted a meaningful "meritocracy"(instead of simply perpetuating the _status quo ante_ and an establishment mentality). But this is precisely what they cannot claim to be or do.[1]

The widespread criticisms of ratings in graduate education were carefully considered in the planning of this study. At the outset consideration was given to whether a national assessment of graduate programs should be undertaken at this time and, if so, what methods should be employed. The next two sections in this chapter examine the background and rationale for the decision by the Conference Board of Associated Research Councils[2] to embark on such a study. The remainder of the chapter describes the selection of disciplines and programs to be covered in the assessment.

The overall study encompasses a total of 2,699 graduate programs in 32 disciplines. In this report--the first of five reports issuing from the study--we examine 596 programs in six disciplines in the mathematical and physical sciences: chemistry, computer sciences, geosciences, mathematics, physics, and statistics/biostatistics. These programs account for more than 90 percent of the research

---

[1] William A. Arrowsmith, "Preface" in _The Ranking Game: The Power of the Academic Elite_, by W. Patrick Dolan, University of Nebraska Printing and Duplicating Service, Lincoln, Nebraska, 1976, p. ix.

[2] The Conference Board includes representatives of the American Council of Learned Societies, American Council on Education, National Research Council, and Social Science Research Council.

doctorates awarded in these six disciplines. It should be emphasized that the selection of disciplines to be covered was determined on the basis of total doctoral awards during the FY1976-78 period (as described later in this chapter), and the exclusion of a particular discipline was in no way based on a judgment of the importance of graduate education or research in that discipline. Also, although the assessment is limited to programs leading to the research-doctorate (Ph.D. or equivalent) degree, the Conference Board and study committee recognize that graduate schools provide many other forms of valuable and needed education.

### PRIOR ATTEMPTS TO ASSESS QUALITY IN GRADUATE EDUCATION

Universities and affiliated organizations have taken the lead in the review of programs in graduate education. At most institutions program reviews are carried out on a regular basis and include a comprehensive examination of the curriculum and educational resources as well as the qualifications of faculty and students. One special form of evaluation is that associated with institutional accreditation:

> The process begins with the institutional or programmatic self-study, a comprehensive effort to measure progress according to previously accepted objectives. The self-study considers the interest of a broad cross-section of constituencies--students, faculty, administrators, alumni, trustees, and in some circumstances the local community. The resulting report is reviewed by the appropriate accrediting commission and serves as the basis for evaluation by a site-visit team from the accrediting group. . . . Public as well as educational needs must be served simultaneously in determining and fostering standards of quality and integrity in the institutions and such specialized programs as they offer. Accreditation, conducted through non-governmental institutional and specialized agencies, provides a major means for meeting those needs.[3]

Although formal accreditation procedures play an important role in higher education, many university administrators do not view such procedures as an adequate means of assessing program quality. Other efforts are being made by universities to evaluate their programs in graduate education. The Educational Testing Service, with the sponsorship of the Council of Graduate Schools in the United States and the Graduate Record Examinations Board, has recently developed a

---

[3]Council on Postsecondary Accreditation, _The Balance Wheel for Accreditation_, Washington, D.C., July 1981, pp. 2-3.

set of procedures to assist institutions in evaluating their own
graduate programs.[4]

While reviews at the institutional (or state) level have proven
useful in assessing the relative strengths and weaknesses of
individual programs, they have not provided the information required
for making national comparisons of graduate programs. Several
attempts have been made at such comparisons. The most widely used of
these have been the studies by Keniston (1959), Cartter (1966), and
Roose and Andersen (1970). All three studies covered a broad range of
disciplines in engineering, the humanities, and the sciences and were
based on the opinions of knowledgeable individuals in the program
areas covered. Keniston[5] surveyed the department chairmen at 25
leading institutions. The Cartter[6] and Roose-Andersen[7] studies
compiled ratings from much larger groups of faculty peers. The stated
motivation for these studies was to increase knowledge concerning the
quality of graduate education:

> A number of reasons can be advanced for undertaking
> such a study. The diversity of the American system of
> higher education has properly been regarded by both
> the professional educator and the layman as a great
> source of strength, since it permits flexibility and
> adaptability and encourages experimentation and
> competing solutions to common problems. Yet diversity
> also poses problems. . . . Diversity can be a costly
> luxury if it is accompanied by ignorance. . . . Just
> as consumer knowledge and honest advertising are
> requisite if a competitive economy is to work satis-
> factorily, so an improved knowledge of opportunities
> and of quality is desirable if a diverse educational
> system is to work effectively.[8]

Although the program ratings from the Cartter and Roose-Andersen
studies are highly correlated, some substantial differences in
successive ratings can be detected for a small number of programs--
suggesting changes in the programs or in the perception of the
programs. For the past decade the Roose-Andersen ratings have

---

[4]For a description of these procedures see M. J. Clark, Graduate
Program Self-Assessment Service: Handbook for Users, Educational
Testing Service, Princeton, New Jersey, 1980.
[5]H. Keniston, Graduate Study in Research in the Arts and Sciences at
the University of Pennsylvania, University of Pennsylvania Press,
Phildelphia, 1959.
[6]A. M. Cartter, An Assessment of Quality in Graduate Education,
American Council on Education, Washington, D.C., 1966.
[7]K. D. Roose and C. J. Andersen, A Rating of Graduate Programs,
American Council on Education, Washington, D.C., 1970.
[8]Cartter, p. 3.

generally been regarded as the best available source of information on the quality of doctoral programs. Although the ratings are now more than 10 years out of date and have been criticized on a variety of grounds, they are still used extensively by individuals within the academic community and by those in federal and state agencies.

A frequently cited criticism of the Cartter and Roose-Andersen studies is their exclusive reliance upon reputational measurement.

> The ACE rankings are but a small part of all the evaluative processes, but they are also the most public, and they are clearly based on the narrow assumptions and elitist structures that so dominate the present direction of higher education in the United States. As long as our most prestigious source of information about post-secondary education is a vague popularity contest, the resultant ignorance will continue to provide a cover for the repetitious aping of a single model. . . . All the attempts to change higher education will ultimately be strangled by the "legitimate" evaluative processes that have already programmed a single set of responses from the start.[9]

A number of other criticisms have been leveled at reputational rankings of graduate programs.[10] First, such studies inherently reflect perceptions that may be several years out of date and do not take into account recent changes in a program. Second, the ratings of individual programs are likely to be influenced by the overall reputation of the university--i.e., an institutional "halo effect." Also, a disproportionately large fraction of the evaluators are graduates of and/or faculty members in the largest programs, which may bias the survey results. Finally, on the basis of such studies it may not be possible to differentiate among many of the lesser known programs in which relatively few faculty members have established national reputations in research.

Despite such criticisms several studies based on methodologies similar to that employed by Cartter and Roose and Andersen have been carried out during the past 10 years. Some of these studies evaluated post-baccalaureate programs in areas not covered in the two earlier reports--including business, religion, educational administration, and medicine. Others have focused exclusively on programs in particular disciplines within the sciences and humanities. A few attempts have been made to assess graduate programs in a broad range of disciplines, many of which were covered in the Roose-Andersen and Cartter ratings, but in the opinion of many each has serious deficiencies in the methods and procedures

---

[9] Dolan, p. 81.
[10] For a discussion of these criticisms, see David S. Webster, "Methods of Assessing Quality," Change, October 1981, pp. 20-24.

employed. In addition to such studies, a myriad of articles have been written on the assessment of graduate programs since the release of the Roose-Andersen report. With the heightening interest in these evaluations, many in the academic community have recognized the need to assess graduate programs, using other criteria in addition to peer judgment.

> Though carefully done and useful in a number of ways, these ratings (Cartter and Roose-Andersen) have been criticized for their failure to reflect the complexity of graduate programs, their tendency to emphasize the traditional values that are highly related to program size and wealth, and their lack of timeliness or currency. Rather than repeat such ratings, many members of the graduate community have voiced a preference for developing ways to assess the quality of graduate programs that would be more comprehensive, sensitive to the different program purposes, and appropriate for use at any time by individual departments or universities.[11]

Several attempts have been made to go beyond the reputational assessment. Clark, Harnett, and Baird, in a pilot study[12] of graduate programs in chemistry, history, and psychology, identified as many as 30 possible measures significant for assessing the quality of graduate education. Glower[13] has ranked engineering schools according to the total amount of research spending and the number of graduates listed in Who's Who in Engineering. House and Yeager[14] rated economics departments on the basis of the total number of pages published by full professors in 45 leading journals in this discipline. Other ratings based on faculty publication records have been compiled for graduate programs in a variety of disciplines, including political science, psychology, and sociology. These and other studies demonstrate the feasibility of a national assessment of graduate programs that is founded on more than reputational standing among faculty peers.

---

[11] Clark, p. 1.

[12] M. J. Clark, R. T. Harnett, and L. L. Baird, Assessing Dimensions of Quality in Doctoral Education: A Technical Report of a National Study in Three Fields, Educational Testing Service, Princeton, New Jersey, 1976.

[13] Donald D. Glower, "A Rational Method for Ranking Engineering Programs," Engineering Education, May 1980.

[14] Donald R. House and James H. Yeager, Jr., "The Distribution of Publication Success Within and Among Top Economics Departments: A Disaggregate View of Recent Evidence," Economic Inquiry, Vol. 16, No. 4, October 1978, pp. 593-598.

## DEVELOPMENT OF STUDY PLANS

In September 1976 the Conference Board, with support from the Carnegie Corporation of New York and the Andrew W. Mellon Foundation, convened a three-day meeting to consider whether a study of programs in graduate education should be undertaken.  The 40 invited participants at this meeting included academic administrators, faculty members, and agency and foundation officials,[15] who represented a variety of institutions, disciplines, and convictions.  In these discussions there was considerable debate concerning whether the potential benefits of such a study outweighed the possible mis-representations of the results.  On the one hand, "a substantial majority of the Conference [participants believed] that the earlier assessments of graduate education have received wide and important use:  by students and their advisors, by the institutions of higher education as aids to planning and the allocation of educational functions, as a check on unwarranted claims of excellence, and in social science research."[16]  On the other hand, the conference participants recognized that a new study assessing the quality of graduate education "would be conducted and received in a very different atmosphere than were the earlier Cartter and Roose-Andersen reports. . . . Where ratings were previously used in deciding where to increase funds and how to balance expanding programs, they might now be used in deciding where to cut off funds and programs."

After an extended debate of these issues, it was the recommendation of this conference that a study with particular emphasis on the effectiveness of doctoral programs in educating research personnel be undertaken.  The recommendation was based principally on four considerations:

(1)  the importance of the study results to national and state bodies,

(2)  the desire to stimulate continuing emphasis on quality in graduate education,

(3)  the need for current evaluations that take into account the many changes that have occurred in programs since the Roose-Andersen study, and

(4)  the value of extending the range of measures used in evaluative studies of graduate programs.

Although many participants expressed interest in an assessment of master's degree and professional degree programs, insurmountable problems prohibited the inclusion of these types of programs in this study.

Following this meeting a 13-member committee,[17] co-chaired by

---

[15] See Appendix G for a list of the participants in this conference.
[16] From a summary of the Woods Hole Conference (see Appendix G).
[17] See Appendix H for a list of members of the planning committee.

Gardner Lindzey and Harriet A. Zuckerman, was formed to develop a detailed plan for a study limited to research-doctorate programs and designed to improve upon the methodologies utilized in earlier studies. In its deliberations the planning committee carefully considered the criticisms of the Roose-Andersen study and other national assessments. Particular attention was paid to the feasibility of compiling a variety of specific measures (e.g., faculty publication records, quality of students, program resources) that were judged to be related to the quality of research-doctorate programs. Attention was also given to making improvements in the survey instrument and procedures used in the Cartter and Roose-Andersen studies. In September 1978 the planning group submitted a comprehensive report describing alternative strategies for an evaluation of the quality and effectiveness of research-doctorate programs.

> The proposed study has its own distinctive features. It is characterized by a sharp focus and a multi-dimensional approach. (1) It will focus only on programs awarding research doctorates; other purposes of doctoral training are acknowledged to be important, but they are outside the scope of the work contemplated. (2) The multidimensional approach represents an explicit recognition of the limitations of studies that make assessments solely in terms of ratings of perceived quality provided by peers--the so-called reputational ratings. Consequently, a variety of quality-related measures will be employed in the proposed study and will be incorporated in the presentation of the results of the study.[18]

This report formed the basis for the decision by the Conference Board to embark on a national assessment of doctorate-level programs in the sciences, engineering, and the humanities.

In June 1980 an 18-member committee was appointed to oversee the study. The committee,[19] made up of individuals from a diverse set of disciplines within the sciences, engineering, and the humanities, includes seven members who had been involved in the planning phase and several members who presently serve or have served as graduate deans at either public or private universities. During the first eight months the committee met three times to review plans for the study activities, make decisions on the selection of disciplines and programs to be covered, and design the survey instruments to be used. Early in the study an effort was made to solicit the views of presidents and graduate deans at more than 250 universities. Their suggestions were most helpful to the committee in drawing up final

---

[18]National Research Council, A Plan to Study the Quality and Effectiveness of Research-Doctorate Programs, 1978 (unpublished report).
[19]See p. iii of this volume for a list of members of the study committee.

plans for the assessment. With the assistance of the Council of Graduate Schools in the United States, the committee and its staff have tried to keep the graduate deans informed about the progress being made in this study. The final section of this chapter describes the procedures followed in determining which research-doctorate programs were to be included in the assessment.

## SELECTION OF DISCIPLINES AND PROGRAMS TO BE EVALUATED

One of the most difficult decisions made by the study committee was the selection of disciplines to be covered in the assessment. Early in the planning stage it was recognized that some important areas of graduate education would have to be left out of the study. Limited financial resources required that efforts be concentrated on a total of no more than about 30 disciplines in the biological sciences, engineering, humanities, mathematical and physical sciences, and social sciences. At its initial meeting the committee decided that the selection of disciplines within each of these five areas should be made primarily on the basis of the total number of doctorates awarded nationally in recent years.

At the time the study was undertaken, aggregate counts of doctoral degrees earned during the FY1976-78 period were available from two independent sources--the Educational Testing Service (ETS) and the National Research Council (NRC). Table 1.1 presents doctoral awards data for 10 disciplines within the mathematical and physical sciences. As alluded to in footnote 1 of the table, discrepancies between the ETS and NRC counts may be explained, in part, by differences in the data collection procedures. The ETS counts, derived from information provided by universities, have been categorized according to the discipline of the department/academic unit in which the degree was earned. The NRC counts were tabulated from the survey responses of FY1976-78 Ph.D. recipients, who had been asked to identify their fields of specialty. Since separate totals for research doctorates in astronomy, atmospheric sciences, environmental sciences, and marine sciences were not available from the ETS manual, the committee made its selection of six disciplines primarily on the basis of the NRC data. In the case of computer sciences, some consideration was given to the fact that the ETS estimate was significantly greater than the NRC estimate.[20]

The selection of the research-doctorate programs to be evaluated in each discipline was made in two stages. Programs meeting _any_ of the following three criteria were initially nominated for inclusion in the study:

> (1) more than a specified number (see below) of research doctorates awarded during the FY1976-78 period,

---

[20] See footnote 4 in Table 1.1.

    (2)  more than one-third of that specified number
        of doctorates awarded in FY1979, or

    (3)  an average rating of 2.0 or higher in the
        Roose-Andersen rating of the scholarly quality
        of departmental faculty.

In each discipline the specified number of doctorates required for
inclusion in the study was determined in such a way that the programs
meeting this criterion accounted for at least 90 percent of the

TABLE 1.1  Number of Research Doctorates Awarded in the Mathematical
and Physical Science Disciplines, FY1976-78

| | Source of Data[1] | |
|---|---|---|
| | ETS | NRC |
| **Disciplines Included in the Assessment** | | |
| Chemistry | 4,624 | 4,739 |
| Physics[2] | 3,139 | 3,033 |
| Mathematics | 1,985 | 1,848 |
| Geosciences[3] | 1,395 | 1,139 |
| Computer Sciences[4] | 728 | 456 |
| Statistics/Biostatistics[5] | 457 | 634 |
| Total | 12,328 | 11,849 |
| **Disciplines Not Included in the Assessment** | | |
| Astronomy | N/A[6] | 408 |
| Marine Sciences | N/A | 406 |
| Atmospheric Sciences | N/A | 246 |
| Environmental Sciences | N/A | 160 |
| Other Physical Sciences | N/A | 132 |
| Total | | 1,352 |

[1] Data on FY1976-78 doctoral awards were derived from two independent
sources:  Educational Testing Service (ETS), Graduate Programs and
Admissions Manual, 1979-81, and NRC's Survey of Earned Doctorates,
1976-78.  Differences in field definitions account for discrepancies
between the ETS and NRC data.
[2] Data from ETS include doctorates in astronomy and astrophysics.
[3] Data from ETS include doctorates in atmospheric sciences and
oceanography.
[4] The ETS data may include some individuals from computer science
departments who earned doctorates in the field of electrical
engineering and consequently are not included in the NRC data.
[5] Data from ETS exclude doctorates in biostatistics.
[6] Not available.

doctorates awarded in that discipline during the FY1976-78 period.  In the mathematical and physical science disciplines, the following numbers of FY1976-78 doctoral awards were required to satisfy the first criterion (above):

> Chemistry--13 or more doctorates
> Computer Sciences--5 or more doctorates
> Geosciences--7 or more doctorates
> Mathematics--7 or more doctorates
> Physics--10 or more doctorates
> Statistics/Biostatistics--5 or more doctorates

A list of the nominated programs at each institution was then sent to a designated individual (usually the graduate dean) who had been appointed by the university president to serve as study coordinator for the institution.  The coordinator was asked to review the list and eliminate any programs no longer offering research doctorates or not belonging in the designated discipline.  The coordinator also was given an opportunity to nominate additional programs that he or she believed should be included in the study.[21]  Coordinators were asked to restrict their nominations to programs that they considered to be "of uncommon distinction" and that had awarded no fewer than two research doctorates during the past two years.  In order to be eligible for inclusion, of course, programs had to belong in one of the disciplines covered in the study.  If the university offered more than one research-doctorate program in a discipline, the coordinator was instructed to provide information on each of them so that these programs could be evaluated separately.

The committee received excellent cooperation from the study co-ordinators at the universities.  Of the 243 institutions that were identified as having one or more research-doctorate programs satisfying the criteria (listed earlier) for inclusion in the study, only 7 declined to participate in the study and another 8 failed to provide the program information requested within the three-month period allotted (despite several reminders).  None of these 15 institutions had doctoral programs that had received strong or distinguished reputational ratings in prior national studies.  Since the information requested had not been provided, the committee decided not to include programs from these institutions in any aspect of the assessment.  In each of the six chapters that follows, a list is given of the universities that met the criteria for inclusion in a particular discipline but that are not represented in the study.

As a result of nominations by institutional coordinators, some programs were added to the original list and others dropped.  Table 1.2 reports the final coverage in each of the six mathematical and physical science disciplines.  The number of programs evaluated varies

---

[21]See Appendix A for the specific instructions given to the coordinators.

TABLE 1.2  Number of Programs Evaluated in Each Discipline and the
Total FY1976-80 Doctoral Awards from These Programs

| Discipline | Programs | FY1976-80 Doctorates* |
|------------|----------|----------------------|
| Chemistry | 145 | 7,304 |
| Computer Sciences | 58 | 1,154 |
| Geosciences | 91 | 1,747 |
| Mathematics | 115 | 2,698 |
| Physics | 123 | 4,271 |
| Statistics/Biostatistics | 64 | 906 |
| TOTAL | 596 | 18,080 |

*The data on doctoral awards were provided by the study coordinator at each of the universities covered in the assessment.

considerably by discipline.  A total of 145 chemistry programs have been included in the study; in computer sciences and statistics/ biostatistics fewer than half this number have been included.  Although the final determination of whether a program should be included in the assessment was left in the hands of the institutional coordinator, it is entirely possible that a few programs meeting the criteria for inclusion in the assessment were overlooked by the coordinators. During the course of the study only two such programs in the mathematical and physical sciences--one in mathematics and one in biostatistics--have been called to the attention of the committee.

In the chapter that follows, a detailed description is given of each of the measures used in the evaluation of research-doctorate programs in the mathematical and physical sciences.  The description includes a discussion of the rationale for using the measure, the source from which data for that measure were derived, and any known limitations that would affect the interpretation of the data reported.  The committee wishes to emphasize that there are limitations associated with each of the measures and that none of the measures should be regarded as a precise indicator of the quality of a program in educating scientists for careers in research.  The reader is strongly urged to consider the descriptive material presented in Chapter II before attempting to interpret the program evaluations reported in subsequent chapters.  In presenting a frank discussion of any shortcomings of each measure, the committee's intent is to reduce the possibility of misuse of the results from this assessment of research-doctorate programs.

# II
# Methodology

Quality . . . you know what it is, yet you don't know
what it is. But that's self-contradictory. But some
things are better than others, that is, they have more
quality. But when you try to say what the quality is,
apart from the things that have it, it all goes poof!
There's nothing to talk about. But if you can't say
what Quality is, how do you know what it is, or how do
you know that it even exists? If no one knows what it
is, then for all practical purposes it doesn't exist
at all. But for all practical purposes it really does
exist. What else are the grades based on? Why else
would people pay fortunes for some things and throw
others in the trash pile? Obviously some things are
better than others . . . but what's the "betterness"?
. . . So round and round you go, spinning mental
wheels and nowhere finding anyplace to get traction.
What the hell is Quality? What is it?

> Robert M. Pirsig
> Zen and the Art of
> Motorcycle Maintenance

Both the planning committee and our own study committee have given
careful consideration to the types of measures to be employed in the
assessment of research-doctorate programs.[1] The committees
recognized that any of the measures that might be used is open to
criticism and that no single measure could be expected to provide an
entirely satisfactory index of the quality of graduate education.
With respect to the use of multiple criteria in educational
assessment, one critic has commented:

---

[1] A description of the measures considered may be found in the third
chapter of the planning committee's report, along with a discussion of
the relative merits of each measure.

> At best each is a partial measure encompassing a
> fraction of the large concept. On occasion its link
> to the real [world] is problematic and tenuous.
> Moreover, each measure [may contain] a load of
> irrelevant superfluities, "extra baggage" unrelated to
> the outcomes under study. By the use of a number of
> such measures, each contributing a different facet of
> information, we can limit the effect of irrelevancies
> and develop a more rounded and truer picture of
> program outcomes.[2]

Although the use of multiple measures alleviates the criticisms
directed at a single dimension or measure, it certainly will not
satisfy those who believe that the quality of graduate programs cannot
be represented by quantitative estimates no matter how many dimensions
they may be intended to represent. Furthermore, the usefulness of the
assessment is dependent on the validity and reliability of the
criteria on which programs are evaluated. The decision concerning
which measures to adopt in the study was made primarily on the basis
of two factors:

> (1) the extent to which a measure was judged to be
>     related to the quality of research-doctorate
>     programs, and
> (2) the feasibility of compiling reliable data for
>     making national comparisons of programs in
>     particular disciplines.

Only measures that were applicable to a majority of the disciplines to
be covered were considered. In reaching a final decision the study
committee found the ETS study,[3] in which 27 separate variables were
examined, especially helpful, even though it was recognized that many
of the measures feasible in institutional self-studies would not be
available in a national study. The committee was aided by the many
suggestions received from university administrators and others within
the academic community.

Although the initial design called for an assessment based on
approximately six measures, the committee concluded that it would be
highly desirable to expand this effort. A total of 16 measures
(listed in Table 2.1) have been utilized in the assessment of
research-doctorate programs in chemistry, computer sciences, geosciences, mathematics, and physics; 15 of these were used in evaluating
programs in statistics/biostatistics. (Data on research expenditures
are unavailable in the latter discipline.) For nine of the measures

---

[2]C. H. Weiss, _Evaluation Research: Methods of Assessing Program
Effectiveness_, Prentice-Hall, Inc., Englewood Cliffs, New Jersey, 1972,
p. 56.
[3]See M. J. Clark et al. (1976) for a description of these variables.

TABLE 2.1  Measures Compiled on Individual Research-Doctorate Programs

Program Size[1]
01  Reported number of faculty members in the program, December 1980.
02  Reported number of program graduates in last five years (July 1975 through June 1980).
03  Reported total number of full-time and part-time graduate students enrolled in the program who intend to earn doctorates, December 1980.

Characteristics of Graduates[2]
04  Fraction of FY1975-79 program graduates who had received some national fellowship or training grant support during their graduate education.
05  Median number of years from first enrollment in graduate school to receipt of the doctorate--FY1975-79 program graduates.[3]
06  Fraction of FY1975-79 program graduates who at the time they completed requirements for the doctorate reported that they had made definite commitments for postgraduation employment.
07  Fraction of FY1975-79 program graduates who at the time they completed requirements for the doctorate reported that they had made definite commitments for postgraduation employment in Ph.D.-granting universities.

Reputational Survey Results[4]
08  Mean rating of the scholarly quality of program faculty.
09  Mean rating of the effectiveness of the program in educating research scholars/scientists
10  Mean rating of the improvement in program quality in the last five years.
11  Mean rating of the evaluators' familiarity with the work of program faculty.

University Library Size[5]
12  Composite index describing the library size in the university in which the program is located, 1979-80.

Research Support
13  Fraction of program faculty members holding research grants from the National Science Foundation; National Institutes of Health; or the Alcohol, Drug Abuse, and Mental Health Administration at any time during the FY1978-80 period.[6]
14  Total expenditures (in thousands of dollars) reported by the university for research and development activities in a specified field, FY1979.[7]

Publication Records[8]
15  Number of published articles attributed to the program, 1978-79.
16  Estimated "overall influence" of published articles attributed to the program, 1978-79.

[1] Based on information provided to the committee by the participating universities.
[2] Based on data compiled in the NRC's Survey of Earned Doctorates.
[3] In reporting standardized scores and correlations with other variables, a shorter time-to-Ph.D. is assigned a higher score.
[4] Based on responses to the committee's survey conducted in April 1981.
[5] Based on data compiled by the Association of Research Libraries.
[6] Based on matching faculty names provided by institutional coordinators with the names of research grant awardees from the three federal agencies.
[7] Based on data provided to the National Science Foundation by universities.
[8] Based on data compiled by the Institute for Scientific Information and developed by Computer Horizons, Inc.

data are available describing most, if not all, of the mathematical and physical science programs included in the assessment. For seven measures the coverage is less complete but encompasses at least a majority of the programs in every discipline. The actual number of programs evaluated on every measure is reported in the second table in each of the next six chapters.

The 16 measures describe a variety of aspects important to the operation and function of research-doctorate programs—and thus are relevant to the quality and effectiveness of programs in educating scientists for careers in research. However, not all of the measures may be viewed as "global indices of quality." Some, such as those relating to program size, are best characterized as "program descriptors" which, although not dimensions of quality per se, are thought to have a significant influence on the effectiveness of programs. Other measures, such as those relating to university library size and support for research and training, describe some of the resources generally recognized as being important in maintaining a vibrant program in graduate education. Measures derived from surveys of faculty peers or from the publication records of faculty members, on the other hand, have traditionally been regarded as indices of the overall quality of graduate programs. Yet these too are not true measures of quality.

> We often settle for an easy-to-gather statistic,
> perfectly legitimate for its own limited purposes, and
> then forget that we haven't measured what we want to
> talk about. Consider, for instance, the reputation
> approach of ranking graduate departments: We ask a
> sample of physics professors (say) which the best
> physics departments are and then tabulate and report
> the results. The "best" departments are those that
> our respondents say are the best. Clearly it is
> useful to know which are the highly regarded depart-
> ments in a given field, but prestige (which is what we
> are measuring here) isn't exactly the same as
> quality.[4]

To be sure, each of the 16 measures reported in this assessment has its own set of limitations. In the sections that follow an explanation is provided of how each measure has been derived and its particular limitations as a descriptor of research-doctorate programs.

### PROGRAM SIZE

Information was collected from the study coordinators at each university on the names and ranks of program faculty, doctoral student

---

[4] John Shelton Reed, "How Not to Measure What a University Does," The Chronicle of Higher Education, Vol 22, No. 12, May 11, 1981, p. 56.

enrollment, and number of Ph.D. graduates in each of the past five years (FY1976-80).  Each coordinator was instructed to include on the faculty list those individuals who, as of December 1, 1980, held academic appointments (typically at the rank of assistant, associate, and full professor) and who participated significantly in doctoral education.  Emeritus and adjunct members generally were not to be included.  Measure 01 represents the number of faculty identified in a program.  Measure 02 is the reported number of graduates who earned Ph.D. or equivalent research doctorates in a program during the period from July 1, 1975, through June 30, 1980.  Measure 03 represents the total number of full-time and part-time students reported to be enrolled in a program in the fall of 1980, who intended to earn research doctorates.  All three of these measures describe different aspects of program size.  In previous studies program size has been shown to be highly correlated with the reputational ratings of a program, and this relationship is examined in detail in this report. It should be noted that since the information was provided by the institutions participating in the study, the data may be influenced by the subjective decisions made by the individuals completing the forms.  For example, some institutional coordinators may be far less restrictive than others in deciding who should be included on the list of program faculty.  To minimize variation in interpretation, detailed instructions were provided to those filling out the forms.[5]  Measure 03 is of particular concern in this regard since the coordinators at some institutions may not have known how many of the students currently enrolled in graduate study intended to earn doctoral degrees.

## CHARACTERISTICS OF GRADUATES

One of the most meaningful measures of the success of a research-doctorate program is the performance of its graduates.  How many go on to lead productive careers in research and/or teaching?  Unfortunately, reliable information on the subsequent employment and career achievements of the graduates of individual programs is not available. In the absence of this directly relevant information, the committee has relied on four indirect measures derived from data compiled in the NRC's Survey of Earned Doctorates.[6]  Although each measure has serious limitations (described below), the committee believes it more desirable to include this information than not to include data about program graduates.

In identifying program graduates who had received their doctorates in the previous five years (FY1975-79),[7] the faculty lists furnished

---

[5] A copy of the survey form and instructions sent to study coordinators is included in Appendix A.
[6] A copy of the questionnaire used in this survey is found in Appendix B.
[7] Survey data for the FY1980 Ph.D. recipients had not yet been compiled at the time this assessment was undertaken.

by the study coordinators at universities were compared with the names of dissertation advisers (available from the NRC survey). The latter source contains records for virtually all individuals who have earned research doctorates from U.S. universities since 1920. The institution, year, and specialty field of Ph.D. recipients were also used in determining the identity of program graduates. It is estimated that this matching process provided information on the graduate training and employment plans of more than 90 percent of the FY1975-79 graduates from the mathematical and physical science programs. In the calculation of each of the four measures derived from the NRC survey, program data are reported only if the survey information is available on at least 10 graduates. Consequently, in the disciplines with smaller programs--computer sciences and statistics/biostatistics--only slightly more than half the programs are included in these measures, whereas more than 90 percent of the chemistry and physics programs are included.

Measure 04 constitutes the fraction of FY1975-79 graduates of a program who had received at least _some_ national fellowship support, including National Institutes of Health (NIH) fellowships or traineeships, National Science Foundation (NSF) fellowships, other federal fellowships, Woodrow Wilson fellowships, or fellowships/ traineeships from other U.S. national organizations. One might expect the more selective programs to have a greater proportion of students with national fellowship support--especially "portable fellowships." Although the committee considered alternative measures of student ability (e.g., Graduate Record Examination scores, undergraduate grade point averages), reliable information of this sort was unavailable for a national assessment. It should be noted that the relevance of the fellowship measure varies considerably among disciplines. In the biomedical sciences a substantial fraction of the graduate students are supported by training grants and fellowships; in the mathematical and physical sciences the majority are supported by research assistantships and teaching assistantships.

Measure 05 is the median number of years elapsed from the time program graduates first enrolled in graduate school to the time they received their doctoral degrees. For purposes of analysis the committee has adopted the conventional wisdom that the most talented students are likely to earn their doctoral degrees in the shortest periods of time--hence, the shorter the median time-to-Ph.D., the higher the standardized score that is assigned. Although this measure has frequently been employed in social science research as a proxy for student ability, one must regard its use here with some skepticism. It is quite possible that the length of time it takes a student to complete requirements for a doctorate may be significantly affected by the explicit or implicit policies of a university or department. For example, in certain cases a short time-to-Ph.D. may be indicative of less stringent requirements for the degree. Furthermore, previous studies have demonstrated that women and members of minority groups, for reasons having nothing to do with their abilities, are more likely than male Caucasians to interrupt their graduate education or to be

enrolled on a part-time basis.[8]  As a consequence, the median
time-to-Ph.D. may be longer for programs with larger fractions of
women and minority students.

Measure 06 represents the fraction of FY 1975-79 program graduates
who reported at the time they had completed requirements for the
doctorate that they had signed contracts or made firm commitments for
postgraduation employment (including postdoctoral appointments as well
as other positions in the academic or nonacademic sectors) and who
provided the names of their prospective employers.  Although this
measure is likely to vary by discipline according to the availability
of employment opportunities, a program's standing relative to other
programs in the same discipline should not be affected by this
variation.  In theory, the graduates with the greatest promise should
have the easiest time finding jobs.  However, the measure is also
influenced by a variety of other factors, such as personal job
preferences and restrictions in geographic mobility, that are
unrelated to the ability of the individual.  It also should be noted
parenthetically that unemployment rates for doctoral recipients are
quite low and that nearly all of the graduates seeking jobs find
positions soon after completing their doctoral programs.[9]
Furthermore, first employment after graduation is by no means a
measure of career achievement, which is what one would like to have if
reliable data were available.

Measure 07, a variant of measure 06, constitutes the fraction of
FY1975-79 program graduates who indicated that they had made firm
commitments for employment in Ph.D.-granting universities and who
provided the names of their prospective employers.  This measure may
be presumed to be an indication of the fraction of graduates likely to
pursue careers in academic research, although there is no evidence
concerning how many of them remain in academic research in the long
term.  In many science disciplines the path from Ph.D. to postdoctoral
apprenticeship to junior faculty has traditionally been regarded as
the road of success for the growth and development of research
talent.  The committee is well aware, of course, that other paths,
such as employment in the major laboratories of industry and
government, provide equally attractive opportunities for growth.
Indeed, in recent years increasing numbers of graduates are entering
the nonacademic sectors.  Unfortunately, the data compiled from the
NRC's Survey of Earned Doctorates do not enable one to distinguish
between employment in the top-flight laboratories of industry and

---

[8] For a detailed analysis of this subject, see Dorothy M. Gilford and
Joan Snyder, Women and Minority Ph.D.'s in the 1970's:  A Data Book,
National Academy of Sciences, Washington, D.C., 1977.
[9] For new Ph.D. recipients in science and engineering the unemployment
rate has been less than 2 percent (see National Research Council,
Postdoctoral Appointments and Disappointments, National Academy Press,
Washington, D.C., 1981, p. 313).

TABLE 2.2  Percentage of FY1975-79 Doctoral Recipients with Definite Commitments for Employment Outside the Academic Sector*

|  |  |
|---|---|
| Chemistry | 45 |
| Computer Sciences | 38 |
| Geosciences | 53 |
| Mathematics | 17 |
| Physics | 42 |
| Statistics/Biostatistics | 29 |

*Percentages are based on responses to the NRC's Survey of Earned Doctorates by those who indicated that they had made firm commitments for postgraduation employment and who provided the names of their prospective employers.  These percentages may be considered lower-bound estimates of the actual percentages of doctoral recipients employed outside the academic sector.

government and employment in other areas of the nonacademic sectors. Accordingly, the committee has relied on a measure that reflects only the academic side and views this measure as a useful and interesting program characteristic rather than a dimension of quality.  In disciplines such as geosciences, chemistry, physics, and computer sciences, in which more than one-third of the graduates take jobs outside the academic environs (see Table 2.2), this limitation is of particular concern.

The inclusion of measures 06 and 07 in this assessment has been an issue much debated by members of the committee; the strenuous objections of three committee members regarding the use of these measures are expressed in the Minority Statement that follows Chapter IX.

## REPUTATIONAL SURVEY RESULTS

In April 1981, survey forms were mailed to a total of 1,788 faculty members in chemistry, computer sciences, geosciences, mathematics, physics, and statistics/biostatistics.  The evaluators were selected from the faculty lists furnished by the study coordinators at the 228 universities covered in the assessment.  These evaluators constituted approximately 13 percent of the total faculty population--13,661 faculty members--in the mathematical and physical science programs being evaluated (see Table 2.3).  The survey sample was chosen on the basis of the number of faculty in a particular program and the number of doctorates awarded in the previous five years (FY1976-80)--with the stipulation that at least one evaluator was selected from every program covered in the assessment.  In selecting the sample each faculty rank was represented in proportion to the total number of individuals holding that rank, and preference was given to those faculty members whom the study coordinators had nominated to serve as evaluators. As shown in Table 2.3, 1,461 individuals, 82 percent of the survey sample in the mathematical and physical sciences, had been recommended by study coordinators.[10]

Each evaluator was asked to consider a stratified random sample of 50 research-doctorate programs in his or her discipline--with programs stratified by the number of faculty members associated with each program. Every program was included on 150 survey forms. The 50 programs to be evaluated appeared on each survey form in random sequence, preceded by an alphabetized list of all programs in that discipline that were being included in the study. No evaluator was asked to consider a program at his or her own institution. Ninety percent of the survey sample group were provided the names of faculty members in each of the 50 programs to be evaluated, along with data on the total number of doctorates awarded in the last five years.[11] The inclusion of this information represents a significant departure from the procedures used in earlier reputational assessments. For purposes of comparison with previous studies, 10 percent (randomly selected in each discipline) were not furnished any information other than the names of the programs.

The survey items were adapted from the form used in the Roose-Andersen study. Prior to mailing, the instrument was pretested using a small sample of faculty members in chemistry and psychology. As a result, two significant improvements were made in the original survey design. A question was added on the extent to which the evaluator was familiar with the work of the faculty in each program. Responses to this question, reported as measure 11, provide some insight into the relationship between faculty recognition and the reputational standing of a program.[12] Also added was a question on the evaluator's field of specialization--thereby making it possible to compare program evaluations in different specialty areas within a particular discipline.

A total of 1,155 faculty members in the mathematical and physical sciences--65 percent of those asked to participate--completed and returned survey forms (see Table 2.3). Two factors probably have contributed to this response rate being approximately 14 percentage points below the rates reported in the Cartter and Roose-Andersen studies. First, because of the considerable expense of printing individualized survey forms (each 25-30 pages), second copies were not sent to sample members not responding to the first mailing[13]--as was

---

[10] A detailed analysis of the survey participants in each discipline is given in subsequent chapters.

[11] This information was furnished to the committee by the study coordinators at the universities participating in the study.

[12] Evidence of the strength of this relationship is provided by correlations presented in Chapters III-VIII, and an analysis of the relationship is provided in Chapter IX.

[13] A follow-up letter was sent to those not responding to the first mailing, and a second copy was distributed to those few evaluators who specifically requested another form.

done in the Cartter and Roose-Andersen efforts.  Second, it is quite apparent that within the academic community there has been a growing dissatisfaction in recent years with educational assessments based on reputational measures.  Indeed, this dissatisfaction was an important factor in the Conference Board's decision to undertake a multidimensional assessment, and some faculty members included in the sample made known to the committee their strong objections to the reputational survey.

TABLE 2.3  Survey Response by Discipline and Characteristics of Evaluator

| | Total Program Faculty N | Survey Sample | | |
| --- | --- | --- | --- | --- |
| | | Total N | Respondents N | % |
| **Discipline of Evaluator** | | | | |
| Chemistry | 3,339 | 435 | 301 | 69 |
| Computer Sciences | 923 | 174 | 108 | 62 |
| Geosciences | 1,419 | 273 | 177 | 65 |
| Mathematics | 3,784 | 348 | 223 | 64 |
| Physics | 3,399 | 369 | 211 | 57 |
| Statistics/Biostatistics | 797 | 189 | 135 | 71 |
| **Faculty Rank** | | | | |
| Professor | 8,133 | 1,090 | 711 | 65 |
| Associate Professor | 3,225 | 471 | 293 | 62 |
| Assistant Professor | 2,120 | 216 | 143 | 66 |
| Other | 183 | 11 | 8 | 73 |
| **Evaluator Selection** | | | | |
| Nominated by Institution | 3,751 | 1,461 | 971 | 66 |
| Other | 9,910 | 327 | 184 | 56 |
| **Survey Form** | | | | |
| With Faculty Names | N/A* | 1,609 | 1,033 | 64 |
| Without Names | N/A | 179 | 122 | 68 |
| Total All Fields | 13,661 | 1,788 | 1,155 | 65 |

*Not applicable.

As can be seen in Table 2.3, there is some variation in the response rates in the six mathematical and physical science disciplines. Of particular interest is the relatively high rate of response from chemists and the low rate from physicists—a result consistent with the findings in the Cartter and Roose-Andersen surveys.[14] It is not surprising to find that the evaluators nominated by study coordinators responded more often than did those who had been selected at random. No appreciable differences were found among the response rates of assistant, associate, and full professors or between the rates of those evaluators who were furnished the abbreviated survey form (without lists of program faculty) and those who were given the longer version.

Each program was considered by an average of approximately 90 survey respondents from other programs in the same discipline. The evaluators were asked to judge programs in terms of scholarly quality of program faculty, effectiveness of the program in educating research scholars/scientists, and change in program quality in the last five years.[15] The mean ratings of a program on these three survey items constitute measures 08, 09, and 10. Evaluators were also asked to indicate the extent to which they were familiar with the work of the program faculty. The average of responses to this item constitutes measure 11.

In making judgments about the quality of faculty, evaluators were instructed to consider the scholarly competence and achievements of the individuals. The ratings were furnished on the following scale:

>5 Distinguished
>4 Strong
>3 Good
>2 Adequate
>1 Marginal
>0 Not sufficient for doctoral education
>X Don't know well enough to evaluate

In assessing the effectiveness of a program, evaluators were asked to consider the accessibility of faculty, the curricula, the instructional and research facilities, the quality of the graduate students, the performance of graduates, and other factors that contribute to a program's effectiveness. This measure was rated accordingly:

>3 Extremely effective
>2 Reasonably effective
>1 Minimally effective
>0 Not effective
>X Don't know well enough to evaluate

---

[14] To compare the response rates obtained in the earlier surveys, see Roose and Andersen, Table 28, p. 29.
[15] A copy of the survey instrument and accompanying instructions are included in Appendix C.

Evaluators were instructed to assess change in program quality on the basis of whether there was an improvement in the last five years in <u>both</u> the scholarly quality of the faculty and the effectiveness in educating research scholars/scientists.  The following alternatives were provided:

> 2  Better than five years ago
> 1  Little or no change in last five years
> 0  Poorer than five years ago
> X  Don't know well enough to evaluate

Evaluators were asked to indicate their familiarity with the work of the program faculty according to the following scale:

> 2  Considerable familiarity
> 1  Some familiarity
> 0  Little or no familiarity

In the computation of mean ratings on measures 08, 09, and 10, the "don't know" responses were ignored.  An average program rating based on fewer than 15 responses (excluding the "don't know" responses) is not reported.

Measures 08, 09, and 10 are subject to many of the same criticisms that have been directed at previous reputational surveys.  Although care has been taken to improve the sampling design and to provide evaluators with some essential information about each program, the survey results merely reflect a consensus of faculty opinions.  As discussed in Chapter I, these opinions may well be based on out-of-date information or be influenced by a variety of factors unrelated to the quality of the program.  In Chapter IX a number of factors that may possibly affect the survey results are examined.  In addition to these limitations, it should be pointed out that the evaluators, on the average, were unfamiliar with almost one-third of the programs they were asked to consider.[16]  As might be expected, the smaller and less prestigious programs were not as well known, and for this reason one might have less confidence in the average ratings of these programs.  For all four survey measures, standard errors of the mean ratings are reported; they tend to be larger for the lesser known programs.  The frequency of response to each of the survey items is discussed in Chapter IX.

Two additional comments should be made regarding the survey activity.  First, it should be emphasized that the ratings derived from the survey reflect a program's standing relative to other programs in the same discipline and provide no basis for making cross-disciplinary comparisons.  For example, the fact that a much larger number of chemistry programs received "distinguished" ratings on measure 08 than did computer science programs indicates nothing

---

[16]See Table 9.6 in Chapter IX.

about the relative quality of faculty in these two disciplines. It may depend, in part, on the total numbers of programs evaluated in these disciplines; in the survey instructions it was suggested to evaluators that no more than 10 percent of the programs listed be designated as "distinguished." Nor is it advisable to compare the rating of a program in one discipline with that of a program in another discipline because the ratings are based on the opinions of different groups of evaluators who were asked to judge entirely different sets of programs. Second, early in the committee's deliberations a decision was made to supplement the ratings obtained from faculty members with ratings from evaluators who hold research-oriented positions in institutions outside the academic sector. These institutions include industrial research laboratories, government research laboratories, and a variety of other research establishments. Over the past 10 years increasing numbers of doctoral recipients have taken positions outside the academic setting. The extensive involvement of these graduates in nonacademic employment is reflected in the percentages reported in Table 2.2: An average of 40 percent of the recent graduates in the mathematical and physical science disciplines who had definite employment plans indicated that they planned to take positions in nonacademic settings. Data from another NRC survey suggest that the actual fraction of scientists employed outside academia may be significantly higher. The committee recognized that the inclusion of nonacademic evaluators would furnish information valuable for assessing nontraditional dimensions of doctoral education and would provide an important new measure not assessed in earlier studies. Results from a survey of this group would provide an interesting comparison with the results obtained from the survey of faculty members. A concentrated effort was made to obtain supplemental funding for adding nonacademic evaluators in selected disciplines to the survey sample, but this effort was unsuccessful. The committee nevertheless remains convinced of the importance of including evaluators from nonacademic research institutions. These institutions are likely to employ increasing fractions of graduates in many disciplines, and it is urged that this group not be overlooked in future assessments of graduate programs.

## UNIVERSITY LIBRARY SIZE

The university library holdings are generally regarded as an important resource for students in graduate (and undergraduate) education. The Association of Research Libraries (ARL) has compiled data from its academic member institutions and developed a composite measure of a university library's size relative to those of other ARL members. The ARL Library Index, as it is called, is based on 10 characteristics: volumes held, volumes added (gross), microform units held, current serials received, expenditures for library materials, expenditures for binding, total salary and wage expenditures, other operating expenditures, number of professional staff, and number of

nonprofessional staff.[17]  The  1979-80 index, which constitutes
measure 12, is available for 89 of the 228 universities included in
the assessment. (These 89 tend to be among the largest institutions.)
The limited coverage of this measure is a major shortcoming.  It
should be noted that the ARL index is a composite description of
library size and not a qualitative evaluation of the collections,
services, or operations of the library.  Also, it is a measure of
aggregate size and does not take into account the library holdings in
a particular department or discipline.  Finally, although universities
with more than one campus were instructed to include figures for the
main campus only, some in fact may have reported library size for the
entire university system.  Whether this misreporting occurred is not
known.

## RESEARCH SUPPORT

Using computerized data files[18] provided by the National Science
Foundation (NSF) and the National Institutes of Health (NIH), it was
possible to identify which faculty members in each program had been
awarded research grants during the FY1978-80 period by either of these
agencies or by the Alcohol, Drug Abuse, and Mental Health Administration (ADAMHA).  The fraction of faculty members in a program who had
received any research grants from these agencies during this three-year
period constitutes measure 13.  Since these awards have been made on
the basis of peer judgment, this measure is considered to reflect the
perceived research competence of program faculty.  However, it should
be noted that significant amounts of support for research in the
mathematical and physical sciences come from other federal agencies as
well, but it was not feasible to compile data from these other
sources.  It is estimated[19] that 55 percent of the university
faculty members in these disciplines who received federal R&D funding
obtained their support from NSF and another 19 percent from NIH.  The
remaining 26 percent received support from the Department of Energy,
Department of Defense, National Aeronautics and Space Administration,
and other federal agencies.  It also should be pointed out that only
those faculty members who served as principal investigators or
co-investigators are counted in the computation of this measure.

Measure 14 describes the total FY1979 expenditures by a university
for R&D in a particular discipline. These data have been furnished to
the NSF[20] by universities and include expenditures of funds from
both federal and nonfederal sources.  If an institution has more than
one program being evaluated in the same discipline, the aggregate
university expenditures for research in that discipline are reported

---

[17] See Appendix D for a description of the calculation of this index.
[18] A description of these files is provided in Appendix E.
[19] Based on special tabulations of data from the NRC's Survey of
Doctorate Recipients, 1979.
[20] A copy of the survey instrument used to collect these data appears
in Appendix E.

# MEASURES COMPILED ON INDIVIDUAL RESEARCH-DOCTORATE PROGRAMS

## Program Size[1]    (See pages 16-17.)

01  Reported <u>number</u> of faculty members in the program, December 1980.

02  Reported <u>number</u> of program graduates in last five years (July 1975 through June 1980).

03  Reported <u>total number</u> of full-time and part-time graduate students enrolled in the program who intend to earn doctorates, December 1980.

## Characteristics of Graduates[2]    (See pages 17-20.)

04  <u>Fraction</u> of FY1975-79 program graduates who had received some national fellowship or training grant support during their graduate education.

05  <u>Median number</u> of years from first enrollment in graduate school to receipt of the doctorate--FY1975-79 program graduates.[3]

06  <u>Fraction</u> of FY1975-79 program graduates who at the time they completed requirements for the doctorate reported that they had made definite commitments for postgraduation employment.

07  <u>Fraction</u> of FY1975-79 program graduates who at the time they completed requirements for the doctorate reported that they had made definite commitments for postgraduation employment in Ph.D.-granting universities.

## Reputational Survey Results[4]    (See pages 20-25.)

08  <u>Mean rating</u> of the scholarly quality of program faculty (scale = 0 to 5).

09  <u>Mean rating</u> of the effectiveness of the program in educating research scholars/scientists (scale = 0 to 3).

10  <u>Mean rating</u> of the improvement in program quality in the last five years (scale = 0 to 2).

11  <u>Mean rating</u> of the evaluators' familiarity with the work of program faculty (scale = 0 to 2).

## University Library Size[5]    (See pages 25-26.)

12  Composite <u>index</u> describing the library size in the university in which the program is located, 1979-80 (scale = -3 to +3).

## Research Support    (See pages 26-27.)

13  <u>Fraction</u> of program faculty members holding research grants from the National Science Foundation, National Institutes of Health, or the Alcohol, Drug Abuse, and Mental Health Administration at any time during the FY1978-80 period.[6]

14  <u>Total expenditures</u> reported by the university for research and development activities in a specified field, FY1979[7] (in thousands of dollars).

## Publication Records[8]    (See pages 27-29.)

15  <u>Number</u> of published articles attributed to the program, 1978-79.

16  Estimated "overall influence" of published articles attributed to the program, 1978-79 (raw values are not reported).

---

[1] Based on information provided to the committee by the participating universities.

[2] Based on data compiled in the NRC's Survey of Earned Doctorates.

[3] In reporting standardized scores and correlations with other variables, a shorter time-to-Ph.D. is assigned a higher score.

[4] Based on responses to the committee's survey conducted in April 1981.

[5] Based on data compiled by the Association of Research Libraries.

[6] Based on matching faculty names provided by institutional coordinators with the names of research grant awardees from the three federal agencies.

[7] Based on data provided to the National Science Foundation by universities.

[8] Based on data compiled by the Institute for Scientific Information and developed by Computer Horizons, Inc.

for each of the programs. In each discipline data are recorded for the 100 universities with the largest R&D expenditures. As already mentioned, these data are not available for statistics and biostatistics programs.

This measure has several limitations related to the procedures by which the data have been collected. The committee notes that there is evidence within the source document[21] that universities employ varying practices for categorizing and reporting expenditures. Apparently, institutional support of research, industrial support of research, and expenditure of indirect costs are reported by different institutions in different categories (or not reported at all). Since measure 14 is based on total expenditures from all sources, the data used here are perturbed only when these types of expenditures are not subsumed under _any_ reporting category. Also, it should be noted that the data being attributed to geosciences programs include university expenditures in all areas of the environmental sciences (geological sciences, atmospheric sciences, and oceanography), and the data for mathematics programs include expenditures in statistics as well as mathematics. In contrast with measure 13, measure 14 is not reported on a scale relative to the number of faculty members and thus reflects the overall level of research activity at an institution in a particular discipline. Although research grants in the sciences and engineering provide some support for graduate students as well, these measures should not be confused with measure 04, which pertains to fellowships and training grants.

## PUBLICATION RECORDS

Data from the 1978 and the 1979 Science Citation Index have been compiled[22] on published articles associated with research-doctorate programs. Publication counts were associated with programs on the basis of the discipline of the journal in which an article appeared and the institution with which the author was affiliated. Coauthored articles were proportionately attributed to the institutions of the individual authors. Articles appearing in multidisciplinary journals (e.g., Science, Nature) were apportioned according to the characteristic mix of subject matter in those journals. For the purposes of assigning publication counts, this mix can be estimated with reasonable accuracy.[23]

---

[21] National Science Foundation, Academic Science:  R and D Funds, Fiscal Year 1979, U.S. Government Printing Office, Washington, D.C., NSF 81-301, 1981.
[22] The publication data have been generated for the committee's use by Computer Horizons, Inc., using source files provided by the Institute for Scientific Information.
[23] Francis Narin, Evaluative Bibliometrics:  The Use of Publications and Citations Analysis in the Evaluation of Scientific Activity, Report to the National Science Foundation, March 1976, p. 203.

Two measures have been derived from the publication records: measure 15--the total number of articles published in the 1978-79 period that have been associated with a research-doctorate program and measure 16--an estimation of the "influence" of these articles. The latter is a product of the number of articles attributed to a program and the estimated influence of the journals in which these articles appeared. The influence of a journal is determined from the weighted number of times, on the average, an article in that journal is cited--with references from frequently cited journals counting more heavily. A more detailed explanation of the derivation of these measures is given in Appendix F. Neither measure 15 nor measure 16 is based on actual counts of articles written only by program faculty. However, extensive analysis of the "influence" index in the fields of physics, chemistry, and biochemistry has demonstrated the stability of this index and the reliability associated with its use.[24] Of course, this does not imply that the measure captures subtle aspects of publication "influence." It is of interest to note that indices similar to measures 15 and 16 have been shown to be highly correlated with the peer ratings of graduate departments compiled in the Roose-Andersen study.[25]

It must be emphasized that these measures encompass articles (published in selected journals) by all authors affiliated with a given university. Included therefore are articles by program faculty members, students and research personnel, and even members of other departments in that university who publish in those journals. Moreover, these measures do not take into account the differing sizes of programs, and the measures clearly do depend on faculty size. Although consideration was given to reporting the number of published articles per faculty member, the committee concluded that since the measure included articles by other individuals besides program faculty members, the aggregate number of articles would be a more reliable measure of overall program quality. It should be noted that if a university had more than one program being evaluated in the same discipline, it is not possible to distinguish the relative contribution of each program. In such cases the aggregate university data in that discipline were assigned to each program.

Since the data are confined to 1978-79, they do not take into account institutional mobility of authors after that period. Thus, articles by authors who have moved from one institution to another since 1979 are credited to the former institution. Also, the publication counts fail to include the contributions of faculty members' publications in journals outside their primary discipline.

---

[24] Narin, pp. 283-307.
[25] Richard C. Anderson, Francis Narin, and Paul McAllister, "Publication Ratings Versus Peer Ratings of Universities," Journal of the American Society for Information Science, March 1978, pp. 91-103, and Lyle V. Jones, "The Assessment of Scholarship," New Directions for Program Evaluation, No. 6, 1980, pp. 1-20.

This point may be especially important for those programs with faculty members whose research is at the intersection of several different disciplines.

The reader should be aware of two additional caveats with regard to the interpretation of measures 15 and 16. First, both measures are based on counts of published articles and do not include books. Since in the mathematical and physical sciences most scholarly contributions are published as journal articles, this may not be a serious limitation. Second, the "influence" measure should not be interpreted as an indicator of the impact of articles by individual authors. Rather it is a measure of the impact of the journals in which articles associated with a particular program have been published. Citation counts, with all their difficulties, would have been preferable since they are attributable to individual authors and they register the impact of books as well as journal articles. However, the difficulty and cost of assembling reliable counts of articles by individual author made their use infeasible.

## ANALYSIS AND PRESENTATION OF THE DATA

The next six chapters present all of the information that has been compiled on individual research-doctorate programs in chemistry, computer sciences, geosciences, mathematics, physics, and statistics/ biostatistics. Each chapter follows a similar format, designed to assist the reader in the interpretation of program data. The first table in each chapter provides a list of the programs evaluated in a discipline--including the names of the universities and departments or academic units in which programs reside--along with the full set of data compiled for individual programs. Programs are listed alphabetically according to name of institution, and both raw and standardized values are given for all but one measure.[26] For the reader's convenience an insert of information from Table 2.1 is provided that identifies each of the 16 measures reported in the table and indicates the raw scale used in reporting values for a particular measure. Standardized values, converted from raw values to have a mean of 50 and a standard deviation of 10, are computed for every measure so that comparisons can easily be made of a program's relative standing on different measures. Thus, a standardized value of 30 corresponds with a raw value that is two standard deviations below the mean for that measure, and a standardized value of 70 represents a raw value two standard deviations above the mean. While the reporting of values in standardized form is convenient for comparing a particular program's standing on different measures, it may be misleading in interpreting actual differences in the values reported for two or more programs--

---

[26] Since the scale used to compute measure 16--the estimated "influence" of published articles--is entirely arbitrary, only standardized values are reported for this measure.

especially when the distribution of the measure being examined is highly skewed. For example, the numbers of published articles (measure 15) associated with four chemistry programs are reported in Table 3.1 as follows:

| Program | Raw Value | Standardized Value |
|---------|-----------|--------------------|
| A | 1 | 37 |
| B | 5 | 38 |
| C | 22 | 41 |
| D | 38 | 43 |

Although programs C and D have many times the number of articles as programs A and B, the differences reported on a standardized scale appear to be small. Thus, the reader is urged to take note of the raw values before attempting to interpret differences in the standardized values given for two or more programs.

The initial table in each chapter also presents estimated standard errors of mean ratings derived from the four survey items (measures 08-11). A standard error is an estimated standard deviation of the sample mean rating and may be used to assess the stability of a mean rating reported for a particular program.[27] For example, one may assert (with .95 confidence) that the population mean rating would lie within two standard errors of the sample mean rating reported in this assessment.

No attempt has been made to establish a composite ranking of programs in a discipline. Indeed, the committee is convinced that no single measure adequately reflects the quality of a research-doctorate program and wishes to emphasize the importance of viewing individual programs from the perspective of multiple indices or dimensions.

The second table in each chapter presents summary statistics (i.e., number of programs evaluated, mean, standard deviation, and decile values) for each of the program measures.[28] The reader should find these statistics helpful in interpreting the data reported on individual programs. Next is a table of the intercorrelations among the various measures for that discipline. This table should be of particular interest to those desiring information about the interrelations of the various measures.

---

[27]The standard error estimate has been computed by dividing the standard deviation of a program's ratings by the square root of the number of ratings. For a more extensive discussion of this topic, see Fred N. Kerlinger, Foundations of Behavioral Research, Holt, Reinhart, and Winston, Inc., New York, 1973, Chapter 12. Readers should note that the estimate is a measure of the variation in response and by no means includes all possible sources of error.

[28]Standardized scores have been computed from precise values of the mean and standard deviation of each measure and not the rounded values reported in the second table of each chapter.

The remainder of each chapter is devoted to an examination of results from the reputational survey. Included are an analysis of the characteristics of survey participants and graphical portrayals of the relationship of mean rating of scholarly quality of faculty (measure 08) with number of faculty (measure 01) and the relationship of mean rating of program effectiveness (measure 09) with number of graduates (measure 02). A frequently mentioned criticism of the Roose-Andersen and Cartter studies is that small but distinguished programs have been penalized in the reputational ratings because they are not as highly visible as larger programs of comparable quality. The comparisons of survey ratings with measures of program size are presented as the first two figures in each chapter and provide evidence about the number of small programs in each discipline that have received high reputational ratings. Since in each case the reputational rating is more highly correlated with the square root of program size than with the size measure itself, measures 01 and 02 are plotted on a square root scale.[29] To assist the reader in interpreting results of the survey evaluations, each chapter concludes with a graphical presentation of the mean rating for every program of the scholarly quality of faculty (measure 08) and an associated "confidence interval" of 1.5 standard errors. In comparing the mean ratings of two programs, if their reported confidence intervals of 1.5 standard errors do not overlap, one may safely conclude that the program ratings are significantly different (at the .05 level of significance)--i.e., the observed difference in mean ratings is too large to be plausibly attributable to sampling error.[30]

The final chapter of this report gives an overview of the evaluation process in the six mathematical and physical science disciplines and includes a summary of general findings. Particular attention is given to some of the extraneous factors that may influence program ratings of individual evaluators and thereby distort the survey results. The chapter concludes with a number of specific suggestions for improving future assessments of research-doctorate programs.

---

[29] For a general discussion of transforming variables to achieve linear fits, see John W. Tukey, Exploring Data Analysis, Addision Wesley, Reading, Massachusetts, 1977.

[30] This rule for comparing nonoverlapping intervals is valid as long as the ratio of the two estimated standard errors does not exceed 2.41. (The exact statistical significance of this criterion then lies between .050 and .034.) Inspection of the standard errors reported in each discipline shows that for programs with mean ratings differing by less than 1.0 (on measure 08), the standard error of one mean very rarely exceeds twice the standard error of another.

# III
# Chemistry Programs

In this chapter 145 research-doctorate programs in chemistry are assessed. These programs, according to the information supplied by their universities, have accounted for 7,304 doctoral degrees awarded during the FY1976-80 period--approximately 93 percent of the aggregate number of chemistry doctorates earned from U.S. universities in this five-year span.[1] On the average, 75 full-time and part-time students intending to earn doctorates were enrolled in a program in December 1980, with an average faculty size of 23 members.[2] The 145 programs, listed in Table 3.1, represent 143 different universities-- the University of Akron and the University of Kansas each have two chemistry programs included in the assessment. All but three of the programs were initiated prior to 1970. In addition to the 143 universities represented in this discipline, another 4 were initially identified as meeting the criteria[3] for inclusion in the assessment:

> University of California--San Francisco
> Fordham University
> Lehigh University
> SUNY at Albany

Chemistry programs at these four institutions have not been included in the evaluations in this discipline since in each case the study coordinator either indicated that the institution did not at that time have a research-doctorate program in chemistry or failed to provide the information requested by the committee.

Before examining individual program results presented in Table 3.1, the reader is urged to refer to Chapter II, in which each of the 16

---

[1] Data from the NRC's Survey of Earned Doctorates indicate that 7,843 research doctorates in chemistry were awarded by U.S. universities between FY1976 and FY1980.
[2] See the reported means for measures 03 and 01 in Table 3.2.
[3] As mentioned in Chapter I, the primary criterion for inclusion was that a university had awarded at least 13 doctorates in chemistry during the FY1976-78 period.

measures used in the assessment is discussed. Summary statistics describing every measure are given in Table 3.2. For all but two measures, data are reported for at least 139 of the 145 chemistry programs. For measure 12, a composite index of the size of a university library, data are available for 87 programs; for measure 14, total university expenditures for research in this discipline, data are available for 95 programs. The programs not evaluated on these two measures are typically smaller--in terms of faculty size and graduate student enrollment--than other chemistry programs. Were data on measures 12 and 14 available for all 145 programs, it is likely that the reported means for these two measures would be appreciably lower (and that some of the correlations of these measures with others would be higher).

Intercorrelations among the 16 measures (Pearson product-moment coefficients) are given in Table 3.3. Of particular note are the high positive correlations of the measures of program size (01-03) with measures of publication records (15, 16) and reputational survey ratings (08, 09, and 11). Figure 3.1 illustrates the relation between the mean rating of the scholarly quality of faculty (measure 08) and the number of faculty members (measure 01) for each of the 145 programs in chemistry. Figure 3.2 plots the mean rating of program effectiveness (measure 09) against the total number of FY1976-80 program graduates (measure 02). Although in both figures there is a significant positive correlation between program size and reputational rating, it is quite apparent that some of the smaller programs received high mean ratings and that some of the larger programs received low mean ratings.

Table 3.4 describes the 301 faculty members who participated in the evaluation of chemistry programs. These individuals constituted 69 percent of those asked to respond to the survey in this discipline and 9 percent of the faculty population in the 145 research-doctorate programs being evaluated.[4] A majority of the survey participants were organic or physical chemists and held the rank of full professor. Almost three-fourths of them had earned their highest degree prior to 1970.

To assist the reader in interpreting results of the survey evaluations, estimated standard errors have been computed for mean ratings of the scholarly quality of the faculty in 145 chemistry programs (and are given in Table 3.1). For each program the mean rating and an associated "confidence interval" of 1.5 standard errors are illustrated in Figure 3.3 (listed in order of highest to lowest mean rating). In comparing two programs, if their confidence intervals do not overlap, one may safely conclude that there is a significant difference in their mean ratings at a .05 level of significance.[5] From this figure it is also apparent that one should

---

[4] See Table 2.3 in Chapter II.
[5] See pp. 29-31 for a discussion of the interpretation of mean ratings and associated confidence intervals.

have somewhat more confidence in the accuracy of the mean ratings of higher-rated programs than lower-rated programs. This generalization results primarily from the fact that evaluators are not as likely to be familiar with the less prestigious programs, and consequently the mean ratings of these programs are usually based on fewer survey responses.

TABLE 3.1  Program Measures (Raw and Standardized Values) in Chemistry

| Prog No. | University - Department/Academic Unit | Program Size (01) | (02) | (03) | Characteristics of Program Graduates (04) | (05) | (06) | (07) |
|---|---|---|---|---|---|---|---|---|
| 001. | Akron, University of | 17 | 18 | 54 | .32 | 7.1 | .87 | .17 |
| | Chemistry | 44 | 42 | 47 | 58 | 34 | 60 | 38 |
| 002. | Akron, University of | 11 | 58 | 102 | .39 | 6.4 | .81 | .05 |
| | Polymer Science | 38 | 52 | 54 | 64 | 44 | 55 | 30 |
| 003. | Alabama, University of-Tuscaloosa | 16 | 20 | 31 | .31 | 6.5 | .60 | .13 |
| | Chemistry | 43 | 43 | 43 | 57 | 42 | 35 | 36 |
| 004. | American University | 11 | 32 | 49 | .11 | 8.5 | .58 | .13 |
| | Chemistry* | 38 | 46 | 46 | 39 | 15 | 33 | 35 |
| 005. | Arizona State University-Tempe | 24 | 45 | 86 | .37 | 6.4 | .75 | .38 |
| | Chemistry | 51 | 49 | 52 | 62 | 43 | 49 | 53 |
| 006. | Arizona, University of-Tucson | 28 | 51 | 92 | .22 | 5.8 | .76 | .31 |
| | Chemistry | 55 | 50 | 53 | 49 | 52 | 50 | 49 |
| 007. | Arkansas, University-Fayetteville | 22 | 32 | 28 | .34 | 6.4 | .84 | .24 |
| | Chemistry | 49 | 46 | 42 | 60 | 44 | 58 | 44 |
| 008. | Atlanta University | 8 | NA | 5 | NA | NA | NA | NA |
| | Chemistry* | 35 | | 39 | | | | |
| 009. | Auburn University | 17 | 10 | 51 | .23 | 6.0 | 1.00 | .33 |
| | Chemistry | 44 | 41 | 46 | 50 | 49 | 73 | 50 |
| 010. | Baylor University-Waco | 12 | 14 | 10 | .33 | 5.5 | .58 | .25 |
| | Chemistry | 39 | 42 | 40 | 59 | 55 | 33 | 44 |
| 011. | Boston College | 19 | 10 | 17 | .33 | 6.2 | .67 | .50 |
| | Chemistry | 46 | 41 | 41 | 59 | 46 | 41 | 62 |
| 012. | Boston University | 22 | 28 | 48 | .39 | 6.4 | .58 | .31 |
| | Chemistry | 49 | 45 | 46 | 64 | 43 | 33 | 49 |
| 013. | Brandeis University | 18 | 39 | 60 | .15 | 5.9 | .71 | .27 |
| | Chemistry | 45 | 47 | 48 | 42 | 50 | 45 | 46 |
| 014. | Brigham Young University | 21 | 21 | 21 | .22 | 5.5 | .77 | .39 |
| | Chemistry | 48 | 43 | 41 | 49 | 55 | 51 | 54 |
| 015. | Brown University | 23 | 59 | 74 | .02 | 5.3 | .80 | .29 |
| | Chemistry | 50 | 52 | 50 | 31 | 58 | 54 | 47 |
| 016. | Bryn Mawr College | 9 | 10 | 14 | NA | NA | NA | NA |
| | Chemistry | 36 | 41 | 40 | | | | |
| 017. | CUNY-Graduate School | 56 | 69 | 90 | .03 | 6.3 | .65 | .32 |
| | Chemistry | 84 | 54 | 53 | 32 | 45 | 39 | 50 |
| 018. | California Institute of Technology | 27 | 112 | 171 | .55 | 5.3 | .91 | .56 |
| | Chemistry and Chemical Engineering | 54 | 64 | 66 | 79 | 58 | 65 | 67 |
| 019. | California, University of-Santa Cruz | 12 | 32 | 50 | .15 | 5.6 | .68 | .42 |
| | Chemistry | 39 | 46 | 46 | 43 | 55 | 42 | 56 |
| 020. | California, University of-Berkeley | 48 | 247 | 359 | .21 | 5.1 | .82 | .38 |
| | Chemistry | 76 | 95 | 96 | 48 | 61 | 56 | 54 |

* indicates program was initiated since 1970.

NOTE: On the first line of data for every program, raw values for each measure are reported; on the second line values are reported in standardized form, with mean = 50 and standard deviation = 10. "NA" indicates that the value for a measure is not available.

TABLE 3.1  Program Measures (Raw and Standardized Values) in Chemistry

| Prog No. | Survey Results (08) | (09) | (10) | (11) | University Library (12) | Research Support (13) | (14) | Published Articles (15) | (16) | Survey Ratings Standard Error (08) | (09) | (10) | (11) |
|---|---|---|---|---|---|---|---|---|---|---|---|---|---|
| 001. | 1.6 | 1.1 | 1.2 | 0.4 | NA | .35 | NA | 69 | | .12 | .11 | .08 | .06 |
|  | 40 | 40 | 55 | 39 | | 44 | | 48 | 42 | | | | |
| 002. | 2.3 | 1.7 | 1.2 | 0.2 | NA | .55 | NA | 69 | | .21 | .15 | .15 | .05 |
|  | 48 | 50 | 55 | 35 | | 53 | | 48 | 42 | | | | |
| 003. | 1.8 | 1.2 | 0.9 | 0.6 | -1.3 | .25 | NA | 57 | | .10 | .08 | .11 | .06 |
|  | 42 | 41 | 45 | 44 | 36 | 39 | | 46 | 44 | | | | |
| 004. | 0.9 | 0.7 | 0.9 | 0.3 | NA | .18 | NA | 11 | | .09 | .11 | .10 | .05 |
|  | 34 | 32 | 43 | 36 | | 36 | | 39 | 40 | | | | |
| 005. | 2.5 | 1.8 | 1.1 | 0.7 | -0.3 | .63 | 1076 | 58 | | .09 | .07 | .08 | .06 |
|  | 50 | 53 | 52 | 47 | 46 | 57 | 44 | 47 | 46 | | | | |
| 006. | 2.9 | 2.0 | 1.3 | 1.0 | 0.9 | .54 | 1269 | 84 | | .08 | .05 | .07 | .07 |
|  | 54 | 56 | 62 | 53 | 58 | 53 | 46 | 51 | 49 | | | | |
| 007. | 2.1 | 1.4 | 1.1 | 0.6 | NA | .55 | 1180 | 31 | | .08 | .08 | .07 | .05 |
|  | 46 | 45 | 51 | 43 | | 53 | 45 | 42 | 43 | | | | |
| 008. | 0.5 | 0.3 | NA | 0.2 | NA | NA | 852 | 1 | | .10 | .10 | NA | .05 |
|  | 29 | 26 | | 34 | | | 42 | 37 | 39 | | | | |
| 009. | 1.4 | 1.0 | 1.0 | 0.4 | NA | .12 | NA | 32 | | .10 | .10 | .10 | .06 |
|  | 39 | 38 | 45 | 38 | | 33 | | 42 | 44 | | | | |
| 010. | 1.3 | 1.0 | 0.8 | 0.4 | NA | .00 | NA | 31 | | .13 | .10 | .10 | .06 |
|  | 38 | 37 | 40 | 38 | | 27 | | 42 | 41 | | | | |
| 011. | 1.6 | 1.0 | 1.1 | 0.5 | NA | .37 | NA | 17 | | .10 | .09 | .06 | .06 |
|  | 40 | 39 | 51 | 42 | | 45 | | 40 | 41 | | | | |
| 012. | 2.0 | 1.3 | 1.0 | 0.7 | -0.4 | .46 | 742 | 23 | | .08 | .08 | .08 | .06 |
|  | 45 | 45 | 46 | 46 | 46 | 49 | 41 | 41 | 43 | | | | |
| 013. | 3.1 | 2.0 | 0.9 | 1.0 | NA | .56 | 769 | 58 | | .08 | .06 | .05 | .06 |
|  | 56 | 57 | 44 | 54 | | 54 | 41 | 47 | 51 | | | | |
| 014. | 2.0 | 1.4 | 1.1 | 0.5 | -0.6 | .14 | 775 | 51 | | .11 | .10 | .08 | .06 |
|  | 45 | 45 | 51 | 42 | 43 | 34 | 41 | 45 | 44 | | | | |
| 015. | 2.9 | 1.9 | 0.9 | 0.9 | -1.1 | .57 | 1036 | 48 | | .07 | .05 | .08 | .07 |
|  | 54 | 54 | 41 | 50 | 39 | 54 | 44 | 45 | 46 | | | | |
| 016. | 1.2 | 0.8 | 0.8 | 0.5 | NA | NA | NA | 4 | | .12 | .10 | .08 | .06 |
|  | 36 | 35 | 39 | 41 | | | | 38 | 40 | | | | |
| 017. | 2.3 | 1.4 | 0.9 | 0.6 | NA | .32 | NA | 112 | | .10 | .10 | .09 | .06 |
|  | 48 | 46 | 43 | 44 | | 43 | | 56 | 52 | | | | |
| 018. | 4.9 | 2.8 | 1.2 | 1.8 | NA | .82 | 5297 | 163 | | .03 | .04 | .05 | .04 |
|  | 74 | 73 | 56 | 73 | | 66 | 80 | 64 | 74 | | | | |
| 019. | 2.4 | 1.6 | 0.6 | 0.7 | NA | .25 | 710 | 38 | | .10 | .08 | .09 | .07 |
|  | 48 | 49 | 30 | 47 | | 39 | 41 | 43 | 45 | | | | |
| 020. | 4.9 | 2.8 | 1.3 | 1.9 | 2.2 | .65 | 3717 | 205 | | .03 | .04 | .05 | .03 |
|  | 74 | 72 | 63 | 74 | 72 | 58 | 66 | 71 | 75 | | | | |

NOTE: On the first line of data for every program, raw values for each measure are reported; on the second line values are reported in standardized form, with mean = 50 and standard deviation = 10. "NA" indicates that the value for a measure is not available. Since the scale used to compute measure (16) is entirely arbitrary, only values in standardized form are reported for this measure.

TABLE 3.1  Program Measures (Raw and Standardized Values) in Chemistry

| Prog No. | University – Department/Academic Unit | Program Size | | | Characteristics of Program Graduates | | | |
|---|---|---|---|---|---|---|---|---|
| | | (01) | (02) | (03) | (04) | (05) | (06) | (07) |
| 021. | California, University of–Davis | 32 | 48 | 97 | .21 | 5.6 | .71 | .41 |
| | *Chemistry* | *59* | *49* | *54* | *48* | *55* | *45* | *56* |
| 022. | California, University of–Irvine | 20 | 39 | 84 | .12 | 5.1 | .78 | .25 |
| | *Chemistry* | *47* | *47* | *52* | *40* | *61* | *52* | *44* |
| 023. | California, University of–Los Angeles | 43 | 142 | 205 | .18 | 5.6 | .80 | .34 |
| | *Chemistry* | *71* | *71* | *71* | *45* | *54* | *54* | *51* |
| 024. | California, University of–Riverside | 18 | 28 | 44 | .27 | 4.9 | .81 | .27 |
| | *Chemistry* | *45* | *45* | *45* | *53* | *64* | *55* | *46* |
| 025. | California, University of–San Diego | 31 | 51 | 88 | .28 | 5.5 | .81 | .58 |
| | *Chemistry* | *58* | *50* | *52* | *54* | *55* | *55* | *68* |
| 026. | California, University of–Santa Barbara | 26 | 63 | 84 | .09 | 6.6 | .79 | .43 |
| | *Chemistry* | *53* | *53* | *52* | *37* | *41* | *53* | *57* |
| 027. | Carnegie-Mellon University | 22 | 32 | 40 | .28 | 6.5 | .87 | .17 |
| | *Chemistry* | *49* | *46* | *44* | *54* | *42* | *60* | *38* |
| 028. | Case Western Reserve University | 19 | 62 | 55 | .19 | 5.6 | .74 | .44 |
| | *Chemistry* | *46* | *53* | *47* | *47* | *54* | *48* | *58* |
| 029. | Catholic University of America | 12 | 23 | 22 | .20 | 6.3 | .65 | .22 |
| | *Chemistry* | *39* | *44* | *41* | *47* | *45* | *40* | *42* |
| 030. | Chicago, University of | 28 | 76 | 119 | .28 | 5.9 | .81 | .47 |
| | *Chemistry* | *55* | *56* | *57* | *54* | *50* | *55* | *60* |
| 031. | Cincinnati, University of | 29 | 39 | 87 | .25 | 5.6 | .81 | .47 |
| | *Chemistry* | *56* | *47* | *52* | *51* | *55* | *55* | *60* |
| 032. | Clark University | 9 | 8 | 15 | NA | NA | NA | NA |
| | *Chemistry* | *36* | *40* | *40* | | | | |
| 033. | Clarkson College of Technology | 15 | 23 | 30 | .20 | 7.0 | .73 | .15 |
| | *Chemistry* | *42* | *44* | *43* | *47* | *35* | *47* | *37* |
| 034. | Clemson University | 17 | 17 | 21 | .27 | 6.0 | .80 | .33 |
| | *Chemistry and Geology* | *44* | *42* | *41* | *53* | *49* | *54* | *50* |
| 035. | Colorado State University–Fort Collins | 24 | 40 | 86 | .11 | 5.6 | .89 | .57 |
| | *Chemistry* | *51* | *48* | *52* | *39* | *55* | *63* | *67* |
| 036. | Colorado, University of | 53 | 59 | 150 | .21 | 5.0 | .86 | .46 |
| | *Chemistry* | *81* | *52* | *62* | *48* | *63* | *60* | *59* |
| 037. | Columbia University | 17 | 94 | 130 | .19 | 5.4 | .86 | .48 |
| | *Chemistry* | *44* | *60* | *59* | *46* | *57* | *59* | *61* |
| 038. | Connecticut, University of–Storrs | 28 | 28 | 32 | .15 | 6.3 | .65 | .15 |
| | *Chemistry* | *55* | *45* | *43* | *43* | *44* | *40* | *37* |
| 039. | Cornell University–Ithaca | 29 | 121 | 153 | .36 | 5.5 | .88 | .45 |
| | *Chemistry* | *56* | *66* | *63* | *62* | *56* | *62* | *58* |
| 040. | Delaware, University of–Newark | 18 | 24 | 98 | .22 | 6.0 | .92 | .58 |
| | *Chemistry* | *45* | *44* | *54* | *49* | *49* | *66* | *68* |

* indicates program was initiated since 1970.

NOTE: On the first line of data for every program, raw values for each measure are reported; on the second line values are reported in standardized form, with mean = 50 and standard deviation = 10. "NA" indicates that the value for a measure is not available.

TABLE 3.1  Program Measures (Raw and Standardized Values) in Chemistry

| Prog No. | Survey Results | | | | University Library | Research Support | | Published Articles | | Survey Ratings Standard Error | | | |
|---|---|---|---|---|---|---|---|---|---|---|---|---|---|
| | (08) | (09) | (10) | (11) | (12) | (13) | (14) | (15) | (16) | (08) | (09) | (10) | (11) |
| 021. | 3.0 | 1.9 | 1.2 | 1.0 | 0.6 | .50 | 838 | 106 | | .07 | .05 | .06 | .06 |
| | 55 | 56 | 59 | 54 | 56 | 51 | 42 | 55 | 54 | | | | |
| 022. | 3.1 | 1.9 | 1.3 | 1.1 | NA | .70 | 1277 | 70 | | .08 | .04 | .07 | .06 |
| | 56 | 56 | 60 | 56 | | 61 | 46 | 49 | 53 | | | | |
| 023. | 4.4 | 2.5 | 1.3 | 1.6 | 2.0 | .84 | 3573 | 161 | | .05 | .05 | .06 | .06 |
| | 69 | 67 | 61 | 68 | 69 | 67 | 65 | 64 | 68 | | | | |
| 024. | 2.6 | 1.6 | 0.9 | 0.9 | -1.0 | .72 | 626 | 59 | | .09 | .08 | .08 | .06 |
| | 51 | 49 | 42 | 52 | 39 | 62 | 40 | 47 | 49 | | | | |
| 025. | 3.7 | 2.1 | 1.2 | 1.2 | -0.0 | .68 | 4225 | 97 | | .07 | .04 | .06 | .07 |
| | 62 | 59 | 56 | 59 | 49 | 59 | 71 | 53 | 54 | | | | |
| 026. | 3.2 | 2.0 | 1.4 | 1.2 | -0.1 | .69 | 1089 | 93 | | .07 | .05 | .06 | .06 |
| | 57 | 56 | 66 | 59 | 48 | 60 | 44 | 52 | 57 | | | | |
| 027. | 2.9 | 1.8 | 0.8 | 1.0 | NA | .55 | 1957 | 62 | | .08 | .06 | .07 | .06 |
| | 54 | 53 | 37 | 52 | | 53 | 51 | 47 | 45 | | | | |
| 028. | 2.8 | 1.8 | 0.5 | 1.0 | -1.3 | .58 | 1413 | 146 | | .07 | .07 | .08 | .07 |
| | 53 | 53 | 26 | 53 | 36 | 55 | 47 | 61 | 54 | | | | |
| 029. | 1.2 | 0.9 | 0.7 | 0.4 | NA | .25 | NA | 9 | | .11 | .11 | .09 | .05 |
| | 36 | 36 | 31 | 39 | | 39 | | 39 | 40 | | | | |
| 030. | 4.4 | 2.5 | 0.9 | 1.6 | 0.9 | .79 | 3418 | 84 | | .06 | .06 | .06 | .06 |
| | 69 | 67 | 44 | 69 | 58 | 65 | 64 | 51 | 55 | | | | |
| 031. | 2.7 | 1.8 | 1.1 | 1.0 | -0.2 | .31 | 814 | 93 | | .08 | .06 | .07 | .06 |
| | 52 | 54 | 53 | 53 | 47 | 42 | 42 | 52 | 48 | | | | |
| 032. | 1.1 | 0.7 | 0.8 | 0.3 | NA | NA | NA | 18 | | .13 | .12 | .09 | .05 |
| | 35 | 33 | 38 | 37 | | | | 40 | 41 | | | | |
| 033. | 1.9 | 1.4 | 1.0 | 0.8 | NA | .67 | 926 | 59 | | .11 | .08 | .06 | .07 |
| | 44 | 46 | 46 | 47 | | 59 | 43 | 47 | 43 | | | | |
| 034. | 1.4 | 1.0 | 1.2 | 0.6 | NA | .35 | NA | 22 | | .10 | .09 | .11 | .07 |
| | 39 | 39 | 56 | 43 | | 44 | | 41 | 41 | | | | |
| 035. | 3.2 | 1.9 | 1.7 | 1.3 | -1.1 | .63 | 1774 | 117 | | .08 | .05 | .07 | .06 |
| | 56 | 55 | 80 | 60 | 38 | 57 | 50 | 56 | 54 | | | | |
| 036. | 3.3 | 2.0 | 1.2 | 1.2 | -0.9 | .68 | 2544 | 95 | | .07 | .04 | .07 | .06 |
| | 58 | 57 | 58 | 58 | 41 | 60 | 56 | 53 | 54 | | | | |
| 037. | 4.6 | 2.7 | 1.3 | 1.6 | 1.7 | .94 | 3531 | 114 | | .06 | .05 | .05 | .06 |
| | 70 | 70 | 60 | 67 | 67 | 72 | 65 | 56 | 62 | | | | |
| 038. | 2.1 | 1.5 | 1.0 | 0.5 | -0.5 | .25 | 1185 | 76 | | .09 | .08 | .08 | .05 |
| | 46 | 47 | 48 | 42 | 44 | 39 | 45 | 50 | 45 | | | | |
| 039. | 4.4 | 2.5 | 1.1 | 1.7 | 1.6 | .72 | 3414 | 197 | | .06 | .05 | .06 | .05 |
| | 68 | 67 | 53 | 70 | 66 | 62 | 64 | 69 | 77 | | | | |
| 040. | 2.5 | 1.8 | 1.1 | 0.8 | NA | .28 | 1015 | 86 | | .09 | .06 | .07 | .06 |
| | 49 | 52 | 50 | 49 | | 41 | 43 | 51 | 50 | | | | |

NOTE: On the first line of data for every program, raw values for each measure are reported; on the second line values are reported in standardized form, with mean = 50 and standard deviation = 10. "NA" indicates that the value for a measure is not available. Since the scale used to compute measure (16) is entirely arbitrary, only values in standardized form are reported for this measure.

TABLE 3.1 Program Measures (Raw and Standardized Values) in Chemistry

| Prog No. | University - Department/Academic Unit | Program Size (01) | (02) | (03) | Characteristics of Program Graduates (04) | (05) | (06) | (07) |
|---|---|---|---|---|---|---|---|---|
| 041. | Denver, University of | 13 | 18 | 10 | NA | NA | NA | NA |
| | *Chemistry* | *40* | *42* | *40* | | | | |
| 042. | Drexel University | 15 | 18 | 35 | .07 | 6.3 | .79 | .07 |
| | *Chemistry* | *42* | *42* | *44* | *36* | *45* | *53* | *32* |
| 043. | Duke University | 22 | 69 | 78 | .22 | 4.9 | .79 | .41 |
| | *Chemistry* | *49* | *54* | *51* | *49* | *64* | *53* | *56* |
| 044. | Emory University | 16 | 25 | 53 | .50 | 5.2 | .44 | .26 |
| | *Chemistry* | *43* | *44* | *46* | *74* | *60* | *19* | *45* |
| 045. | Florida State University-Tallahassee | 31 | 54 | 86 | .26 | 5.7 | .83 | .45 |
| | *Chemistry* | *58* | *51* | *52* | *53* | *52* | *57* | *59* |
| 046. | Florida, University of-Gainesville | 35 | 82 | 130 | .14 | 6.0 | .71 | .31 |
| | *Chemistry* | *62* | *57* | *59* | *42* | *49* | *45* | *49* |
| 047. | Georgetown University | 14 | 33 | 50 | .35 | 5.9 | .79 | .35 |
| | *Chemistry* | *41* | *46* | *46* | *61* | *50* | *53* | *52* |
| 048. | Georgia Institute of Technology | 24 | 63 | 67 | .18 | 5.9 | .73 | .17 |
| | *Chemistry* | *51* | *53* | *49* | *46* | *50* | *47* | *39* |
| 049. | Georgia, University of-Athens | 21 | 61 | 64 | .16 | 5.6 | .76 | .29 |
| | *Chemistry* | *48* | *52* | *48* | *43* | *54* | *51* | *47* |
| 050. | Harvard University | 24 | 144 | 174 | .57 | 5.7 | .86 | .51 |
| | *Chemistry/Chemical Physics* | *51* | *72* | *66* | *81* | *53* | *59* | *63* |
| 051. | Hawaii, University of | 17 | 33 | 26 | .07 | 6.1 | .78 | .44 |
| | *Chemistry* | *44* | *46* | *42* | *36* | *47* | *52* | *58* |
| 052. | Houston, University of | 23 | 62 | 83 | .21 | 5.2 | .65 | .35 |
| | *Chemistry* | *50* | *53* | *51* | *48* | *60* | *40* | *51* |
| 053. | Howard University | 20 | 41 | 40 | .54 | 6.5 | .65 | .10 |
| | *Chemistry* | *47* | *48* | *44* | *78* | *42* | *39* | *33* |
| 054. | Idaho, University of-Moscow | 12 | 18 | 17 | .15 | 5.5 | .65 | .10 |
| | *Chemistry* | *39* | *42* | *41* | *43* | *55* | *40* | *34* |
| 055. | Illinois Institute of Technology | 15 | 12 | 17 | .50 | 5.7 | .73 | .27 |
| | *Chemistry* | *42* | *41* | *41* | *74* | *53* | *47* | *46* |
| 056. | Illinois, University of-Chicago Circle | 20 | 50 | 52 | .05 | 6.8 | .68 | .45 |
| | *Chemistry* | *47* | *50* | *46* | *34* | *38* | *43* | *58* |
| 057. | Illinois, University-Urbana/Champaign | 43 | 227 | 294 | .21 | 5.0 | .83 | .30 |
| | *Chemistry* | *71* | *91* | *86* | *48* | *62* | *57* | *48* |
| 058. | Indiana University-Bloomington | 36 | 114 | 184 | .24 | 5.2 | .82 | .36 |
| | *Chemistry* | *63* | *65* | *68* | *51* | *60* | *56* | *52* |
| 059. | Institute of Paper Chemistry-Appleton, Wi | 12 | 31 | 16 | .18 | 6.0 | .77 | .00 |
| | *Chemistry* | *39* | *45* | *40* | *45* | *49* | *51* | *26* |
| 060. | Iowa State University-Ames | 33 | 114 | 192 | .16 | 5.7 | .85 | .23 |
| | *Chemistry* | *60* | *65* | *69* | *44* | *53* | *59* | *43* |

\* indicates program was initiated since 1970.

NOTE: On the first line of data for every program, raw values for each measure are reported; on the second line values are reported in standardized form, with mean = 50 and standard deviation = 10. "NA" indicates that the value for a measure is not available.

TABLE 3.1 Program Measures (Raw and Standardized Values) in Chemistry

| Prog No. | Survey Results (08) | (09) | (10) | (11) | University Library (12) | Research Support (13) | (14) | Published Articles (15) | (16) | Survey Ratings Standard Error (08) | (09) | (10) | (11) |
|---|---|---|---|---|---|---|---|---|---|---|---|---|---|
| 041. | 1.4 | 0.9 | 1.1 | 0.4 | NA | .46 | 636 | 24 | | .10 | .09 | .12 | .06 |
| | *38* | *37* | *53* | *40* | | *49* | *40* | *41* | *42* | | | | |
| 042. | 1.5 | 1.1 | 1.1 | 0.6 | NA | .40 | NA | 30 | | .10 | .10 | .09 | .07 |
| | *39* | *40* | *51* | *43* | | *46* | | *42* | *43* | | | | |
| 043. | 2.9 | 1.9 | 1.2 | 0.9 | 0.3 | .64 | 743 | 88 | | .08 | .05 | .07 | .07 |
| | *54* | *55* | *58* | *50* | *53* | *58* | *41* | *52* | *48* | | | | |
| 044. | 2.3 | 1.6 | 1.5 | 0.7 | -0.6 | .50 | NA | 33 | | .10 | .08 | .09 | .07 |
| | *47* | *49* | *70* | *47* | *43* | *51* | | *43* | *44* | | | | |
| 045. | 3.3 | 2.0 | 1.1 | 1.2 | -0.4 | .74 | 2478 | 60 | | .08 | .05 | .06 | .06 |
| | *57* | *57* | *51* | *58* | *45* | *63* | *56* | *47* | *47* | | | | |
| 046. | 3.2 | 2.0 | 1.2 | 1.1 | 0.8 | .34 | 1968 | 130 | | .09 | .04 | .07 | .07 |
| | *57* | *57* | *56* | *55* | *57* | *44* | *52* | *58* | *52* | | | | |
| 047. | 1.9 | 1.5 | 1.0 | 0.5 | -0.6 | .64 | 811 | 27 | | .11 | .08 | .06 | .06 |
| | *44* | *47* | *49* | *41* | *43* | *58* | *42* | *42* | *43* | | | | |
| 048. | 3.1 | 1.9 | 1.2 | 1.1 | NA | .50 | 3577 | 74 | | .08 | .05 | .07 | .06 |
| | *55* | *55* | *56* | *55* | | *51* | *65* | *49* | *47* | | | | |
| 049. | 3.0 | 1.9 | 1.0 | 1.1 | 0.4 | .38 | 1065 | 126 | | .07 | .05 | .08 | .07 |
| | *55* | *54* | *47* | *55* | *54* | *45* | *44* | *58* | *54* | | | | |
| 050. | 4.9 | 2.8 | 0.8 | 1.7 | 3.0 | .75 | 4283 | 114 | | .03 | .05 | .06 | .05 |
| | *74* | *71* | *40* | *71* | *80* | *63* | *71* | *56* | *64* | | | | |
| 051. | 2.2 | 1.5 | 1.1 | 0.7 | -0.1 | .41 | NA | 64 | | .08 | .07 | .07 | .06 |
| | *47* | *48* | *52* | *46* | *48* | *47* | | *48* | *51* | | | | |
| 052. | 2.6 | 1.8 | 1.3 | 0.9 | -0.9 | .39 | 2091 | 114 | | .07 | .06 | .09 | .07 |
| | *51* | *52* | *63* | *51* | *40* | *46* | *53* | *56* | *54* | | | | |
| 053. | 1.8 | 1.2 | 1.3 | 0.5 | -0.4 | .35 | 756 | 30 | | .08 | .09 | .08 | .06 |
| | *43* | *42* | *62* | *41* | *45* | *44* | *41* | *42* | *41* | | | | |
| 054. | 1.6 | 1.1 | 1.0 | 0.6 | NA | .33 | NA | 22 | | .10 | .10 | .07 | .06 |
| | *41* | *39* | *49* | *43* | | *43* | | *41* | *42* | | | | |
| 055. | 1.9 | 1.2 | 0.6 | 0.8 | NA | .53 | NA | 36 | | .09 | .09 | .08 | .06 |
| | *43* | *42* | *28* | *47* | | *53* | | *43* | *44* | | | | |
| 056. | 2.2 | 1.5 | 1.1 | 0.8 | NA | .60 | 705 | 43 | | .08 | .08 | .07 | .06 |
| | *47* | *48* | *52* | *49* | | *56* | *41* | *44* | *47* | | | | |
| 057. | 4.5 | 2.7 | 1.0 | 1.7 | 2.0 | .79 | 3963 | 222 | | .06 | .05 | .07 | .04 |
| | *69* | *69* | *49* | *71* | *69* | *65* | *68* | *74* | *78* | | | | |
| 058. | 3.7 | 2.2 | 1.3 | 1.4 | 0.9 | .61 | 2611 | 124 | | .06 | .06 | .08 | .05 |
| | *62* | *60* | *61* | *64* | *59* | *56* | *57* | *57* | *60* | | | | |
| 059. | 1.2 | 1.2 | 0.8 | 0.2 | NA | .00 | 617 | 3 | | .18 | .12 | .10 | .04 |
| | *36* | *41* | *38* | *34* | | *27* | *40* | *38* | *39* | | | | |
| 060. | 3.6 | 2.3 | 1.0 | 1.3 | -0.5 | .33 | 1288 | 154 | | .08 | .06 | .07 | .06 |
| | *60* | *63* | *47* | *61* | *44* | *43* | *46* | *62* | *59* | | | | |

NOTE: On the first line of data for every program, raw values for each measure are reported; on the second line values are reported in standardized form, with mean = 50 and standard deviation = 10. "NA" indicates that the value for a measure is not available. Since the scale used to compute measure (16) is entirely arbitrary, only values in standardized form are reported for this measure.

TABLE 3.1  Program Measures (Raw and Standardized Values) in Chemistry

| Prog No. | University - Department/Academic Unit | Program Size (01) | (02) | (03) | Characteristics of Program Graduates (04) | (05) | (06) | (07) |
|---|---|---|---|---|---|---|---|---|
| 061. | Iowa, University of-Iowa City | 23 | 50 | 63 | .19 | 5.4 | .69 | .37 |
|  | *Chemistry* | *50* | *50* | *48* | *46* | *57* | *44* | *53* |
| 062. | Johns Hopkins University | 18 | 55 | 50 | .31 | 5.3 | .84 | .53 |
|  | *Chemistry* | *45* | *51* | *46* | *57* | *59* | *58* | *64* |
| 063. | Kansas State University-Manhattan | 18 | 22 | 50 | .33 | 6.4 | .85 | .05 |
|  | *Chemistry* | *45* | *43* | *46* | *59* | *43* | *59* | *30* |
| 064. | Kansas, University of | 20 | 41 | 49 | .10 | 5.7 | .88 | .35 |
|  | *Chemistry* | *47* | *48* | *46* | *38* | *53* | *62* | *51* |
| 065. | Kansas, University of | 9 | 14 | 28 | .07 | 5.5 | .93 | .07 |
|  | *Pharmaceutical Chemistry* | *36* | *42* | *42* | *35* | *55* | *66* | *32* |
| 066. | Kent State University | 24 | 25 | 35 | .15 | 7.2 | .76 | .40 |
|  | *Chemistry* | *51* | *44* | *44* | *43* | *33* | *50* | *55* |
| 067. | Kentucky, University of | 24 | 14 | 37 | .12 | 6.0 | .81 | .38 |
|  | *Chemistry* | *51* | *42* | *44* | *40* | *49* | *55* | *53* |
| 068. | Louisiana State University-Baton Rouge | 25 | 41 | 71 | .25 | 6.0 | .63 | .21 |
|  | *Chemistry* | *52* | *48* | *49* | *51* | *49* | *37* | *41* |
| 069. | Louisville, University of | 17 | 15 | 27 | .35 | 7.9 | .69 | .19 |
|  | *Chemistry* | *44* | *42* | *42* | *61* | *23* | *43* | *40* |
| 070. | Loyola University of Chicago | 17 | 9 | NA | .20 | 8.3 | .90 | .40 |
|  | *Chemistry* | *44* | *40* |  | *47* | *18* | *64* | *55* |
| 071. | Maryland, University of-College Park | 54 | 96 | 180 | .17 | 5.7 | .74 | .16 |
|  | *Chemistry* | *82* | *60* | *67* | *45* | *53* | *49* | *38* |
| 072. | Massachusetts Institute of Technology | 31 | 163 | 175 | .32 | 4.4 | .70 | .22 |
|  | *Chemistry* | *58* | *76* | *66* | *58* | *71* | *45* | *42* |
| 073. | Massachusetts, University of-Amherst | 23 | 81 | 90 | .10 | 5.5 | .79 | .20 |
|  | *Chemistry* | *50* | *57* | *53* | *38* | *55* | *53* | *40* |
| 074. | Miami, University of-Florida | 16 | 17 | 18 | .24 | 5.4 | .43 | .14 |
|  | *Chemistry* | *43* | *42* | *41* | *51* | *57* | *18* | *37* |
| 075. | Michigan State University-East Lansing | 40 | 142 | 174 | .12 | 5.4 | .78 | .32 |
|  | *Chemistry* | *67* | *71* | *66* | *40* | *57* | *52* | *50* |
| 076. | Michigan, University of-Ann Arbor | 43 | 82 | 103 | .22 | 5.5 | .76 | .38 |
|  | *Chemistry* | *71* | *57* | *55* | *49* | *55* | *51* | *54* |
| 077. | Minnesota, University of | 47 | 80 | 153 | .26 | 5.7 | .81 | .36 |
|  | *Chemistry* | *75* | *57* | *63* | *52* | *53* | *55* | *52* |
| 078. | Missouri, University of-Columbia | 16 | 17 | 13 | .22 | 4.5 | .61 | .39 |
|  | *Chemistry* | *43* | *42* | *40* | *49* | *69* | *36* | *54* |
| 079. | Missouri, University of-Kansas City | 16 | 19 | 38 | .25 | 7.2 | NA | NA |
|  | *Chemistry* | *43* | *43* | *44* | *52* | *33* |  |  |
| 080. | Missouri, University of-Rolla | 23 | 11 | 18 | .07 | 6.3 | .57 | .29 |
|  | *Chemistry* | *50* | *41* | *41* | *35* | *44* | *32* | *47* |

* indicates program was initiated since 1970.

NOTE: On the first line of data for every program, raw values for each measure are reported; on the second line values are reported in standardized form, with mean = 50 and standard deviation = 10. "NA" indicates that the value for a measure is not available.

TABLE 3.1  Program Measures (Raw and Standardized Values) in Chemistry

| Prog No. | Survey Results | | | | University Library | Research Support | | Published Articles | | Survey Ratings Standard Error | | | |
|---|---|---|---|---|---|---|---|---|---|---|---|---|---|
| | (08) | (09) | (10) | (11) | (12) | (13) | (14) | (15) | (16) | (08) | (09) | (10) | (11) |
| 061. | 2.5 | 1.7 | 0.9 | 0.8 | 0.3 | .48 | 594 | 51 | | .10 | .07 | .07 | .07 |
| | 49 | 51 | 41 | 49 | 52 | 50 | 40 | 45 | 47 | | | | |
| 062. | 3.1 | 1.9 | 0.7 | 1.1 | -0.4 | .61 | 1138 | 54 | | .08 | .06 | .06 | .07 |
| | 56 | 56 | 35 | 55 | 45 | 56 | 45 | 46 | 47 | | | | |
| 063. | 2.4 | 1.6 | 1.3 | 0.8 | NA | .61 | NA | 49 | | .09 | .07 | .08 | .06 |
| | 49 | 49 | 61 | 50 | | 56 | | 45 | 47 | | | | |
| 064. | 2.6 | 1.8 | 0.9 | 0.8 | 0.1 | .40 | 957 | 45 | | .09 | .07 | .06 | .06 |
| | 50 | 53 | 40 | 47 | 50 | 46 | 43 | 45 | 46 | | | | |
| 065. | 1.9 | 1.6 | NA | 0.2 | 0.1 | NA | 957 | 42 | | .27 | .22 | NA | .04 |
| | 44 | 50 | | 33 | 50 | | 43 | 44 | 44 | | | | |
| 066. | 1.8 | 1.4 | 0.9 | 0.5 | -1.8 | .25 | NA | 45 | | .09 | .09 | .09 | .05 |
| | 43 | 45 | 41 | 40 | 31 | 39 | | 45 | 44 | | | | |
| 067. | 1.8 | 1.3 | 1.1 | 0.7 | -0.1 | .08 | NA | 68 | | .09 | .08 | .09 | .06 |
| | 43 | 43 | 51 | 45 | 49 | 31 | | 48 | 44 | | | | |
| 068. | 2.8 | 1.7 | 1.0 | 0.9 | -0.3 | .40 | 1161 | 89 | | .08 | .07 | .06 | .07 |
| | 53 | 52 | 46 | 51 | 46 | 46 | 45 | 52 | 54 | | | | |
| 069. | 1.4 | 0.9 | 0.9 | 0.3 | NA | .29 | NA | 23 | | .12 | .11 | .09 | .05 |
| | 38 | 37 | 44 | 37 | | 41 | | 41 | 42 | | | | |
| 070. | 1.4 | 1.1 | 0.9 | 0.5 | NA | .06 | NA | 16 | | .12 | .10 | .07 | .06 |
| | 38 | 39 | 41 | 41 | | 30 | | 40 | 41 | | | | |
| 071. | 2.9 | 1.9 | 1.1 | 1.0 | 0.2 | .52 | 1907 | 66 | | .08 | .06 | .08 | .06 |
| | 54 | 54 | 51 | 53 | 51 | 52 | 51 | 48 | 46 | | | | |
| 072. | 4.8 | 2.8 | 1.2 | 1.7 | -0.3 | .97 | 5324 | 235 | | .06 | .05 | .06 | .05 |
| | 73 | 71 | 54 | 71 | 46 | 73 | 80 | 76 | 76 | | | | |
| 073. | 2.6 | 1.9 | 1.1 | 0.9 | -0.7 | .61 | 2621 | 201 | | .08 | .06 | .05 | .06 |
| | 51 | 55 | 50 | 50 | 42 | 56 | 57 | 70 | 54 | | | | |
| 074. | 1.2 | 0.9 | 0.9 | 0.4 | NA | .19 | NA | 27 | | .12 | .10 | .11 | .06 |
| | 37 | 35 | 45 | 39 | | 36 | | 42 | 43 | | | | |
| 075. | 3.5 | 2.1 | 1.0 | 1.4 | 0.3 | .50 | 1491 | 109 | | .07 | .05 | .06 | .06 |
| | 60 | 59 | 45 | 62 | 53 | 51 | 47 | 55 | 57 | | | | |
| 076. | 3.3 | 2.1 | 0.8 | 1.4 | 1.8 | .56 | 1406 | 150 | | .08 | .05 | .07 | .06 |
| | 58 | 58 | 40 | 63 | 67 | 54 | 47 | 62 | 56 | | | | |
| 077. | 3.6 | 2.1 | 1.4 | 1.3 | 1.2 | .51 | 2162 | 159 | | .06 | .04 | .06 | .05 |
| | 61 | 59 | 68 | 61 | 61 | 52 | 53 | 63 | 61 | | | | |
| 078. | 1.8 | 1.1 | 0.9 | 0.7 | -0.2 | .25 | NA | 54 | | .09 | .08 | .09 | .06 |
| | 42 | 40 | 45 | 45 | 47 | 39 | | 46 | 44 | | | | |
| 079. | 1.6 | 1.1 | 1.0 | 0.4 | NA | .19 | NA | NA | | .11 | .10 | .08 | .06 |
| | 40 | 40 | 49 | 40 | | 36 | | | NA | | | | |
| 080. | 1.4 | 1.0 | 1.0 | 0.5 | NA | .26 | NA | 74 | | .11 | .10 | .07 | .06 |
| | 39 | 39 | 47 | 42 | | 40 | | 49 | 48 | | | | |

NOTE: On the first line of data for every program, raw values for each measure are reported; on the second line values are reported in standardized form, with mean = 50 and standard deviation = 10. "NA" indicates that the value for a measure is not available. Since the scale used to compute measure (16) is entirely arbitrary, only values in standardized form are reported for this measure.

TABLE 3.1  Program Measures (Raw and Standardized Values) in Chemistry

| Prog No. | University – Department/Academic Unit | Program Size | | | Characteristics of Program Graduates | | | |
|---|---|---|---|---|---|---|---|---|
| | | (01) | (02) | (03) | (04) | (05) | (06) | (07) |
| 081. | Montana State University–Bozeman | 21 | 13 | 20 | .44 | 6.6 | .80 | .53 |
| | *Chemistry* | *48* | *41* | *41* | *69* | *40* | *54* | *65* |
| 082. | Nebraska, University of–Lincoln | 25 | 61 | 94 | .21 | 6.0 | .85 | .52 |
| | *Chemistry* | *52* | *52* | *53* | *48* | *48* | *59* | *64* |
| 083. | New Hampshire, University of | 15 | 25 | 23 | .24 | 5.1 | .81 | .43 |
| | *Chemistry* | *42* | *44* | *42* | *51* | *61* | *55* | *57* |
| 084. | New Mexico, University of–Albuquerque | 17 | 32 | 56 | .16 | 5.9 | .84 | .39 |
| | *Chemistry* | *44* | *46* | *47* | *44* | *50* | *58* | *54* |
| 085. | New Orleans, University of | 19 | 23 | 33 | .14 | 6.0 | .82 | .14 |
| | *Chemistry* | *46* | *44* | *43* | *41* | *49* | *56* | *36* |
| 086. | New York University | 27 | 43 | 60 | .24 | 6.7 | .71 | .46 |
| | *Chemistry* | *54* | *48* | *48* | *51* | *39* | *46* | *60* |
| 087. | North Carolina State University–Raleigh | 26 | 21 | 65 | .07 | 6.5 | .79 | .28 |
| | *Chemistry* | *53* | *43* | *48* | *35* | *42* | *53* | *46* |
| 088. | North Carolina, University of–Chapel Hill | 30 | 102 | 143 | .11 | 4.9 | .85 | .43 |
| | *Chemistry* | *57* | *62* | *61* | *39* | *63* | *59* | *58* |
| 089. | North Dakota State University–Fargo | 14 | 18 | 16 | .11 | 5.2 | .89 | .06 |
| | *Chemistry/Polymers Coatings* | *41* | *42* | *40* | *39* | *59* | *63* | *30* |
| 090. | North Dakota, University of–Grand Forks | 15 | 8 | 20 | .36 | 5.5 | .80 | .10 |
| | *Chemistry* | *42* | *40* | *41* | *62* | *55* | *54* | *34* |
| 091. | North Texas State University–Denton | 18 | 18 | 37 | .05 | 5.5 | .43 | .38 |
| | *Chemistry* | *45* | *42* | *44* | *33* | *55* | *18* | *54* |
| 092. | Northeastern University | 13 | 25 | 45 | .19 | 6.1 | .89 | .15 |
| | *Chemistry* | *40* | *44* | *45* | *46* | *47* | *63* | *37* |
| 093. | Northern Illinois University–De Kalb | 21 | 18 | 15 | .16 | 6.7 | .89 | .39 |
| | *Chemistry* | *48* | *42* | *40* | *43* | *39* | *63* | *54* |
| 094. | Northwestern University | 28 | 126 | 124 | .37 | 5.2 | .82 | .33 |
| | *Chemistry* | *55* | *67* | *58* | *62* | *60* | *56* | *50* |
| 095. | Notre Dame, University of | 16 | 23 | 42 | .22 | 6.0 | .73 | .46 |
| | *Chemistry* | *43* | *44* | *45* | *49* | *49* | *47* | *60* |
| 096. | Ohio State University–Columbus | 39 | 142 | 220 | .11 | 5.4 | .76 | .29 |
| | *Chemistry* | *66* | *71* | *74* | *39* | *57* | *51* | *47* |
| 097. | Ohio University–Athens | 16 | 19 | 47 | .06 | 5.9 | .88 | .44 |
| | *Chemistry* | *43* | *43* | *46* | *35* | *50* | *61* | *58* |
| 098. | Oklahoma State University–Stillwater | 17 | 33 | 47 | .36 | 6.2 | .70 | .21 |
| | *Chemistry* | *44* | *46* | *46* | *62* | *46* | *44* | *42* |
| 099. | Oklahoma, University of–Norman | 26 | 27 | 28 | .23 | 6.3 | .81 | .36 |
| | *Chemistry* | *53* | *45* | *42* | *50* | *45* | *55* | *52* |
| 100. | Oregon State University–Corvallis | 25 | 44 | 75 | .32 | 5.6 | .83 | .36 |
| | *Chemistry* | *52* | *48* | *50* | *58* | *55* | *57* | *52* |

\* indicates program was initiated since 1970.

NOTE: On the first line of data for every program, raw values for each measure are reported; on the second line values are reported in standardized form, with mean = 50 and standard deviation = 10. "NA" indicates that the value for a measure is not available.

TABLE 3.1 Program Measures (Raw and Standardized Values) in Chemistry

| Prog No. | Survey Results (08) | (09) | (10) | (11) | University Library (12) | Research Support (13) | (14) | Published Articles (15) | (16) | Survey Ratings Standard Error (08) | (09) | (10) | (11) |
|---|---|---|---|---|---|---|---|---|---|---|---|---|---|
| 081. | 1.4 | 1.0 | 1.2 | 0.4 | NA | .24 | 947 | 45 | | .10 | .11 | .13 | .06 |
| | 39 | 38 | 56 | 38 | | 39 | 43 | 45 | 44 | | | | |
| 082. | 2.8 | 1.9 | 1.5 | 1.0 | -0.5 | .52 | 1745 | 72 | | .09 | .06 | .07 | .08 |
| | 53 | 54 | 71 | 53 | 44 | 52 | 50 | 49 | 49 | | | | |
| 083. | 1.7 | 1.2 | 0.9 | 0.5 | NA | .47 | NA | 22 | | .10 | .09 | .06 | .06 |
| | 42 | 42 | 41 | 41 | | 49 | | 41 | 41 | | | | |
| 084. | 1.7 | 1.4 | 1.1 | 0.4 | -1.0 | .29 | NA | 29 | | .09 | .09 | .10 | .06 |
| | 42 | 45 | 52 | 40 | 39 | 41 | | 42 | 42 | | | | |
| 085. | 2.1 | 1.4 | 0.9 | 0.8 | NA | .26 | NA | 43 | | .09 | .08 | .08 | .07 |
| | 46 | 45 | 41 | 48 | | 40 | | 44 | 43 | | | | |
| 086. | 2.4 | 1.5 | 0.8 | 0.8 | 0.5 | .37 | NA | 17 | | .08 | .09 | .07 | .06 |
| | 48 | 48 | 36 | 48 | 54 | 45 | | 40 | 41 | | | | |
| 087. | 2.1 | 1.5 | 1.2 | 0.7 | NA | .19 | NA | 85 | | .10 | .08 | .07 | .06 |
| | 45 | 48 | 58 | 45 | | 36 | | 51 | 44 | | | | |
| 088. | 3.7 | 2.3 | 1.5 | 1.3 | 1.0 | .60 | 1685 | 139 | | .08 | .06 | .06 | .06 |
| | 62 | 63 | 69 | 60 | 59 | 56 | 49 | 60 | 59 | | | | |
| 089. | 1.3 | 0.9 | 0.9 | 0.3 | NA | .00 | NA | 29 | | .11 | .12 | .10 | .05 |
| | 37 | 36 | 43 | 37 | | 27 | | 42 | 41 | | | | |
| 090. | 1.4 | 0.8 | 0.9 | 0.4 | NA | .33 | NA | 21 | | .13 | .11 | .11 | .06 |
| | 39 | 34 | 41 | 39 | | 43 | | 41 | 41 | | | | |
| 091. | 1.5 | 0.9 | 1.0 | 0.4 | NA | .33 | NA | 35 | | .12 | .10 | .10 | .06 |
| | 40 | 36 | 46 | 39 | | 43 | | 43 | 43 | | | | |
| 092. | 1.6 | 1.3 | 1.1 | 0.6 | NA | .62 | NA | 44 | | .11 | .11 | .08 | .06 |
| | 41 | 43 | 52 | 43 | | 57 | | 44 | 44 | | | | |
| 093. | 1.4 | 1.0 | 1.0 | 0.4 | NA | .24 | NA | 33 | | .10 | .10 | .09 | .06 |
| | 38 | 38 | 45 | 40 | | 39 | | 43 | 43 | | | | |
| 094. | 4.1 | 2.4 | 1.0 | 1.6 | 0.3 | .89 | 1851 | 160 | | .07 | .05 | .07 | .05 |
| | 65 | 64 | 45 | 67 | 52 | 70 | 51 | 63 | 64 | | | | |
| 095. | 2.7 | 1.6 | 0.9 | 1.1 | -1.3 | .50 | 2746 | 136 | | .09 | .06 | .07 | .07 |
| | 51 | 50 | 44 | 55 | 36 | 51 | 58 | 59 | 63 | | | | |
| 096. | 3.9 | 2.3 | 1.1 | 1.5 | 0.9 | .59 | 2039 | 231 | | .06 | .05 | .06 | .06 |
| | 63 | 62 | 52 | 66 | 58 | 55 | 52 | 75 | 75 | | | | |
| 097. | 1.6 | 1.0 | 0.7 | 0.5 | NA | .31 | NA | 14 | | .10 | .09 | .08 | .06 |
| | 40 | 39 | 35 | 41 | | 42 | | 39 | 41 | | | | |
| 098. | 1.9 | 1.4 | 1.2 | 0.6 | -1.9 | .53 | NA | 35 | | .10 | .10 | .08 | .06 |
| | 44 | 45 | 57 | 44 | 30 | 52 | | 43 | 43 | | | | |
| 099. | 2.3 | 1.5 | 1.2 | 0.6 | -0.6 | .46 | 869 | 62 | | .10 | .08 | .07 | .06 |
| | 47 | 48 | 55 | 44 | 44 | 49 | 42 | 47 | 45 | | | | |
| 100. | 2.8 | 1.9 | 1.1 | 0.9 | NA | .56 | 1065 | 61 | | .09 | .06 | .07 | .07 |
| | 52 | 54 | 52 | 51 | | 54 | 44 | 47 | 47 | | | | |

NOTE: On the first line of data for every program, raw values for each measure are reported; on the second line values are reported in standardized form, with mean = 50 and standard deviation = 10. "NA" indicates that the value for a measure is not available. Since the scale used to compute measure (16) is entirely arbitrary, only values in standardized form are reported for this measure.

TABLE 3.1  Program Measures (Raw and Standardized Values) in Chemistry

| Prog No. | University – Department/Academic Unit | Program Size | | | Characteristics of Program Graduates | | | |
|---|---|---|---|---|---|---|---|---|
| | | (01) | (02) | (03) | (04) | (05) | (06) | (07) |
| 101. | Oregon, University of-Eugene | 23 | 59 | 64 | .34 | 5.4 | .71 | .41 |
| | *Chemistry* | *50* | *52* | *48* | *60* | *57* | *45* | *56* |
| 102. | Pennsylvania State University | 25 | 99 | 174 | .26 | 5.8 | .71 | .24 |
| | *Chemistry* | *52* | *61* | *66* | *53* | *51* | *45* | *43* |
| 103. | Pennsylvania, University of | 25 | 69 | 137 | .33 | 6.0 | .73 | .35 |
| | *Chemistry* | *52* | *54* | *60* | *59* | *49* | *47* | *51* |
| 104. | Pittsburgh, University of | 23 | 83 | 136 | .24 | 5.1 | .76 | .42 |
| | *Chemistry* | *50* | *57* | *60* | *50* | *61* | *51* | *56* |
| 105. | Polytech Institute of New York | 18 | 34 | 36 | .47 | 7.3 | .52 | .23 |
| | *Chemistry* | *45* | *46* | *44* | *71* | *32* | *27* | *43* |
| 106. | Princeton University | 17 | 77 | 80 | .37 | 5.4 | .80 | .36 |
| | *Chemistry* | *44* | *56* | *51* | *62* | *57* | *54* | *52* |
| 107. | Purdue University-West Lafayette | 47 | 170 | 218 | .17 | 5.6 | .88 | .27 |
| | *Chemistry* | *75* | *78* | *73* | *44* | *55* | *62* | *46* |
| 108. | Rensselaer Polytechnic Institute | 30 | 39 | 74 | .29 | 6.3 | .71 | .25 |
| | *Chemistry* | *57* | *47* | *50* | *55* | *45* | *45* | *44* |
| 109. | Rhode Island, University of | 15 | 21 | 17 | .25 | 6.6 | .70 | .40 |
| | *Chemistry* | *42* | *43* | *41* | *52* | *41* | *44* | *55* |
| 110. | Rice University | 19 | 46 | 48 | .20 | 4.5 | .65 | .21 |
| | *Chemistry* | *46* | *49* | *46* | *48* | *68* | *39* | *41* |
| 111. | Rochester, University of | 22 | 47 | 85 | .15 | 5.4 | .78 | .33 |
| | *Chemistry* | *49* | *49* | *52* | *43* | *57* | *52* | *50* |
| 112. | Rutgers, The State University-New Brunswick | 42 | 48 | 191 | .23 | 6.7 | .75 | .31 |
| | *Chemistry* | *70* | *49* | *69* | *50* | *39* | *49* | *49* |
| 113. | Rutgers, The State University-Newark | 14 | 31 | 101 | .21 | 8.5 | .66 | .21 |
| | *Chemistry* | *41* | *45* | *54* | *48* | *15* | *40* | *41* |
| 114. | SUNY at Binghamton | 16 | 20 | 36 | .10 | 7.3 | .63 | .21 |
| | *Chemistry* | *43* | *43* | *44* | *38* | *31* | *38* | *42* |
| 115. | SUNY at Buffalo | 25 | 73 | 87 | .18 | 6.5 | .83 | .32 |
| | *Chemistry* | *52* | *55* | *52* | *46* | *43* | *57* | *50* |
| 116. | SUNY at Stony Brook | 32 | 40 | 56 | .15 | 6.3 | .69 | .41 |
| | *Chemistry* | *59* | *48* | *47* | *43* | *45* | *44* | *56* |
| 117. | SUNY-College of Environ Science & Forestry | 14 | 21 | 17 | .25 | 7.3 | .69 | .13 |
| | *Chemistry* | *41* | *43* | *41* | *52* | *31* | *43* | *35* |
| 118. | South Carolina, University of-Columbia | 25 | 55 | 109 | .13 | 4.6 | .75 | .46 |
| | *Chemistry* | *52* | *51* | *56* | *41* | *67* | *49* | *59* |
| 119. | South Florida, University of-Tampa | 17 | 19 | 20 | .06 | 5.5 | .77 | .59 |
| | *Chemistry* | *44* | *43* | *41* | *34* | *55* | *51* | *69* |
| 120. | Southern California, University of | 26 | 48 | 84 | .17 | 5.2 | .89 | .59 |
| | *Chemistry* | *53* | *49* | *52* | *44* | *60* | *63* | *69* |

\* indicates program was initiated since 1970.

NOTE: On the first line of data for every program, raw values for each measure are reported; on the second line values are reported in standardized form, with mean = 50 and standard deviation = 10. "NA" indicates that the value for a measure is not available.

TABLE 3.1  Program Measures (Raw and Standardized Values) in Chemistry

| Prog No. | Survey Results (08) | (09) | (10) | (11) | University Library (12) | Research Support (13) | (14) | Published Articles (15) | (16) | Survey Ratings Standard Error (08) | (09) | (10) | (11) |
|---|---|---|---|---|---|---|---|---|---|---|---|---|---|
| 101. | 3.4 | 2.0 | 1.2 | 1.0 | -0.9 | .74 | 1074 | 46 | | .08 | .06 | .06 | .07 |
| | 58 | 57 | 58 | 53 | 40 | 62 | 44 | 45 | 48 | | | | |
| 102. | 3.5 | 2.1 | 1.2 | 1.4 | 0.7 | .80 | 2303 | 169 | | .07 | .05 | .07 | .06 |
| | 60 | 60 | 54 | 64 | 56 | 65 | 54 | 65 | 59 | | | | |
| 103. | 3.4 | 2.1 | 1.2 | 1.2 | 0.7 | .76 | 2211 | 105 | | .07 | .05 | .06 | .06 |
| | 59 | 59 | 55 | 59 | 56 | 63 | 54 | 54 | 54 | | | | |
| 104. | 3.2 | 2.0 | 0.9 | 1.2 | 0.1 | .65 | 1728 | 157 | | .07 | .05 | .09 | .07 |
| | 57 | 57 | 41 | 58 | 50 | 58 | 49 | 63 | 62 | | | | |
| 105. | 2.0 | 1.5 | 0.7 | 0.5 | NA | .67 | 720 | 50 | | .10 | .10 | .10 | .06 |
| | 44 | 48 | 35 | 40 | | 59 | 41 | 45 | 43 | | | | |
| 106. | 4.0 | 2.4 | 0.8 | 1.6 | 0.9 | .77 | 2242 | 133 | | .07 | .05 | .06 | .05 |
| | 64 | 64 | 39 | 67 | 58 | 64 | 54 | 59 | 60 | | | | |
| 107. | 3.9 | 2.3 | 1.2 | 1.6 | -0.5 | .47 | 3438 | 292 | | .08 | .06 | .06 | .05 |
| | 64 | 62 | 56 | 69 | 44 | 50 | 64 | 85 | 79 | | | | |
| 108. | 2.5 | 1.8 | 1.1 | 0.7 | NA | .40 | 968 | 77 | | .08 | .07 | .07 | .06 |
| | 50 | 53 | 52 | 45 | | 46 | 43 | 50 | 45 | | | | |
| 109. | 1.7 | 1.2 | 1.0 | 0.4 | NA | .27 | NA | 33 | | .12 | .09 | .09 | .05 |
| | 41 | 42 | 47 | 38 | | 40 | | 43 | 41 | | | | |
| 110. | 3.3 | 2.1 | 1.1 | 1.1 | -1.4 | .63 | 1227 | 64 | | .07 | .05 | .08 | .06 |
| | 58 | 59 | 52 | 57 | 35 | 57 | 45 | 48 | 50 | | | | |
| 111. | 3.5 | 2.2 | 1.5 | 1.1 | -0.6 | .68 | 2536 | 50 | | .08 | .05 | .08 | .07 |
| | 59 | 60 | 68 | 56 | 43 | 60 | 56 | 45 | 48 | | | | |
| 112. | 2.7 | 1.8 | 1.1 | 0.9 | 0.8 | .31 | 991 | 109 | | .08 | .05 | .07 | .06 |
| | 52 | 53 | 50 | 51 | 57 | 42 | 43 | 55 | 55 | | | | |
| 113. | 1.7 | 1.2 | 0.9 | 0.5 | NA | .21 | 991 | 13 | | .12 | .09 | .09 | .06 |
| | 41 | 42 | 43 | 42 | | 37 | 43 | 39 | 41 | | | | |
| 114. | 2.0 | 1.3 | 1.0 | 0.7 | NA | .44 | NA | 38 | | .09 | .08 | .10 | .06 |
| | 44 | 45 | 47 | 47 | | 48 | | 43 | 43 | | | | |
| 115. | 3.0 | 1.9 | 1.1 | 1.1 | 0.3 | .64 | 1143 | 173 | | .08 | .05 | .08 | .06 |
| | 55 | 55 | 50 | 56 | 52 | 58 | 45 | 66 | 59 | | | | |
| 116. | 3.1 | 1.9 | 1.0 | 1.1 | -0.6 | .50 | 1801 | 91 | | .08 | .05 | .06 | .05 |
| | 56 | 54 | 49 | 57 | 43 | 51 | 50 | 52 | 54 | | | | |
| 117. | 2.0 | 1.2 | 0.6 | 0.4 | NA | .57 | 1404 | 44 | | .15 | .13 | .11 | .06 |
| | 44 | 42 | 30 | 38 | | 54 | 47 | 44 | 44 | | | | |
| 118. | 2.7 | 1.8 | 1.6 | 0.8 | -0.4 | .68 | 933 | 81 | | .08 | .06 | .07 | .07 |
| | 52 | 54 | 73 | 49 | 46 | 60 | 43 | 50 | 49 | | | | |
| 119. | 1.3 | 0.8 | 0.9 | 0.5 | NA | .29 | NA | 30 | | .10 | .10 | .10 | .06 |
| | 37 | 35 | 45 | 41 | | 41 | | 42 | 42 | | | | |
| 120. | 3.4 | 2.0 | 1.6 | 1.2 | 0.4 | .81 | 2181 | 144 | | .07 | .05 | .06 | .06 |
| | 59 | 56 | 73 | 59 | 53 | 66 | 53 | 61 | 62 | | | | |

NOTE: On the first line of data for every program, raw values for each measure are reported; on the second line values are reported in standardized form, with mean = 50 and standard deviation = 10. "NA" indicates that the value for a measure is not available. Since the scale used to compute measure (16) is entirely arbitrary, only values in standardized form are reported for this measure.

TABLE 3.1  Program Measures (Raw and Standardized Values) in Chemistry

| Prog No. | University - Department/Academic Unit | Program Size | | | Characteristics of Program Graduates | | | |
|---|---|---|---|---|---|---|---|---|
| | | (01) | (02) | (03) | (04) | (05) | (06) | (07) |
| 121. | Southern Illinois University-Carbondale | 16 | 23 | 22 | .08 | 6.1 | .88 | .63 |
| | *Chemistry and Biochemistry* | *43* | *44* | *41* | *36* | *47* | *61* | *71* |
| 122. | Southern Mississippi, Univ of-Hattiesburg | 14 | 17 | 14 | .36 | 5.3 | .86 | .50 |
| | *Chemistry* | *41* | *42* | *40* | *61* | *59* | *60* | *62* |
| 123. | Stanford University | 18 | 108 | 109 | .27 | 5.1 | .88 | .43 |
| | *Chemistry* | *45* | *63* | *56* | *54* | *60* | *61* | *57* |
| 124. | Syracuse University | 18 | 23 | 36 | .28 | 6.8 | .67 | .17 |
| | *Chemistry* | *45* | *44* | *44* | *54* | *39* | *41* | *38* |
| 125. | Temple University | 18 | 17 | 35 | .30 | 7.3 | .68 | .26 |
| | *Chemistry* | *45* | *42* | *44* | *56* | *32* | *42* | *45* |
| 126. | Tennessee, University of-Knoxville | 28 | 43 | 50 | .20 | 5.9 | .96 | .20 |
| | *Chemistry* | *55* | *48* | *46* | *47* | *50* | *69* | *41* |
| 127. | Texas A & M University | 46 | 107 | 149 | .33 | 5.4 | .66 | .31 |
| | *Chemistry* | *74* | *63* | *62* | *59* | *57* | *41* | *49* |
| 128. | Texas Tech University-Lubbock | 13 | 28 | 59 | .24 | 6.3 | .76 | .62 |
| | *Chemistry* | *40* | *45* | *47* | *51* | *44* | *50* | *71* |
| 129. | Texas, University of-Austin | 36 | 89 | 230 | .30 | 5.8 | .66 | .29 |
| | *Chemistry* | *63* | *59* | *75* | *56* | *51* | *41* | *47* |
| 130. | Tulane University | 12 | 11 | 11 | .25 | 5.8 | .67 | .25 |
| | *Chemistry* | *39* | *41* | *40* | *52* | *51* | *41* | *44* |
| 131. | Utah State University-Logan | 16 | 22 | 21 | .32 | 7.3 | .68 | .26 |
| | *Chemistry and Biochemistry* | *43* | *43* | *41* | *58* | *31* | *43* | *45* |
| 132. | Utah, University of-Salt Lake City | 27 | 57 | 90 | .27 | 5.8 | .75 | .39 |
| | *Chemistry* | *54* | *51* | *53* | *53* | *52* | *49* | *54* |
| 133. | Vanderbilt University | 19 | 31 | 37 | .37 | 5.0 | .71 | .37 |
| | *Chemistry* | *46* | *45* | *44* | *63* | *62* | *46* | *53* |
| 134. | Vermont, University of | 13 | 24 | 26 | .36 | 5.4 | .95 | .67 |
| | *Chemistry* | *40* | *44* | *42* | *62* | *57* | *69* | *74* |
| 135. | Virginia Polytechnic Institute & State Univ | 30 | 55 | 93 | .22 | 5.9 | .65 | .15 |
| | *Chemistry* | *57* | *51* | *53* | *49* | *50* | *40* | *37* |
| 136. | Virginia, University of | 26 | 36 | 73 | .39 | 5.2 | .71 | .43 |
| | *Chemistry* | *53* | *47* | *50* | *64* | *60* | *46* | *57* |
| 137. | Washington State University-Pullman | 29 | 17 | 40 | .15 | 5.8 | .78 | .33 |
| | *Chemistry* | *56* | *42* | *44* | *43* | *52* | *52* | *50* |
| 138. | Washington University-Saint Louis | 19 | 35 | 52 | .38 | 6.1 | .83 | .58 |
| | *Chemistry* | *46* | *46* | *46* | *63* | *48* | *56* | *68* |
| 139. | Washington, University of-Seattle | 34 | 74 | 56 | .28 | 6.2 | .66 | .30 |
| | *Chemistry* | *61* | *55* | *47* | *54* | *46* | *41* | *48* |
| 140. | Wayne State University | 30 | 64 | 52 | .16 | 6.5 | .74 | .26 |
| | *Chemistry* | *57* | *53* | *46* | *43* | *43* | *48* | *45* |

* indicates program was initiated since 1970.


NOTE: On the first line of data for every program, raw values for each measure are reported; on the second line values are reported in standardized form, with mean = 50 and standard deviation = 10. "NA" indicates that the value for a measure is not available.

TABLE 3.1  Program Measures (Raw and Standardized Values) in Chemistry

| Prog No. | Survey Results (08) | (09) | (10) | (11) | University Library (12) | Research Support (13) | (14) | Published Articles (15) | (16) | Survey Ratings Standard Error (08) | (09) | (10) | (11) |
|---|---|---|---|---|---|---|---|---|---|---|---|---|---|
| 121. | 1.7 | 1.1 | 0.7 | 0.6 | -0.2 | .13 | NA | 30 | | .10 | .09 | .07 | .06 |
| | 41 | 40 | 33 | 44 | 47 | 33 | | 42 | 43 | | | | |
| 122. | 0.8 | 0.6 | 1.0 | 0.2 | NA | .29 | NA | 23 | | .08 | .10 | .13 | .05 |
| | 32 | 30 | 47 | 35 | | 41 | | 41 | 41 | | | | |
| 123. | 4.5 | 2.6 | 0.7 | 1.6 | 2.0 | .83 | 4159 | 177 | | .06 | .05 | .07 | .05 |
| | 70 | 68 | 35 | 69 | 70 | 67 | 70 | 66 | 65 | | | | |
| 124. | 2.3 | 1.6 | 1.0 | 0.9 | -0.3 | .44 | 691 | 40 | | .07 | .08 | .07 | .06 |
| | 48 | 49 | 46 | 50 | 46 | 48 | 41 | 44 | 44 | | | | |
| 125. | 1.7 | 1.1 | 1.0 | 0.5 | -0.4 | .22 | NA | 27 | | .11 | .09 | .07 | .06 |
| | 41 | 40 | 47 | 41 | 45 | 38 | | 42 | 43 | | | | |
| 126. | 2.2 | 1.6 | 1.1 | 0.5 | -0.4 | .32 | 929 | 76 | | .10 | .09 | .09 | .06 |
| | 47 | 49 | 52 | 41 | 45 | 43 | 43 | 50 | 45 | | | | |
| 127. | 3.7 | 2.0 | 1.4 | 1.3 | -0.5 | .50 | 3715 | 246 | | .07 | .04 | .07 | .06 |
| | 61 | 58 | 67 | 61 | 45 | 51 | 66 | 78 | 74 | | | | |
| 128. | 2.2 | 1.4 | 1.3 | 0.8 | NA | .54 | 736 | 38 | | .10 | .08 | .08 | .07 |
| | 46 | 46 | 59 | 49 | | 53 | 41 | 43 | 44 | | | | |
| 129. | 3.8 | 2.2 | 1.4 | 1.3 | 1.6 | .58 | 3373 | 261 | | .07 | .05 | .06 | .06 |
| | 63 | 61 | 64 | 61 | 66 | 55 | 63 | 80 | 80 | | | | |
| 130. | 1.6 | 1.1 | 0.7 | 0.7 | -1.0 | .50 | NA | 1 | | .08 | .08 | .09 | .07 |
| | 40 | 40 | 35 | 46 | 39 | 51 | | 37 | 39 | | | | |
| 131. | 1.6 | 1.1 | 1.1 | 0.3 | NA | .44 | NA | 22 | | .13 | .12 | .09 | .05 |
| | 40 | 41 | 52 | 37 | | 48 | | 41 | 41 | | | | |
| 132. | 3.7 | 2.1 | 1.6 | 1.4 | -0.6 | .74 | 2290 | 140 | | .07 | .05 | .06 | .06 |
| | 62 | 59 | 73 | 63 | 43 | 62 | 54 | 60 | 63 | | | | |
| 133. | 2.6 | 1.6 | 1.0 | 0.8 | -0.7 | .42 | NA | 72 | | .09 | .08 | .06 | .07 |
| | 50 | 50 | 49 | 48 | 42 | 47 | | 49 | 46 | | | | |
| 134. | 2.2 | 1.4 | 1.1 | 0.7 | NA | .54 | NA | 38 | | .09 | .09 | .08 | .07 |
| | 46 | 46 | 51 | 45 | | 53 | | 43 | 44 | | | | |
| 135. | 2.4 | 1.7 | 1.2 | 0.8 | -0.0 | .30 | 1281 | 73 | | .11 | .07 | .08 | .07 |
| | 48 | 51 | 57 | 48 | 49 | 42 | 46 | 49 | 48 | | | | |
| 136. | 2.8 | 1.8 | 1.4 | 0.9 | 0.7 | .46 | 813 | 98 | | .08 | .06 | .08 | .06 |
| | 52 | 53 | 67 | 51 | 57 | 49 | 42 | 53 | 57 | | | | |
| 137. | 2.5 | 1.8 | 1.2 | 0.7 | -0.3 | .69 | 785 | 61 | | .09 | .08 | .06 | .06 |
| | 50 | 54 | 57 | 46 | 46 | 60 | 42 | 47 | 45 | | | | |
| 138. | 2.7 | 1.8 | 1.1 | 0.8 | -0.4 | .47 | 741 | 28 | | .08 | .06 | .08 | .07 |
| | 52 | 52 | 52 | 48 | 45 | 50 | 41 | 42 | 43 | | | | |
| 139. | 3.5 | 2.0 | 1.0 | 1.2 | 1.5 | .44 | 921 | 98 | | .07 | .06 | .06 | .05 |
| | 59 | 57 | 49 | 58 | 64 | 48 | 43 | 53 | 55 | | | | |
| 140. | 3.0 | 1.9 | 1.0 | 1.1 | -0.4 | .70 | 2022 | 79 | | .08 | .05 | .08 | .07 |
| | 54 | 54 | 45 | 56 | 46 | 61 | 52 | 50 | 52 | | | | |

NOTE: On the first line of data for every program, raw values for each measure are reported; on the second line values are reported in standardized form, with mean = 50 and standard deviation = 10. "NA" indicates that the value for a measure is not available. Since the scale used to compute measure (16) is entirely arbitrary, only values in standardized form are reported for this measure.

TABLE 3.1   Program Measures (Raw and Standardized Values) in Chemistry

| Prog No. | University - Department/Academic Unit | Program Size | | | Characteristics of Program Graduates | | | |
|---|---|---|---|---|---|---|---|---|
| | | (01) | (02) | (03) | (04) | (05) | (06) | (07) |
| 141. | Western Michigan University | 17 | 7 | 12 | NA | NA | NA | NA |
| | *Chemistry* | 44 | 40 | 40 | | | | |
| 142. | Wisconsin, University of-Madison | 39 | 200 | 226 | .15 | 5.6 | .90 | .40 |
| | *Chemistry* | 66 | 85 | 75 | 43 | 54 | 64 | 55 |
| 143. | Wisconsin, University of-Milwaukee | 17 | 19 | 35 | .05 | 6.0 | .62 | .24 |
| | *Chemistry** | 44 | 43 | 44 | 34 | 49 | 37 | 43 |
| 144. | Wyoming, University of | 16 | 18 | 21 | .26 | 5.5 | .58 | .26 |
| | *Chemistry* | 43 | 42 | 41 | 53 | 55 | 33 | 45 |
| 145. | Yale University | 24 | 72 | 118 | .25 | 4.9 | .83 | .56 |
| | *Chemistry* | 51 | 55 | 57 | 52 | 64 | 57 | 66 |

\* indicates program was initiated since 1970.


NOTE: On the first line of data for every program, raw values for each measure are reported; on the second line values are reported in standardized form, with mean = 50 and standard deviation = 10. "NA" indicates that the value for a measure is not available.

TABLE 3.1  Program Measures (Raw and Standardized Values) in Chemistry

| Prog No. | Survey Results (08) | (09) | (10) | (11) | University Library (12) | Research Support (13) | (14) | Published Articles (15) | (16) | Survey Ratings Standard Error (08) | (09) | (10) | (11) |
|---|---|---|---|---|---|---|---|---|---|---|---|---|---|
| 141. | 0.9 | 0.6 | 0.8 | 0.4 | NA | .00 | NA | 5 | | .10 | .09 | .07 | .06 |
| | _34_ | _31_ | _38_ | _39_ | | _27_ | | _38_ | _40_ | | | | |
| 142. | 4.4 | 2.6 | 1.0 | 1.7 | 1.6 | .77 | 5450 | 253 | | .06 | .05 | .06 | .05 |
| | _69_ | _67_ | _49_ | _70_ | _65_ | _64_ | _81_ | _79_ | _81_ | | | | |
| 143. | 1.9 | 1.4 | 1.3 | 0.6 | NA | .53 | 669 | NA | | .11 | .09 | .07 | .06 |
| | _43_ | _45_ | _61_ | _44_ | | _52_ | _41_ | | _NA_ | | | | |
| 144. | 1.4 | 0.8 | 0.9 | 0.6 | NA | .31 | NA | 17 | | .11 | .09 | .10 | .06 |
| | _38_ | _35_ | _42_ | _43_ | | _42_ | | _40_ | _41_ | | | | |
| 145. | 3.9 | 2.4 | 1.1 | 1.4 | 2.1 | .79 | 2107 | 98 | | .06 | .05 | .08 | .06 |
| | _64_ | _64_ | _52_ | _64_ | _70_ | _65_ | _53_ | _53_ | _57_ | | | | |

NOTE: On the first line of data for every program, raw values for each measure are reported; on the second line values are reported in standardized form, with mean = 50 and standard deviation = 10. "NA" indicates that the value for a measure is not available. Since the scale used to compute measure (16) is entirely arbitrary, only values in standardized form are reported for this measure.

TABLE 3.2  Summary Statistics Describing Each Program Measure--Chemistry

| Measure | Number of Programs Evaluated | Mean | Standard Deviation | DECILES | | | | | | | | |
|---|---|---|---|---|---|---|---|---|---|---|---|---|
| | | | | 1 | 2 | 3 | 4 | 5 | 6 | 7 | 8 | 9 |
| **Program Size** | | | | | | | | | | | | |
| 01 Raw Value | 145 | 23 | 10 | 13 | 16 | 17 | 18 | 21 | 23 | 26 | 29 | 36 |
| Std Value | 145 | 50 | 10 | 40 | 43 | 44 | 45 | 48 | 50 | 53 | 56 | 63 |
| 02 Raw Value | 144 | 51 | 43 | 14 | 18 | 23 | 28 | 36 | 46 | 59 | 73 | 108 |
| Std Value | 144 | 50 | 10 | 42 | 42 | 44 | 45 | 47 | 49 | 52 | 55 | 63 |
| 03 Raw Value | 144 | 75 | 62 | 17 | 23 | 35 | 47 | 53 | 72 | 87 | 109 | 173 |
| Std Value | 144 | 50 | 10 | 41 | 42 | 44 | 46 | 47 | 50 | 52 | 56 | 66 |
| **Program Graduates** | | | | | | | | | | | | |
| 04 Raw Value | 140 | .23 | .11 | .09 | .14 | .16 | .20 | .22 | .25 | .28 | .33 | .37 |
| Std Value | 140 | 50 | 10 | 37 | 42 | 44 | 47 | 49 | 52 | 55 | 59 | 63 |
| 05 Raw Value | 140 | 5.9 | .7 | 6.8 | 6.4 | 6.2 | 6.0 | 5.8 | 5.6 | 5.5 | 5.4 | 5.1 |
| Std Value | 140 | 50 | 10 | 38 | 43 | 46 | 49 | 51 | 54 | 56 | 57 | 61 |
| 06 Raw Value | 139 | .76 | .10 | .63 | .67 | .71 | .74 | .77 | .80 | .82 | .85 | .88 |
| Std Value | 139 | 50 | 10 | 37 | 41 | 45 | 48 | 51 | 54 | 56 | 59 | 62 |
| 07 Raw Value | 139 | .33 | .14 | .14 | .21 | .25 | .29 | .33 | .37 | .40 | .44 | .52 |
| Std Value | 139 | 50 | 10 | 36 | 41 | 44 | 47 | 50 | 53 | 55 | 58 | 64 |
| **Survey Results** | | | | | | | | | | | | |
| 08 Raw Value | 145 | 2.5 | 1.0 | 1.4 | 1.6 | 1.9 | 2.1 | 2.5 | 2.7 | 3.1 | 3.4 | 3.9 |
| Std Value | 145 | 50 | 10 | 38 | 41 | 44 | 46 | 50 | 52 | 56 | 59 | 64 |
| 09 Raw Value | 145 | 1.6 | .5 | .9 | 1.1 | 1.3 | 1.5 | 1.6 | 1.8 | 1.9 | 2.0 | 2.3 |
| Std Value | 145 | 50 | 10 | 36 | 40 | 44 | 48 | 49 | 53 | 55 | 57 | 62 |
| 10 Raw Value | 143 | 1.1 | .2 | .8 | .9 | .9 | 1.0 | 1.1 | 1.1 | 1.2 | 1.2 | 1.3 |
| Std Value | 143 | 50 | 10 | 38 | 42 | 42 | 47 | 52 | 52 | 57 | 57 | 61 |
| 11 Raw Value | 145 | .9 | .4 | .4 | .5 | .6 | .7 | .8 | .9 | 1.1 | 1.2 | 1.5 |
| Std Value | 145 | 50 | 10 | 39 | 41 | 44 | 46 | 49 | 51 | 56 | 58 | 66 |
| **University Library** | | | | | | | | | | | | |
| 12 Raw Value | 87 | .1 | 1.0 | -1.1 | -.6 | -.5 | -.4 | -.2 | .1 | .4 | .9 | 1.6 |
| Std Value | 87 | 50 | 10 | 38 | 43 | 44 | 45 | 47 | 50 | 53 | 58 | 65 |
| **Research Support** | | | | | | | | | | | | |
| 13 Raw Value | 141 | .48 | .21 | .22 | .29 | .34 | .43 | .50 | .54 | .61 | .67 | .75 |
| Std Value | 141 | 50 | 10 | 38 | 41 | 43 | 48 | 51 | 53 | 56 | 59 | 63 |
| 14 Raw Value | 95 | 1788 | 1186 | 728 | 813 | 952 | 1074 | 1273 | 1745 | 2099 | 2544 | 3575 |
| Std Value | 95 | 50 | 10 | 41 | 42 | 43 | 44 | 46 | 50 | 53 | 56 | 65 |
| **Publication Records** | | | | | | | | | | | | |
| 15 Raw Value | 143 | 78 | 61 | 19 | 29 | 38 | 48 | 61 | 74 | 93 | 120 | 161 |
| Std Value | 143 | 50 | 10 | 40 | 42 | 43 | 45 | 47 | 49 | 52 | 57 | 64 |
| 16 Std Value | 143 | 50 | 10 | 41 | 42 | 43 | 44 | 46 | 48 | 54 | 57 | 63 |

NOTE: Standardized values reported in the preceding table have been computed from exact values of the mean and standard deviation and not the rounded values reported here. Since the scale used to compute measure 16 is entirely arbitrary, only data in standardized form are reported for this measure.

TABLE 3.3  Intercorrelations Among Program Measures on 145 Programs in Chemistry

Measure

| | 01 | 02 | 03 | 04 | 05 | 06 | 07 | 08 | 09 | 10 | 11 | 12 | 13 | 14 | 15 | 16 |
|---|---|---|---|---|---|---|---|---|---|---|---|---|---|---|---|---|
| **Program Size** | | | | | | | | | | | | | | | | |
| 01 | | .68 | .75 | -.11 | .24 | .11 | .14 | .64 | .62 | .35 | .63 | .39 | .31 | .43 | .68 | .65 |
| 02 | | | .92 | .02 | .38 | .23 | .13 | .83 | .81 | .23 | .83 | .61 | .57 | .72 | .83 | .86 |
| 03 | | | | .01 | .32 | .24 | .15 | .81 | .79 | .35 | .80 | .61 | .51 | .66 | .81 | .84 |
| **Program Graduates** | | | | | | | | | | | | | | | | |
| 04 | | | | | .00 | -.07 | .05 | .11 | .08 | .01 | .07 | .10 | .20 | .18 | -.03 | .03 |
| 05 | | | | | | .17 | .26 | .47 | .46 | .23 | .46 | .19 | .39 | .35 | .38 | .41 |
| 06 | | | | | | | .32 | .28 | .30 | .10 | .25 | .28 | .20 | .31 | .21 | .22 |
| 07 | | | | | | | | .30 | .27 | .16 | .32 | .20 | .23 | .20 | .15 | .23 |
| **Survey Results** | | | | | | | | | | | | | | | | |
| 08 | | | | | | | | | .98 | .35 | .96 | .66 | .77 | .79 | .80 | .86 |
| 09 | | | | | | | | | | .36 | .92 | .65 | .77 | .74 | .78 | .82 |
| 10 | | | | | | | | | | | .31 | .03 | .32 | .14 | .33 | .33 |
| 11 | | | | | | | | | | | | .62 | .74 | .77 | .81 | .88 |
| **University Library** | | | | | | | | | | | | | | | | |
| 12 | | | | | | | | | | | | | .37 | .45 | .46 | .56 |
| **Research Support** | | | | | | | | | | | | | | | | |
| 13 | | | | | | | | | | | | | | .55 | .52 | .60 |
| 14 | | | | | | | | | | | | | | | .70 | .78 |
| **Publication Records** | | | | | | | | | | | | | | | | |
| 15 | | | | | | | | | | | | | | | | .95 |
| 16 | | | | | | | | | | | | | | | | |

NOTE: Since in computing correlation coefficients program data must be available for both of the measures
being correlated, the actual number of programs on which each coefficient is based varies.

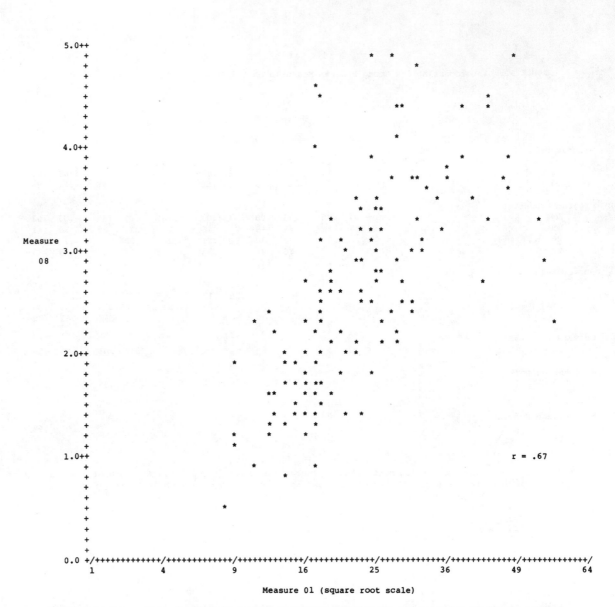

FIGURE 3.1 Mean rating of scholarly quality of faculty (measure 08) versus number of faculty members (measure 01)--145 programs in chemistry.

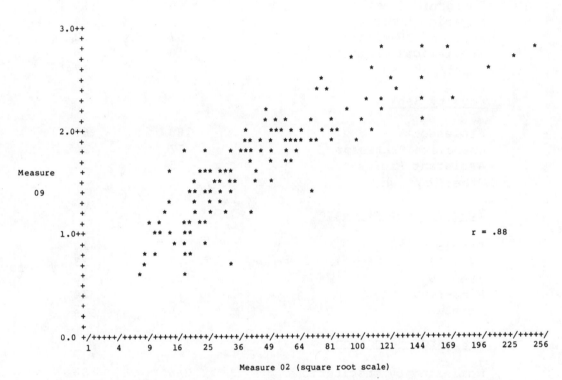

FIGURE 3.2  Mean rating of program effectiveness in educating research scholars/scientists (measure 09) versus number of graduates in last five years (measure 02)--144 programs in chemistry.

TABLE 3.4  Characteristics of Survey Participants in Chemistry

|  | Respondents | |
|---|---|---|
|  | N | % |
| **Field of Specialization** | | |
| Analytical Chemistry | 39 | 13 |
| Biochemistry | 10 | 3 |
| Inorganic Chemistry | 46 | 15 |
| Organic Chemistry | 101 | 34 |
| Physical Chemistry | 67 | 22 |
| Theoretical Chemistry | 18 | 6 |
| Other/Unknown | 20 | 7 |
| **Faculty Rank** | | |
| Professor | 188 | 63 |
| Associate Professor | 77 | 26 |
| Assistant Professor | 35 | 12 |
| Other/Unknown | 1 | 0 |
| **Year of Highest Degree** | | |
| Pre-1950 | 31 | 10 |
| 1950-59 | 70 | 23 |
| 1960-69 | 121 | 40 |
| Post-1969 | 73 | 24 |
| Unknown | 6 | 2 |
| **Evaluator Selection** | | |
| Nominated by Institution | 266 | 88 |
| Other | 35 | 12 |
| **Survey Form** | | |
| With Faculty Names | 271 | 90 |
| Without Names | 30 | 10 |
| **Total Evaluators** | 301 | 100 |

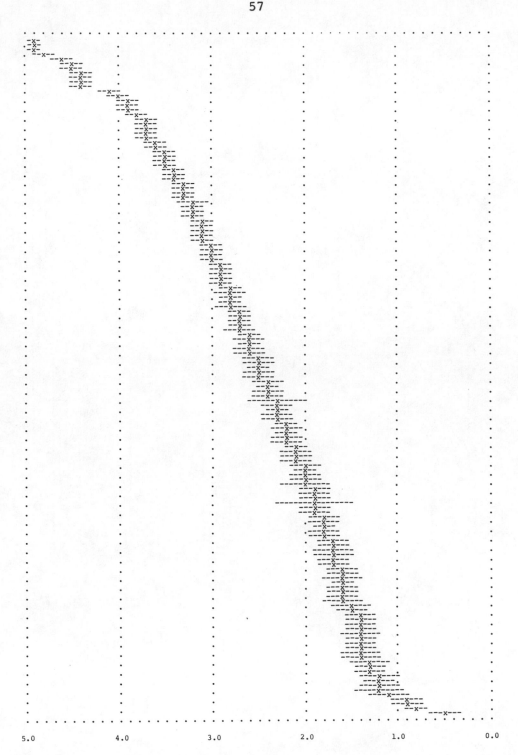

Mean Survey Rating (Measure 08)

FIGURE 3.3  Mean rating of scholarly quality of faculty in 145 programs in chemistry.

NOTE:  Programs are listed in sequence of mean rating, with the highest-rated program appearing at the
top of the page.  The broken lines (---) indicate a confidence interval of ±1.5 standard errors
around the reported mean (x) of each program.

# IV
# Computer Science Programs

In this chapter 58 research-doctorate programs in computer sciences are assessed. These programs, according to the information supplied by their universities, have accounted for 1,154 doctoral degrees awarded during the FY1976-80 period--approximately 86 percent of the aggregate number of computer science and computer engineering doctorates earned from U.S. universities in this five-year span.[1] Because computer sciences is a younger discipline than the other five mathematical and physical sciences covered in this assessment and because computer science programs may be found in a variety of settings within universities, the committee encountered some difficulty in identifying research-doctorate programs that have produced graduates in this discipline. On the average, 41 full-time and part-time students intending to earn doctorates were enrolled in a program in December 1980, with an average faculty size of 16 members.[2] Most of the 58 programs, listed in Table 4.1, are located in computer science or computer and information science departments. Approximately 20 percent are found in departments of electrical engineering. Fifteen programs were initiated since 1970, and no two programs are located in the same university. In addition to the 58 institutions represented in this discipline, another 7 were initially identified as meeting the criteria[3] for inclusion in the assessment:

> University of Chicago
> George Washington University
> Harvard University
> Northeastern University

---

[1]Data from the NRC's Survey of Earned Doctorates indicate that 889 research doctorates in computer sciences and another 458 research doctorates in computer engineering were awarded by U.S. universities between FY1976 and FY1980.
[2]See the reported means for measures 03 and 01 in Table 4.2.
[3]As mentioned in Chapter I, the primary criterion for inclusion was that a university had awarded at least 5 doctorates in computer sciences during the FY1976-78 period.

Purdue University
University of Southwest Louisiana
University of Texas, Health Science
Center--Dallas

The latter two institutions chose not to participate in the assessment in any discipline. Computer science programs at the other five institutions have not been included in the evaluations in this discipline, since in each case the study coordinator either indicated that the institution did not at that time have a research-doctorate program in computer sciences or failed to provide the information requested by the committee.

Before examining individual program results presented in Table 4.1, the reader is urged to refer to Chapter II, in which each of the 16 measures used in the assessment is discussed. Summary statistics describing every measure are given in Table 4.2. For nine of the measures, data are reported for at least 56 of the 58 computer science programs. For measures 04-07, which pertain to characteristics of the program graduates, data are presented for only approximately half of the programs; the other half had too few graduates on which to base statistics.[*] For measure 12, a composite index of the size of a university library, data are available for 49 programs; for measure 14, total university expenditures for research in this discipline, data are available for 44 programs. The programs not evaluated on measures 12 and 14 are typically smaller--in terms of faculty size and graduate student enrollment--than other computer science programs. Were data on these two measures available for all 58 programs, it is likely that their reported means would be appreciably lower (and that some of the correlations of these measures with others would be higher). With respect to measure 13, the fraction of faculty with research support from the National Science Foundation, the National Institutes of Health, and the Alcohol, Drug Abuse, and Mental Health Administration, data are reported for 45 programs that had at least 10 faculty members.

Intercorrelations among the 16 measures (Pearson product-moment coefficients) are given in Table 4.3. Of particular note are the high positive correlations of the measures of program size (01-03) with measures of publication records (15, 16) and reputational survey ratings (08 and 09). Figure 4.1 illustrates the relation between the mean rating of the scholarly quality of faculty (measure 08) and the number of faculty members (measure 01) for each of 57 programs in computer sciences. Figure 4.2 plots the mean rating of program effectiveness (measure 09) against the total number of FY1976-80 program graduates (measure 02). Although in both figures there is a significant positive correlation between program size and reputa-

---

[*]As mentioned in Chapter II, data for measures 04-07 are not reported if they are based on the survey responses of fewer than 10 FY1975-79 program graduates.

tional rating, it is quite apparent that some of the smaller programs received high mean ratings and that some of the larger programs received low mean ratings.

Table 4.4 describes the 108 faculty members who participated in the evaluation of computer science programs. These individuals constituted 62 percent of those asked to respond to the survey in this discipline and 12 percent of the faculty population in the 58 research-doctorate programs being evaluated.[5] A majority of the survey participants had earned their highest degree since 1970, and almost one-third held the rank of assistant professor. Two exceptions should be noted with regard to the survey evaluations in this discipline. Regretably, ratings are unavailable for the program in the Department of Computer and Communications Sciences at the University of Michigan since an entirely inaccurate list of its faculty members was included on the survey form. Also, it has been called to the attention of the committee that the faculty list (used in the survey) for the Department of Computer Science at Columbia University was missing the names of four members. The committee has decided to report the survey results for this program but cautions that the reputational ratings may have been influenced by the omission of these names.

To assist the reader in interpreting results of the survey evaluations, estimated standard errors have been computed for mean ratings of the scholarly quality of faculty in 57 computer science programs (and are given in Table 4.1). For each program the mean rating and an associated "confidence interval" of 1.5 standard errors are illustrated in Figure 4.3 (listed in order of highest to lowest mean rating). In comparing two programs, if their confidence intervals do not overlap, one may conclude that there is a significant difference in their mean ratings at a .05 level of significance.[6] From this figure it is also apparent that one should have somewhat more confidence in the accuracy of the mean ratings of higher-rated programs than lower-rated programs. This generalization results primarily from the fact that evaluators are not as likely to be familiar with the less prestigious programs, and consequently the mean ratings of these programs are usually based on fewer survey responses.

---

[5] See Table 2.3 in Chapter II.
[6] See pp. 29-31 for a discussion of the interpretation of mean ratings and associated confidence intervals.

TABLE 4.1  Program Measures (Raw and Standardized Values) in Computer Sciences

| Prog No. | University - Department/Academic Unit | Program Size | | | Characteristics of Program Graduates | | | |
|---|---|---|---|---|---|---|---|---|
| | | (01) | (02) | (03) | (04) | (05) | (06) | (07) |
| 001. | Arizona, University of-Tucson | 7 | 4 | 13 | NA | NA | NA | NA |
| | *Computer Sciences* | 40 | 42 | 43 | | | | |
| 002. | Brown University | 8 | 4 | 21 | NA | NA | NA | NA |
| | *Computer Science\** | 41 | 42 | 45 | | | | |
| 003. | California Institute of Technology | 5 | 5 | 22 | NA | NA | NA | NA |
| | *Computer Science\** | 37 | 42 | 45 | | | | |
| 004. | California, University of-Berkeley | 30 | 43 | 53 | .14 | 6.3 | .79 | .24 |
| | *Electrical Engineering & Computer Sciences* | 66 | 62 | 53 | 47 | 51 | 50 | 41 |
| 005. | California, University of-Irvine | 12 | 15 | 46 | .09 | 5.3 | .64 | .27 |
| | *Information and Computer Science* | 45 | 47 | 51 | 43 | 60 | 32 | 43 |
| 006. | California, University of-Los Angeles | 36 | 55 | 103 | .00 | 7.9 | .65 | .25 |
| | *Computer Science* | 73 | 68 | 65 | 35 | 37 | 34 | 42 |
| 007. | California, University of-San Diego | 9 | 7 | 17 | NA | NA | NA | NA |
| | *Electrical Engineering & Computer Science* | 42 | 43 | 44 | | | | |
| 008. | California, University of-Santa Barbara | 8 | 10 | 9 | NA | NA | NA | NA |
| | *Electrical and Computer Engineering* | 41 | 45 | 42 | | | | |
| 009. | Carnegie-Mellon University | 31 | 41 | 83 | .22 | 6.8 | .81 | .50 |
| | *Computer Science* | 68 | 61 | 60 | 54 | 47 | 51 | 57 |
| 010. | Case Western Reserve University | 7 | 9 | 10 | NA | NA | NA | NA |
| | *Computer Engin/Computing & Information Sci* | 40 | 44 | 42 | | | | |
| 011. | Columbia University | 11 | 2 | 17 | NA | NA | NA | NA |
| | *Computer Science\** | 44 | 41 | 44 | | | | |
| 012. | Connecticut, University of-Storrs | 11 | 8 | 12 | NA | NA | NA | NA |
| | *Electrical Engineering & Computer Science* | 44 | 44 | 43 | | | | |
| 013. | Cornell University-Ithaca | 14 | 34 | 48 | .27 | 5.5 | .97 | .63 |
| | *Computer Science* | 48 | 57 | 52 | 58 | 59 | 69 | 65 |
| 014. | Duke University | 13 | 10 | 29 | NA | NA | NA | NA |
| | *Computer Science\** | 47 | 45 | 47 | | | | |
| 015. | Georgia Institute of Technology | 11 | 8 | 30 | .10 | NA | .80 | .10 |
| | *Information and Computer Science* | 44 | 44 | 47 | 43 | | 50 | 33 |
| 016. | Illinois, University-Urbana/Champaign | 30 | 112 | 125 | .13 | 6.1 | .85 | .28 |
| | *Computer Science* | 66 | 97 | 71 | 46 | 53 | 56 | 44 |
| 017. | Indiana University-Bloomington | 15 | NA | 16 | NA | NA | NA | NA |
| | *Computer Science\** | 49 | | 44 | | | | |
| 018. | Iowa State University-Ames | 15 | 18 | 17 | .07 | 5.8 | .73 | .33 |
| | *Computer Science* | 49 | 49 | 44 | 41 | 56 | 43 | 47 |
| 019. | Iowa, University of-Iowa City | 12 | 11 | 20 | .36 | 5.3 | .80 | .70 |
| | *Computer Science* | 45 | 45 | 45 | 67 | 61 | 50 | 69 |
| 020. | Kansas State University-Manhattan | 8 | 7 | 12 | NA | NA | NA | NA |
| | *Computer Science\** | 41 | 43 | 43 | | | | |

\* indicates program was initiated since 1970.

NOTE: On the first line of data for every program, raw values for each measure are reported; on the second line values are reported in standardized form, with mean = 50 and standard deviation = 10. "NA" indicates that the value for a measure is not available.

TABLE 4.1  Program Measures (Raw and Standardized Values) in Computer Sciences

| Prog No. | Survey Results (08) | (09) | (10) | (11) | University Library (12) | Research Support (13) | (14) | Published Articles (15) | (16) | Survey Ratings Standard Error (08) | (09) | (10) | (11) |
|---|---|---|---|---|---|---|---|---|---|---|---|---|---|
| 001. | 2.4 | 1.4 | 1.3 | 1.0 | 0.9 | NA | 323 | 15 | | .11 | .10 | .11 | .07 |
| | 48 | 48 | 57 | 54 | 55 | | 44 | 44 | 46 | | | | |
| 002. | 2.9 | 1.7 | 1.5 | 1.2 | -1.1 | NA | 417 | 25 | | .10 | .09 | .08 | .06 |
| | 54 | 53 | 63 | 57 | 35 | | 45 | 47 | 47 | | | | |
| 003. | 2.5 | 1.5 | 0.8 | 0.9 | NA | NA | 871 | 26 | | .17 | .11 | .12 | .08 |
| | 50 | 50 | 40 | 50 | | | 48 | 48 | 47 | | | | |
| 004. | 4.5 | 2.6 | 1.3 | 1.6 | 2.2 | .60 | NA | 134 | | .08 | .06 | .07 | .05 |
| | 70 | 69 | 57 | 68 | 69 | 63 | | 83 | 82 | | | | |
| 005. | 2.4 | 1.4 | 0.9 | 0.8 | NA | .17 | 98 | 12 | | .09 | .09 | .13 | .07 |
| | 49 | 47 | 42 | 50 | | 40 | 43 | 43 | 44 | | | | |
| 006. | 3.8 | 2.2 | 1.3 | 1.3 | 2.0 | .61 | 126 | 77 | | .08 | .05 | .08 | .05 |
| | 63 | 62 | 57 | 60 | 66 | 63 | 43 | 64 | 61 | | | | |
| 007. | 2.6 | 1.2 | 1.1 | 0.8 | -0.0 | NA | 376 | 21 | | .13 | .12 | .11 | .07 |
| | 51 | 45 | 49 | 49 | 45 | | 45 | 46 | 48 | | | | |
| 008. | 2.1 | 1.2 | 1.1 | 0.8 | -0.1 | NA | 305 | 27 | | .11 | .11 | .12 | .07 |
| | 46 | 43 | 50 | 48 | 45 | | 44 | 48 | 49 | | | | |
| 009. | 4.8 | 2.7 | 1.1 | 1.8 | NA | .26 | 3649 | 53 | | .05 | .05 | .07 | .05 |
| | 73 | 71 | 50 | 72 | | 44 | 67 | 56 | 61 | | | | |
| 010. | 1.3 | 0.8 | 0.4 | 0.4 | -1.3 | NA | NA | 24 | | .11 | .13 | .10 | .06 |
| | 37 | 36 | 24 | 40 | 32 | | | 47 | 44 | | | | |
| 011. | 2.5 | 1.2 | 1.6 | 0.8 | 1.7 | .36 | NA | 23 | | .12 | .11 | .08 | .07 |
| | 50 | 45 | 67 | 49 | 64 | 50 | | 47 | 46 | | | | |
| 012. | 1.7 | 1.0 | 1.2 | 0.5 | -0.5 | .36 | 435 | 12 | | .13 | .12 | .09 | .06 |
| | 41 | 41 | 54 | 43 | 41 | 50 | 45 | 43 | 43 | | | | |
| 013. | 4.3 | 2.5 | 1.1 | 1.6 | 1.6 | .57 | 987 | 52 | | .07 | .06 | .07 | .05 |
| | 68 | 68 | 49 | 67 | 62 | 61 | 49 | 56 | 54 | | | | |
| 014. | 2.4 | 1.5 | 1.3 | 0.7 | 0.3 | .46 | 218 | 12 | | .10 | .10 | .10 | .07 |
| | 49 | 50 | 56 | 46 | 49 | 55 | 44 | 43 | 43 | | | | |
| 015. | 2.7 | 1.6 | 1.8 | 0.8 | NA | .27 | 4056 | 30 | | .10 | .08 | .06 | .07 |
| | 52 | 51 | 75 | 50 | | 45 | 69 | 49 | 48 | | | | |
| 016. | 3.8 | 2.3 | 1.0 | 1.4 | 2.0 | .53 | 3357 | 155 | | .09 | .07 | .07 | .06 |
| | 63 | 63 | 46 | 62 | 66 | 59 | 65 | 89 | 83 | | | | |
| 017. | 2.3 | 1.3 | 1.6 | 0.8 | 0.9 | .53 | 67 | 20 | | .11 | .12 | .08 | .07 |
| | 48 | 46 | 66 | 49 | 55 | 59 | 43 | 46 | 46 | | | | |
| 018. | 1.7 | 1.2 | 1.2 | 0.4 | -0.5 | .27 | NA | 14 | | .12 | .13 | .09 | .06 |
| | 42 | 45 | 51 | 40 | 40 | 45 | | 44 | 43 | | | | |
| 019. | 1.7 | 1.1 | 1.0 | 0.5 | 0.3 | .25 | 355 | 10 | | .12 | .10 | .11 | .07 |
| | 41 | 43 | 44 | 41 | 49 | 44 | 45 | 42 | 42 | | | | |
| 020. | 0.9 | 0.6 | 0.9 | 0.2 | NA | NA | 153 | 13 | | .13 | .11 | .10 | .04 |
| | 33 | 33 | 41 | 34 | | | 43 | 43 | 42 | | | | |

NOTE: On the first line of data for every program, raw values for each measure are reported; on the second line values are reported in standardized form, with mean = 50 and standard deviation = 10. "NA" indicates that the value for a measure is not available. Since the scale used to compute measure (16) is entirely arbitrary, only values in standardized form are reported for this measure.

TABLE 4.1  Program Measures (Raw and Standardized Values) in Computer Sciences

| Prog No. | University – Department/Academic Unit | Program Size | | | Characteristics of Program Graduates | | | |
|---|---|---|---|---|---|---|---|---|
| | | (01) | (02) | (03) | (04) | (05) | (06) | (07) |
| 021. | Kansas, University of<br>*Computer Science\** | 16<br>*50* | 6<br>*43* | 6<br>*41* | NA | NA | NA | NA |
| 022. | Maryland, University of-College Park<br>*Computer Science* | 28<br>*64* | 35<br>*58* | 50<br>*52* | .08<br>*42* | 8.0<br>*37* | .76<br>*46* | .24<br>*41* |
| 023. | Massachusetts Institute of Technology<br>*Electrical Engineering and Computer Science* | 34<br>*71* | 62<br>*71* | 135<br>*73* | .23<br>*55* | 6.4<br>*51* | .77<br>*47* | .39<br>*50* |
| 024. | Massachusetts, University of-Amherst<br>*Computer and Information Sciences\** | 16<br>*50* | 16<br>*48* | 70<br>*57* | .00<br>*35* | 5.4<br>*59* | .82<br>*52* | .36<br>*49* |
| 025. | Michigan State University-East Lansing<br>*Computer Science* | 14<br>*48* | 8<br>*44* | 20<br>*45* | NA | NA | NA | NA |
| 026. | Michigan, University of-Ann Arbor<br>*Computer and Communication Sciences* | 10<br>*43* | 18<br>*49* | 38<br>*49* | .30<br>*61* | 6.4<br>*50* | .79<br>*49* | .63<br>*65* |
| 027. | Minnesota, University of<br>*Computer Science* | 21<br>*56* | 13<br>*46* | 28<br>*47* | .25<br>*57* | 6.5<br>*50* | .81<br>*52* | .50<br>*57* |
| 028. | Missouri, University of-Rolla<br>*Computer Science\** | 11<br>*44* | 9<br>*44* | 11<br>*42* | NA | NA | NA | NA |
| 029. | New York University<br>*Computer Science* | 13<br>*47* | 19<br>*49* | 63<br>*55* | .21<br>*53* | 8.0<br>*37* | .80<br>*50* | .20<br>*39* |
| 030. | North Carolina, University of-Chapel Hill<br>*Computer Science* | 8<br>*41* | 17<br>*48* | 23<br>*45* | .07<br>*41* | 7.5<br>*41* | .88<br>*59* | .19<br>*38* |
| 031. | Northwestern University<br>*Electrical Engineering & Computer Sciences* | 24<br>*59* | 41<br>*61* | 16<br>*44* | .08<br>*41* | 6.5<br>*50* | .92<br>*63* | .33<br>*47* |
| 032. | Ohio State University-Columbus<br>*Computer and Information Science* | 21<br>*56* | 43<br>*62* | 90<br>*62* | .09<br>*43* | 6.3<br>*51* | .76<br>*46* | .48<br>*56* |
| 033. | Oklahoma, University of-Norman<br>*Electrical Engineering and Computer Sci* | 15<br>*49* | 1<br>*40* | 26<br>*46* | NA | NA | NA | NA |
| 034. | Pennsylvania State University<br>*Computer Sciences* | 16<br>*50* | 10<br>*45* | 32<br>*48* | .15<br>*48* | 6.3<br>*52* | .77<br>*47* | .39<br>*50* |
| 035. | Pennsylvania, University of<br>*Computer and Information Science* | 29<br>*65* | 25<br>*52* | 54<br>*53* | .14<br>*47* | 6.4<br>*51* | .79<br>*50* | .41<br>*52* |
| 036. | Pittsburgh, University of<br>*Computer Science\** | 12<br>*45* | 10<br>*45* | 20<br>*45* | NA | NA | NA | NA |
| 037. | Polytech Institute of New York<br>*Electrical Engineering and Computer Science* | 8<br>*41* | 6<br>*43* | 33<br>*48* | NA | NA | NA | NA |
| 038. | Princeton University<br>*Electrical Engineering and Computer Science* | 9<br>*42* | 21<br>*50* | 26<br>*46* | .23<br>*55* | 4.3<br>*69* | 1.00<br>*72* | .60<br>*63* |
| 039. | Rice University<br>*Mathematical Sciences* | 22<br>*57* | 19<br>*49* | 23<br>*45* | .29<br>*60* | 4.9<br>*63* | .52<br>*20* | .14<br>*35* |
| 040. | Rochester, University of<br>*Computer Science\** | 11<br>*44* | 5<br>*42* | 36<br>*49* | NA | NA | NA | NA |

\* indicates program was initiated since 1970.

NOTE: On the first line of data for every program, raw values for each measure are reported; on the second line values are reported in standardized form, with mean = 50 and standard deviation = 10. "NA" indicates that the value for a measure is not available.

TABLE 4.1  Program Measures (Raw and Standardized Values) in Computer Sciences

| Prog No. | Survey Results (08) | (09) | (10) | (11) | University Library (12) | Research Support (13) | (14) | Published Articles (15) | (16) | Survey Ratings Standard Error (08) | (09) | (10) | (11) |
|---|---|---|---|---|---|---|---|---|---|---|---|---|---|
| 021. | 1.9 | 1.1 | 0.8 | 0.7 | 0.1 | .06 | 51 | 29 | | .10 | .09 | .09 | .07 |
| | 44 | 43 | 38 | 47 | 47 | 34 | 43 | 48 | 48 | | | | |
| 022. | 3.1 | 1.9 | 1.2 | 1.3 | 0.2 | .21 | 3942 | 58 | | .08 | .06 | .07 | .06 |
| | 56 | 58 | 54 | 61 | 47 | 42 | 68 | 58 | 57 | | | | |
| 023. | 4.9 | 2.8 | 1.1 | 1.8 | -0.3 | .32 | 6646 | 108 | | .03 | .05 | .05 | .04 |
| | 74 | 73 | 50 | 72 | 42 | 48 | 86 | 74 | 79 | | | | |
| 024. | 2.8 | 1.7 | 1.3 | 1.0 | -0.7 | .56 | 790 | 21 | | .09 | .09 | .09 | .07 |
| | 53 | 52 | 55 | 53 | 38 | 61 | 47 | 46 | 45 | | | | |
| 025. | 1.5 | 0.9 | 0.9 | 0.3 | 0.3 | .36 | 295 | 12 | | .13 | .12 | .08 | .05 |
| | 39 | 39 | 42 | 36 | 49 | 50 | 44 | 43 | 43 | | | | |
| 026. | NA | NA | NA | NA | 1.8 | .60 | 1710 | 41 | | NA | NA | NA | NA |
| | | | | | 64 | 63 | 54 | 52 | 53 | | | | |
| 027. | 2.7 | 1.6 | 1.2 | 0.8 | 1.2 | .38 | 126 | 56 | | .11 | .08 | .09 | .07 |
| | 52 | 52 | 52 | 49 | 58 | 51 | 43 | 57 | 60 | | | | |
| 028. | 1.2 | 0.8 | 1.1 | 0.3 | NA | .18 | 60 | 13 | | .14 | .11 | .09 | .05 |
| | 37 | 37 | 50 | 37 | | 40 | 43 | 43 | 43 | | | | |
| 029. | 2.8 | 1.7 | 1.0 | 0.9 | 0.5 | .39 | 1192 | 9 | | .11 | .07 | .06 | .07 |
| | 53 | 53 | 45 | 51 | 51 | 51 | 50 | 42 | 45 | | | | |
| 030. | 2.7 | 1.7 | 1.1 | 1.0 | 1.0 | NA | 461 | 24 | | .10 | .07 | .08 | .06 |
| | 52 | 54 | 49 | 53 | 56 | | 45 | 47 | 47 | | | | |
| 031. | 2.4 | 1.5 | 1.1 | 0.6 | 0.3 | .33 | NA | 33 | | .10 | .10 | .09 | .06 |
| | 49 | 50 | 50 | 45 | 49 | 48 | | 50 | 53 | | | | |
| 032. | 2.4 | 1.6 | 1.1 | 0.9 | 0.9 | .19 | 567 | 62 | | .10 | .07 | .07 | .07 |
| | 49 | 51 | 49 | 51 | 55 | 41 | 46 | 59 | 58 | | | | |
| 033. | 0.8 | 0.3 | NA | 0.1 | -0.6 | .00 | NA | 9 | | .14 | .09 | NA | .03 |
| | 32 | 28 | | 33 | 40 | 31 | | 42 | 42 | | | | |
| 034. | 2.1 | 1.3 | 0.4 | 0.9 | 0.7 | .63 | 3707 | 37 | | .11 | .09 | .09 | .08 |
| | 46 | 46 | 24 | 51 | 53 | 64 | 67 | 51 | 52 | | | | |
| 035. | 2.7 | 1.8 | 1.1 | 1.0 | 0.7 | .24 | 1586 | 42 | | .12 | .08 | .09 | .08 |
| | 52 | 54 | 49 | 53 | 53 | 43 | 53 | 53 | 49 | | | | |
| 036. | 1.9 | 1.2 | 0.9 | 0.7 | 0.1 | .33 | 839 | 32 | | .12 | .10 | .09 | .06 |
| | 44 | 45 | 41 | 46 | 46 | 48 | 48 | 49 | 47 | | | | |
| 037. | 1.2 | 0.9 | 0.7 | 0.2 | NA | NA | NA | 11 | | .16 | .14 | .15 | .04 |
| | 37 | 38 | 35 | 36 | | | | 43 | 43 | | | | |
| 038. | 3.0 | 1.9 | 0.7 | 1.1 | 0.9 | NA | 422 | 21 | | .10 | .08 | .10 | .07 |
| | 55 | 57 | 35 | 57 | 55 | | 45 | 46 | 50 | | | | |
| 039. | 2.4 | 1.6 | 1.1 | 0.6 | -1.4 | .41 | NA | 13 | | .12 | .10 | .11 | .06 |
| | 49 | 51 | 50 | 43 | 31 | 52 | | 43 | 44 | | | | |
| 040. | 2.7 | 1.7 | 1.7 | 1.1 | -0.6 | .36 | 365 | 10 | | .09 | .09 | .07 | .07 |
| | 52 | 54 | 70 | 55 | 39 | 50 | 45 | 42 | 44 | | | | |

NOTE: On the first line of data for every program, raw values for each measure are reported; on the second line values are reported in standardized form, with mean = 50 and standard deviation = 10. "NA" indicates that the value for a measure is not available. Since the scale used to compute measure (16) is entirely arbitrary, only values in standardized form are reported for this measure.

TABLE 4.1 Program Measures (Raw and Standardized Values) in Computer Sciences

| Prog No. | University - Department/Academic Unit | Program Size | | | Characteristics of Program Graduates | | | |
|---|---|---|---|---|---|---|---|---|
| | | (01) | (02) | (03) | (04) | (05) | (06) | (07) |
| 041. | Rutgers, The State University-New Brunswick *Computer Science** | 29 65 | 6 43 | 243 99 | NA | NA | NA | NA |
| 042. | SUNY at Buffalo *Computer Science* | 11 44 | 8 44 | 62 55 | NA | NA | NA | NA |
| 043. | SUNY at Stony Brook *Computer Science** | 16 50 | 21 50 | 22 45 | .05 39 | 5.5 58 | .85 56 | .65 66 |
| 044. | Southern California, University of *Computer Science* | 10 43 | 18 49 | 7 42 | NA | NA | NA | NA |
| 045. | Southern Methodist University *Computer Science and Engineering* | 10 43 | 11 45 | 25 46 | NA | NA | NA | NA |
| 046. | Stanford University *Computer Science* | 21 56 | 74 77 | 95 63 | .53 81 | 6.8 47 | .87 58 | .51 58 |
| 047. | Stevens Institute of Technology *Electrical Engineering/ Computer Science* | 10 43 | 10 45 | 12 43 | NA | 9.0 28 | NA | NA |
| 048. | Syracuse University *Computer Sciences* | 44 83 | 22 51 | 68 57 | .17 49 | 6.0 54 | .83 54 | .17 37 |
| 049. | Texas A & M University *Industrial Engineering* | 10 43 | 31 55 | 22 45 | .08 41 | 9.4 25 | .74 44 | .23 41 |
| 050. | Texas, University of-Austin *Computer Sciences* | 27 63 | 22 51 | 100 65 | .23 55 | 7.0 45 | .77 47 | .42 53 |
| 051. | Utah, University of-Salt Lake City *Computer Science** | 13 47 | 25 52 | 32 48 | .21 53 | 5.4 59 | .87 58 | .40 51 |
| 052. | Vanderbilt University *Computer Science* | 9 42 | 9 44 | 19 44 | NA | NA | NA | NA |
| 053. | Virginia, University of *Applied Mathematics and Computer Science** | 9 42 | 9 44 | 12 43 | NA | NA | NA | NA |
| 054. | Washington State University-Pullman *Computer Science* | 12 45 | 8 44 | 15 43 | NA | NA | NA | NA |
| 055. | Washington University-Saint Louis *Computer Science* | 8 41 | 13 46 | 14 43 | NA | NA | NA | NA |
| 056. | Washington, University of-Seattle *Computer Sciences* | 12 45 | 14 47 | 25 46 | .36 67 | 7.7 40 | NA | NA |
| 057. | Wisconsin, University of-Madison *Computer Sciences* | 25 61 | 39 60 | 77 59 | .16 49 | 6.6 49 | .82 52 | .32 46 |
| 058. | Yale University *Computer Science* | 16 50 | 27 53 | 45 51 | .13 46 | 5.3 60 | .74 44 | .57 61 |

* indicates program was initiated since 1970.

NOTE: On the first line of data for every program, raw values for each measure are reported; on the second line values are reported in standardized form, with mean = 50 and standard deviation = 10. "NA" indicates that the value for a measure is not available.

TABLE 4.1 Program Measures (Raw and Standardized Values) in Computer Sciences

| Prog No. | Survey Results (08) | (09) | (10) | (11) | University Library (12) | Research Support (13) | (14) | Published Articles (15) | (16) | Survey Ratings Standard Error (08) | (09) | (10) | (11) |
|---|---|---|---|---|---|---|---|---|---|---|---|---|---|
| 041. | 2.4 | 1.4 | 1.4 | 0.7 | 0.8 | .17 | 1043 | 27 | | .11 | .10 | .09 | .07 |
| | 49 | 47 | 60 | 47 | 54 | 40 | 49 | 48 | 48 | | | | |
| 042. | 2.3 | 1.4 | 0.9 | 0.9 | 0.3 | .46 | 556 | 31 | | .11 | .08 | .09 | .06 |
| | 48 | 49 | 43 | 51 | 48 | 55 | 46 | 49 | 48 | | | | |
| 043. | 2.7 | 1.8 | 1.2 | 0.9 | -0.6 | .56 | 312 | 14 | | .10 | .05 | .09 | .07 |
| | 52 | 56 | 52 | 52 | 39 | 61 | 44 | 44 | 45 | | | | |
| 044. | 3.2 | 1.8 | 1.3 | 1.3 | 0.4 | .20 | NA | 75 | | .09 | .06 | .09 | .06 |
| | 57 | 56 | 57 | 60 | 49 | 41 | | 63 | 61 | | | | |
| 045. | 1.6 | 1.0 | 0.8 | 0.5 | NA | .20 | 77 | 16 | | .14 | .12 | .13 | .06 |
| | 41 | 40 | 37 | 41 | | 41 | 43 | 44 | 44 | | | | |
| 046. | 5.0 | 2.8 | 1.1 | 1.9 | 2.0 | .71 | 5008 | 106 | | .02 | .04 | .06 | .03 |
| | 75 | 74 | 48 | 74 | 67 | 69 | 76 | 73 | 80 | | | | |
| 047. | 1.2 | 0.9 | 1.0 | 0.2 | NA | .00 | NA | 2 | | .16 | .16 | .18 | .04 |
| | 36 | 38 | 45 | 34 | | 31 | | 40 | 40 | | | | |
| 048. | 2.4 | 1.4 | 1.1 | 0.6 | -0.3 | .25 | 918 | 38 | | .13 | .09 | .09 | .06 |
| | 49 | 48 | 50 | 44 | 42 | 44 | 48 | 51 | 48 | | | | |
| 049. | 1.1 | 0.7 | 1.0 | 0.3 | -0.5 | .00 | NA | 32 | | .13 | .10 | .08 | .05 |
| | 35 | 36 | 44 | 37 | 41 | 31 | | 49 | 46 | | | | |
| 050. | 3.2 | 2.1 | 1.3 | 1.3 | 1.6 | .48 | 1380 | 53 | | .10 | .06 | .09 | .06 |
| | 57 | 60 | 55 | 60 | 62 | 56 | 51 | 56 | 57 | | | | |
| 051. | 2.8 | 1.9 | 1.0 | 1.0 | -0.6 | .54 | 606 | 21 | | .08 | .06 | .10 | .07 |
| | 53 | 57 | 45 | 54 | 39 | 59 | 46 | 46 | 46 | | | | |
| 052. | 1.8 | 1.1 | 1.5 | 0.6 | -0.7 | NA | NA | 6 | | .11 | .12 | .11 | .06 |
| | 43 | 42 | 64 | 44 | 38 | | | 41 | 41 | | | | |
| 053. | 1.7 | 1.2 | 1.3 | 0.4 | 0.7 | NA | 263 | 23 | | .14 | .12 | .12 | .05 |
| | 42 | 44 | 55 | 40 | 53 | | 44 | 47 | 44 | | | | |
| 054. | 1.5 | 1.0 | 1.2 | 0.3 | -0.3 | .25 | NA | 9 | | .15 | .10 | .11 | .06 |
| | 40 | 40 | 52 | 38 | 43 | 44 | | 42 | 42 | | | | |
| 055. | 1.4 | 1.0 | 1.0 | 0.4 | -0.4 | NA | NA | 10 | | .15 | .11 | .10 | .06 |
| | 39 | 41 | 45 | 39 | 42 | | | 42 | 44 | | | | |
| 056. | 3.4 | 2.0 | 1.7 | 1.3 | 1.5 | .75 | 473 | 17 | | .09 | .07 | .08 | .07 |
| | 59 | 59 | 69 | 59 | 61 | 71 | 45 | 45 | 44 | | | | |
| 057. | 3.2 | 1.9 | 1.3 | 1.1 | 1.6 | .56 | 672 | 55 | | .10 | .05 | .08 | .07 |
| | 57 | 58 | 56 | 57 | 62 | 60 | 47 | 57 | 56 | | | | |
| 058. | 3.5 | 2.1 | 1.1 | 1.3 | 2.1 | .44 | 1672 | 22 | | .08 | .05 | .09 | .07 |
| | 60 | 61 | 49 | 61 | 67 | 54 | 53 | 46 | 47 | | | | |

NOTE: On the first line of data for every program, raw values for each measure are reported; on the second line values are reported in standardized form, with mean = 50 and standard deviation = 10. "NA" indicates that the value for a measure is not available. Since the scale used to compute measure (16) is entirely arbitrary, only values in standardized form are reported for this measure.

TABLE 4.2  Summary Statistics Describing Each Program Measure—Computer Sciences

| Measure | Number of Programs Evaluated | Mean | Standard Deviation | DECILES | | | | | | | | |
|---|---|---|---|---|---|---|---|---|---|---|---|---|
| | | | | 1 | 2 | 3 | 4 | 5 | 6 | 7 | 8 | 9 |
| **Program Size** | | | | | | | | | | | | |
| 01 Raw Value | 58 | 16 | 9 | 8 | 9 | 10 | 11 | 12 | 15 | 16 | 23 | 29 |
| Std Value | 58 | 50 | 10 | 41 | 42 | 43 | 44 | 45 | 49 | 50 | 58 | 65 |
| 02 Raw Value | 57 | 20 | 20 | 5 | 7 | 9 | 10 | 13 | 18 | 21 | 29 | 42 |
| Std Value | 57 | 50 | 10 | 42 | 43 | 44 | 45 | 46 | 49 | 50 | 54 | 61 |
| 03 Raw Value | 58 | 41 | 40 | 12 | 15 | 18 | 22 | 25 | 32 | 46 | 62 | 91 |
| Std Value | 58 | 50 | 10 | 43 | 43 | 44 | 45 | 46 | 48 | 51 | 55 | 62 |
| **Program Graduates** | | | | | | | | | | | | |
| 04 Raw Value | 31 | .17 | .11 | .05 | .08 | .09 | .13 | .15 | .20 | .22 | .25 | .30 |
| Std Value | 31 | 50 | 10 | 39 | 42 | 43 | 46 | 48 | 53 | 55 | 57 | 62 |
| 05 Raw Value | 31 | 6.5 | 1.2 | 8.0 | 7.6 | 6.8 | 6.5 | 6.4 | 6.3 | 5.8 | 5.4 | 5.3 |
| Std Value | 31 | 50 | 10 | 37 | 40 | 47 | 50 | 51 | 51 | 56 | 59 | 60 |
| 06 Raw Value | 30 | .80 | .09 | .65 | .74 | .77 | .79 | .80 | .81 | .82 | .85 | .88 |
| Std Value | 30 | 50 | 10 | 33 | 43 | 47 | 49 | 50 | 51 | 52 | 56 | 59 |
| 07 Raw Value | 30 | .38 | .16 | .17 | .23 | .25 | .32 | .36 | .40 | .48 | .51 | .63 |
| Std Value | 30 | 50 | 10 | 37 | 41 | 42 | 46 | 49 | 51 | 56 | 58 | 66 |
| **Survey Results** | | | | | | | | | | | | |
| 08 Raw Value | 57 | 2.5 | 1.0 | 1.2 | 1.6 | 1.9 | 2.3 | 2.4 | 2.7 | 2.8 | 3.2 | 3.8 |
| Std Value | 57 | 50 | 10 | 37 | 41 | 44 | 48 | 49 | 52 | 53 | 57 | 63 |
| 09 Raw Value | 57 | 1.5 | .6 | .8 | 1.0 | 1.2 | 1.3 | 1.5 | 1.6 | 1.7 | 1.9 | 2.2 |
| Std Value | 57 | 50 | 10 | 37 | 41 | 44 | 46 | 50 | 52 | 53 | 57 | 63 |
| 10 Raw Value | 56 | 1.1 | .3 | .8 | .9 | 1.0 | 1.1 | 1.1 | 1.1 | 1.2 | 1.3 | 1.5 |
| Std Value | 56 | 50 | 10 | 38 | 41 | 45 | 49 | 49 | 49 | 53 | 56 | 64 |
| 11 Raw Value | 57 | .9 | .4 | .3 | .4 | .6 | .7 | .8 | .9 | 1.0 | 1.2 | 1.4 |
| Std Value | 57 | 50 | 10 | 37 | 40 | 44 | 47 | 49 | 51 | 53 | 58 | 63 |
| **University Library** | | | | | | | | | | | | |
| 12 Raw Value | 49 | .4 | 1.0 | -.7 | -.6 | -.3 | .1 | .3 | .7 | .9 | 1.5 | 1.8 |
| Std Value | 49 | 50 | 10 | 38 | 39 | 42 | 47 | 49 | 53 | 55 | 61 | 64 |
| **Research Support** | | | | | | | | | | | | |
| 13 Raw Value | 45 | .36 | .19 | .12 | .20 | .25 | .27 | .36 | .39 | .47 | .56 | .60 |
| Std Value | 45 | 50 | 10 | 37 | 42 | 44 | 45 | 50 | 52 | 56 | 61 | 63 |
| 14 Raw Value | 44 | 1171 | 1501 | 85 | 205 | 314 | 401 | 473 | 719 | 973 | 1603 | 3684 |
| Std Value | 44 | 50 | 10 | 43 | 44 | 44 | 45 | 45 | 47 | 49 | 53 | 67 |
| **Publication Records** | | | | | | | | | | | | |
| 15 Raw Value | 58 | 34 | 31 | 10 | 12 | 14 | 21 | 23 | 27 | 33 | 52 | 65 |
| Std Value | 58 | 50 | 10 | 42 | 43 | 44 | 46 | 47 | 48 | 50 | 56 | 60 |
| 16 Std Value | 58 | 50 | 10 | 42 | 43 | 44 | 45 | 47 | 48 | 49 | 55 | 61 |

NOTE: Standardized values reported in the preceding table have been computed from exact values of the mean and standard deviation and not the rounded values reported here. Since the scale used to compute measure 16 is entirely arbitrary, only data in standardized form are reported for this measure.

TABLE 4.3  Intercorrelations Among Program Measures on 58 Programs in Computer Sciences

|  | 01 | 02 | 03 | 04 | 05 | 06 | 07 | 08 | 09 | 10 | 11 | 12 | 13 | 14 | 15 | 16 |
|---|---|---|---|---|---|---|---|---|---|---|---|---|---|---|---|---|
| **Program Size** | | | | | | | | | | | | | | | | |
| 01 | | .62 | .67 | -.11 | -.03 | -.17 | -.26 | .54 | .54 | .13 | .45 | .28 | .12 | .44 | .62 | .61 |
| 02 | | | .52 | .05 | -.07 | .12 | -.05 | .66 | .68 | -.02 | .61 | .44 | .34 | .58 | .85 | .84 |
| 03 | | | | .04 | -.05 | -.06 | -.07 | .50 | .49 | .12 | .46 | .33 | .13 | .43 | .51 | .52 |
| **Program Graduates** | | | | | | | | | | | | | | | | |
| 04 | | | | | .12 | .17 | .43 | .35 | .34 | -.07 | .30 | .23 | .34 | .22 | .09 | .20 |
| 05 | | | | | | .13 | .44 | .14 | .17 | -.22 | .10 | -.17 | .29 | -.21 | -.07 | -.04 |
| 06 | | | | | | | .41 | .21 | .26 | -.11 | .25 | .23 | .26 | -.03 | .10 | .14 |
| 07 | | | | | | | | .17 | .23 | -.31 | .23 | .24 | .26 | -.16 | -.08 | -.01 |
| **Survey Results** | | | | | | | | | | | | | | | | |
| 08 | | | | | | | | | .98 | .29 | .97 | .58 | .59 | .63 | .70 | .77 |
| 09 | | | | | | | | | | .26 | .95 | .54 | .61 | .61 | .69 | .75 |
| 10 | | | | | | | | | | | .26 | .16 | .18 | -.02 | .04 | .05 |
| 11 | | | | | | | | | | | | .56 | .57 | .64 | .69 | .74 |
| **University Library** | | | | | | | | | | | | | | | | |
| 12 | | | | | | | | | | | | | .49 | .16 | .52 | .52 |
| **Research Support** | | | | | | | | | | | | | | | | |
| 13 | | | | | | | | | | | | | | .10 | .32 | .35 |
| 14 | | | | | | | | | | | | | | | .66 | .73 |
| **Publication Records** | | | | | | | | | | | | | | | | |
| 15 | | | | | | | | | | | | | | | | .98 |
| 16 | | | | | | | | | | | | | | | | |

NOTE: Since in computing correlation coefficients program data must be available for both of the measures being correlated, the actual number of programs on which each coefficient is based varies.

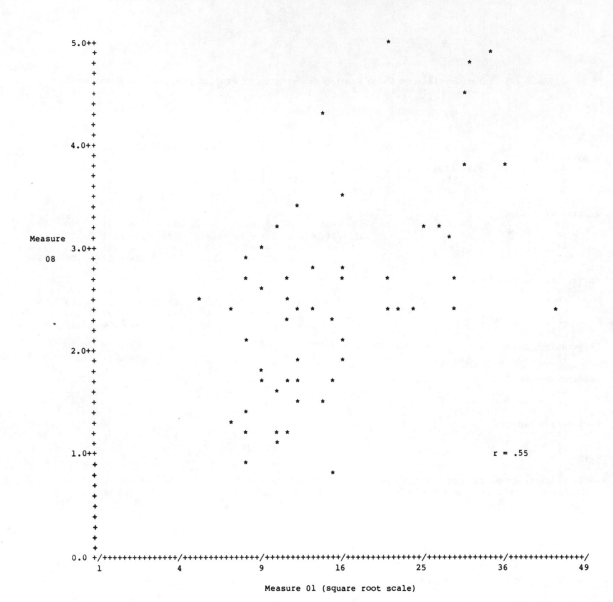

FIGURE 4.1  Mean rating of scholarly quality of faculty (measure 08) versus number of faculty members (measure 01)--57 programs in computer sciences.

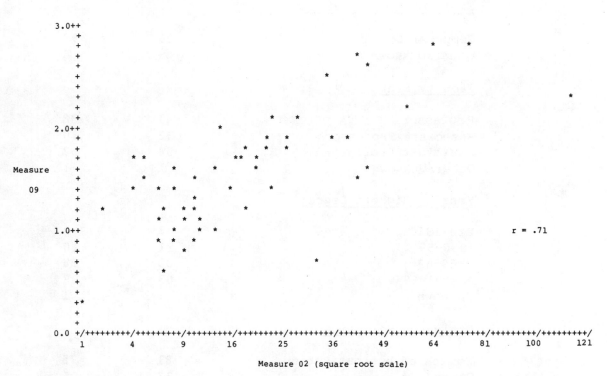

FIGURE 4.2  Mean rating of program effectiveness in educating research scholars/scientists (measure 09) versus number of graduates in last five years (measure 02)--56 programs in computer sciences.

TABLE 4.4  Characteristics of Survey Participants in Computer Sciences

|  | Respondents | |
|---|---|---|
|  | N | % |
| **Field of Specialization** | | |
| Computer Sciences | 99 | 92 |
| Other/Unknown | 9 | 8 |
| **Faculty Rank** | | |
| Professor | 41 | 38 |
| Associate Professor | 32 | 30 |
| Assistant Professor | 34 | 32 |
| Other/Unknown | 1 | 1 |
| **Year of Highest Degree** | | |
| Pre-1950 | 3 | 3 |
| 1950-59 | 11 | 10 |
| 1960-69 | 30 | 28 |
| Post-1969 | 63 | 58 |
| Unknown | 1 | 1 |
| **Evaluator Selection** | | |
| Nominated by Institution | 81 | 75 |
| Other | 27 | 25 |
| **Survey Form** | | |
| With Faculty Names | 97 | 90 |
| Without Names | 11 | 10 |
| **Total Evaluators** | 108 | 100 |

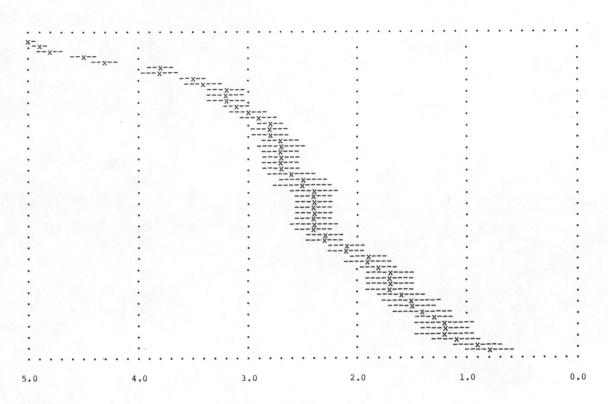

5.0                4.0                3.0                2.0                1.0                0.0

Mean Survey Rating (Measure 08)

FIGURE 4.3  Mean rating of scholarly quality of faculty in 57 programs in computer sciences.

NOTE:  Programs are listed in sequence of mean rating, with the highest-rated program appearing at the
        top of the page.  The broken lines (---) indicate a confidence interval of ±1.5 standard errors
        around the reported mean (x) of each program.

# V
# Geoscience Programs

In this chapter 91 research-doctorate programs in the geosciences--
including geology, geochemistry, geophysics, and general earth
sciences--are assessed.  These programs, according to the information
supplied by their universities, have accounted for 1,747 doctoral
degrees awarded during the FY1976-80 period--approximately 93 percent
of the aggregate number of geoscience doctorates earned from U.S.
universities in this five-year span.[1]  On the average, 25 full-time
and part-time students intending to earn doctorates were enrolled in a
program in December 1980, with an average faculty size of 16
members.[2]  The 91 programs, listed in Table 5.1, represent 82
different universities.  The University of California (Berkeley),
University of Missouri (Rolla), Ohio State University (Columbus),
Princeton University, and Texas A&M University each have two
geoscience programs included in the assessment, and Pennsylvania State
University and Stanford University each have three.  All but 5 of the
91 geoscience programs were initiated prior to 1970.  In addition to
the 82 universities represented in this discipline, another 5 were
initially identified as meeting the criteria[3] for inclusion in the
assessment:

> University of California--San Diego
> Colorado School of Mines
> Colorado State University
> North Carolina State University--Raleigh
> University of Rochester

The Colorado School of Mines chose not to participate in the study in
any of the disciplines.  Geoscience programs at the other four

---

[1] Data from the NRC's Survey of Earned Doctorates indicate that 1,871
research doctorates in geosciences were awarded by U.S. universities
between FY1976 and FY1980.
[2] See the reported means for measures 03 and 01 in Table 5.2.
[3] As mentioned in Chapter I, the primary criterion for inclusion was
that a university had awarded at least 7 doctorates in the geosciences
during the FY1976-78 period.

institutions have not been included in the evaluations in this
discipline, since in each case the study coordinator either indicated
that the institution did not at that time have a research-doctorate
program in geosciences or failed to provide the information requested
by the committee.

Before examining individual program results presented in Table 5.1,
the reader is urged to refer to Chapter II, in which each of the 16
measures used in the assessment is discussed. Summary statistics
describing every measure are given in Table 5.2. For nine of the
measures, data are reported for at least 89 of the 91 geoscience
programs. For measures 04-07, which pertain to characteristics of the
program graduates, data are presented for only two-thirds of the
programs; the other one-third had too few graduates on which to base
statistics.[4] For measure 12, a composite index of the size of a
university library, data are available for 69 programs; for measure
14, total university expenditures for research in this discipline,
data are available for 73 programs. With respect to the measure 14,
it should be noted that the reported data include expenditures for
research in atmospheric sciences and oceanography as well as in the
geosciences. The programs not evaluated on measures 12 and 14 are
typically smaller--in terms of faculty size and graduate student
enrollment--than other geoscience programs. Were data on measures 12
and 14 available for all 91 programs, it is likely that the reported
means for these two measures would be appreciably lower (and that some
of the correlations of these measures with others would be higher).
With respect to measure 13, the fraction of faculty with research
support from the National Science Foundation, the National Institutes
of Health, and the Alcohol, Drug Abuse, and Mental Health
Adminstration, data are reported for 72 programs that had at least 10
faculty members.

Intercorrelations among the 16 measures (Pearson product-moment
coefficients) are given in Table 5.3. Of particular note are the high
positive correlations of the measures of the numbers of doctoral
graduates and students (02, 03) with measures of publication records
(15-16) and reputational survey ratings (08, 09, and 11). Figure 5.1
illustrates the relationship between the mean rating of the scholarly
quality of faculty (measure 08) and the number of faculty members
(measure 01) for each of the 91 geoscience programs. Figure 5.2 plots
the mean rating of program effectiveness (measure 09) against the
total number of FY1976-80 program graduates (measure 02). Although in
both figures there is a significant positive correlation between
program size and reputational rating, it is quite apparent that some
of the smaller programs received high mean ratings and some of the
larger programs received low mean ratings.

Table 5.4 describes the 177 faculty members who participated in
the evaluation of geoscience programs. These individuals constituted

---

[4]As mentioned in Chapter II, data for measures 04-07 are not
reported if they are based on the survey responses of fewer than 10
FY1975-79 program graduates.
[5]See Table 2.3 in Chapter II.

65 percent of those asked to respond to the survey in this discipline and 12 percent of the faculty population in the 91 research-doctorate programs being evaluated.[5]  More than one-third of the survey participants were geologists, and approximately two-thirds held the rank of full professor.  Almost three-fourths of them had earned their highest degree prior to 1970.

To assist the reader in interpreting results of the survey evaluations, estimated standard errors have been computed for mean ratings of the scholarly quality of faculty in 91 geoscience programs (and are given in Table 5.1).  For each program the mean rating and an associated "confidence interval" of 1.5 standard errors are illustrated in Figure 5.3 (listed in order of highest to lowest mean rating).  In comparing two programs, if their confidence intervals do not overlap, one may conclude that there is a significant difference in their mean ratings at a .05 level of significance.[6]  From this figure it is also apparent that one should have somewhat more confidence in the accuracy of the mean ratings of higher-rated programs than lower-rated programs.  This generalization results primarily from the fact that evaluators are not as likely to be familiar with the less prestigious programs, and consequently the mean ratings of these programs are usually based on fewer survey responses.

---

[6]See pp. 29-31 for a discussion of the interpretation of mean ratings and associated confidence intervals.

TABLE 5.1 Program Measures (Raw and Standardized Values) in Geosciences

| Prog No. | University - Department/Academic Unit | Program Size (01) | (02) | (03) | Characteristics of Program Graduates (04) | (05) | (06) | (07) |
|----------|----------------------------------------|------|------|------|------|------|------|------|
| 001. | Alaska, University of *Geophysical Institute* | 52 / 95 | 6 / 41 | 11 / 43 | NA | NA | NA | NA |
| 002. | Arizona State University-Tempe *Geology** | 19 / 54 | 8 / 42 | 15 / 45 | .27 / 51 | 6.8 / 52 | .64 / 39 | .46 / 67 |
| 003. | Arizona, University of-Tucson *Geosciences** | 26 / 63 | 27 / 55 | 40 / 57 | .30 / 53 | 8.8 / 30 | .52 / 29 | .13 / 43 |
| 004. | Boston University *Geology* | 6 / 38 | 9 / 43 | 1 / 38 | .00 / 30 | 9.8 / 18 | .58 / 34 | .00 / 33 |
| 005. | Brown University *Geological Sciences* | 16 / 51 | 20 / 51 | 43 / 59 | .32 / 54 | 6.5 / 55 | .81 / 54 | .38 / 62 |
| 006. | California Institute of Technology *Geological and Planetary Sciences* | 28 / 65 | 60 / 78 | 69 / 71 | .33 / 55 | 6.5 / 55 | .64 / 39 | .36 / 60 |
| 007. | California, University of-Santa Cruz *Earth Sciences* | 10 / 43 | 16 / 48 | 31 / 53 | .06 / 35 | 5.8 / 63 | .94 / 65 | .29 / 55 |
| 008. | California, University of-Berkeley *Geology* | 11 / 44 | 25 / 54 | 29 / 52 | .16 / 42 | 6.1 / 59 | .74 / 47 | .26 / 53 |
| 009. | California, University of-Berkeley *Geophysics* | 5 / 37 | 16 / 48 | 21 / 48 | .21 / 46 | 6.3 / 57 | .79 / 52 | .29 / 55 |
| 010. | California, University of-Davis *Geology* | 12 / 46 | 14 / 46 | 21 / 48 | NA | NA | NA | NA |
| 011. | California, University of-Los Angeles *Earth and Space Sciences* | 39 / 79 | 58 / 77 | 68 / 71 | .24 / 48 | 6.9 / 51 | .70 / 44 | .30 / 56 |
| 012. | California, University of-Riverside *Earth Sciences* | 8 / 41 | 7 / 42 | 7 / 41 | NA | NA | NA | NA |
| 013. | California, University of-Santa Barbara *Geological Sciences* | 18 / 53 | 18 / 49 | 42 / 58 | .17 / 43 | 7.7 / 42 | .92 / 63 | .33 / 58 |
| 014. | Case Western Reserve University *eological Sciences* | 7 / 39 | 12 / 45 | 7 / 41 | .62 / 77 | 6.8 / 52 | .69 / 44 | .23 / 51 |
| 015. | Chicago, University of *Geophysical Sciences* | 20 | 16 | 24 | .36 | 5.7 | .86 | .43 |
| 016. | Cincinnati, University of *Geology* | 11 / 44 | 13 / 46 | 11 / 43 | .15 / 42 | 7.3 / 46 | .92 / 63 | .08 / 40 |
| 017. | Colorado, University of *Geological Sciences* | 19 / 54 | 35 / 61 | 39 / 57 | .25 / 49 | 7.5 / 44 | .81 / 54 | .16 / 45 |
| 018. | Columbia University *Geological Sciences* | 28 / 65 | 55 / 75 | 108 / 90 | .22 / 47 | 6.7 / 53 | .85 / 57 | .37 / 61 |
| 019. | Cornell University-Ithaca *Geological Sciences* | 18 / 53 | 16 / 48 | 42 / 58 | .36 / 58 | 5.8 / 63 | .82 / 54 | .46 / 67 |
| 020. | Delaware, University of-Newark *Geology* | 10 / 43 | 14 / 46 | 9 / 42 | .36 / 57 | 7.5 / 44 | .79 / 52 | .00 / 33 |

* indicates program was initiated since 1970.

NOTE: On the first line of data for every program, raw values for each measure are reported; on the second line values are reported in standardized form, with mean = 50 and standard deviation = 10. "NA" indicates that the value for a measure is not available.

TABLE 5.1  Program Measures (Raw and Standardized Values) in Geosciences

| Prog No. | Survey Results (08) | (09) | (10) | (11) | University Library (12) | Research Support (13) | (14) | Published Articles (15) | (16) | Survey Ratings Standard Error (08) | (09) | (10) | (11) |
|---|---|---|---|---|---|---|---|---|---|---|---|---|---|
| 001. | 2.5 | 1.4 | 1.4 | 0.3 | NA | .58 | 25987 | 56 | | .18 | .16 | .14 | .05 |
| | 45 | 41 | 58 | 35 | | 55 | 99 | 53 | 48 | | | | |
| 002. | 3.1 | 1.9 | 1.6 | 1.1 | -0.3 | .58 | 898 | 32 | | .08 | .06 | .08 | .07 |
| | 52 | 53 | 66 | 55 | 44 | 55 | 43 | 47 | 48 | | | | |
| 003. | 3.7 | 2.2 | 1.8 | 1.2 | 0.9 | .62 | 10111 | 72 | | .08 | .06 | .05 | .07 |
| | 59 | 60 | 74 | 60 | 55 | 57 | 64 | 57 | 63 | | | | |
| 004. | 1.3 | 0.9 | 0.8 | 0.4 | -0.4 | NA | NA | 11 | | .12 | .09 | .10 | .06 |
| | 30 | 29 | 39 | 38 | 43 | | | 42 | 41 | | | | |
| 005. | 3.7 | 2.1 | 1.4 | 1.4 | -1.1 | .88 | 1479 | 35 | | .09 | .06 | .08 | .07 |
| | 59 | 59 | 59 | 65 | 37 | 69 | 44 | 48 | 50 | | | | |
| 006. | 4.9 | 2.8 | 1.2 | 1.7 | NA | .64 | 6414 | 145 | | .04 | .05 | .06 | .05 |
| | 74 | 73 | 54 | 73 | | 58 | 56 | 76 | 77 | | | | |
| 007. | 3.1 | 1.9 | 1.3 | 1.0 | NA | .70 | 855 | 17 | | .08 | .06 | .07 | .07 |
| | 52 | 53 | 54 | 54 | | 61 | 43 | 43 | 44 | | | | |
| 008. | 4.1 | 2.2 | 0.8 | 1.4 | 2.2 | .73 | 2232 | 110 | | .07 | .06 | .07 | .07 |
| | 64 | 61 | 36 | 65 | 68 | 62 | 46 | 67 | 71 | | | | |
| 009. | 3.5 | 2.1 | 0.9 | 0.8 | 2.2 | NA | 2232 | 110 | | .12 | .07 | .10 | .09 |
| | 57 | 57 | 40 | 49 | 68 | | 46 | 67 | 71 | | | | |
| 010. | 2.9 | 1.8 | 1.2 | 1.0 | 0.6 | .67 | 597 | 34 | | .09 | .06 | .09 | .07 |
| | 50 | 52 | 53 | 53 | 53 | 59 | 42 | 48 | 48 | | | | |
| 011. | 4.5 | 2.4 | 1.4 | 1.5 | 2.0 | .54 | 5359 | 171 | | .06 | .06 | .06 | .06 |
| | 69 | 65 | 59 | 69 | 66 | 53 | 53 | 83 | 85 | | | | |
| 012. | 2.0 | 1.4 | 0.7 | 0.8 | -1.0 | NA | 716 | 27 | | .12 | .07 | .09 | .07 |
| | 38 | 41 | 35 | 47 | 37 | | 42 | 46 | 45 | | | | |
| 013. | 3.7 | 2.2 | 1.6 | 1.3 | -0.1 | .50 | 2883 | 35 | | .08 | .06 | .07 | .07 |
| | 59 | 60 | 67 | 63 | 46 | 51 | 47 | 48 | 49 | | | | |
| 014. | 2.2 | 1.3 | 0.3 | 0.7 | -1.3 | NA | NA | 14 | | .11 | .09 | .08 | .07 |
| | 41 | 41 | 20 | 46 | 34 | | | 42 | 44 | | | | |
| 015. | 4.3 | 2.3 | 1.2 | 1.4 | 0.9 | .85 | 2186 | 80 | | .08 | .07 | .06 | .07 |
| | 66 | 61 | 54 | 63 | 55 | 68 | 46 | 59 | 63 | | | | |
| 016. | 2.8 | 1.7 | 1.3 | 0.9 | -0.2 | .27 | NA | 10 | | .10 | .07 | .09 | .07 |
| | 48 | 48 | 56 | 51 | 44 | 41 | | 41 | 42 | | | | |
| 017. | 3.1 | 2.0 | 1.3 | 1.0 | -0.9 | .32 | 3977 | 79 | | .09 | .05 | .08 | .06 |
| | 52 | 55 | 54 | 54 | 38 | 43 | 50 | 59 | 56 | | | | |
| 018. | 4.3 | 2.4 | 1.0 | 1.5 | 1.7 | .71 | 13637 | 167 | | .08 | .06 | .07 | .06 |
| | 67 | 65 | 44 | 68 | 63 | 61 | 73 | 82 | 78 | | | | |
| 019. | 4.0 | 2.3 | 1.7 | 1.4 | 1.6 | .39 | 2887 | 84 | | .07 | .06 | .05 | .06 |
| | 63 | 62 | 70 | 66 | 62 | 46 | 47 | 60 | 63 | | | | |
| 020. | 1.8 | 1.3 | 1.2 | 0.5 | NA | .30 | 3086 | 33 | | .14 | .11 | .14 | .07 |
| | 37 | 39 | 51 | 39 | | 42 | 48 | 47 | 47 | | | | |

NOTE: On the first line of data for every program, raw values for each measure are reported; on the second line values are reported in standardized form, with mean = 50 and standard deviation = 10. "NA" indicates that the value for a measure is not available. Since the scale used to compute measure (16) is entirely arbitrary, only values in standardized form are reported for this measure.

TABLE 5.1  Program Measures (Raw and Standardized Values) in Geosciences

| Prog No. | University - Department/Academic Unit | Program Size | | | Characteristics of Program Graduates | | | |
|---|---|---|---|---|---|---|---|---|
| | | (01) | (02) | (03) | (04) | (05) | (06) | (07) |
| 021. | Florida State University-Tallahassee | 10 | 11 | 16 | NA | NA | NA | NA |
| | *Geology* | 43 | 44 | 46 | | | | |
| 022. | George Washington University | 10 | 11 | 14 | .09 | 8.5 | NA | NA |
| | *Geology* | 43 | 44 | 45 | 37 | 33 | | |
| 023. | Harvard University | 13 | 50 | 36 | .42 | 6.2 | .76 | .36 |
| | *Geological Sciences* | 47 | 71 | 55 | 62 | 59 | 49 | 60 |
| 024. | Hawaii, University of | 33 | 21 | 22 | .31 | 8.3 | .56 | .19 |
| | *Geology and Geophysics* | 72 | 51 | 49 | 54 | 35 | 33 | 47 |
| 025. | Houston, University of | 14 | 2 | 26 | NA | NA | NA | NA |
| | *Geology\** | 48 | 38 | 50 | | | | |
| 026. | Idaho, University of-Moscow | 13 | 21 | NA | .19 | 7.3 | .65 | .15 |
| | *Geology* | 47 | 51 | | 45 | 47 | 40 | 45 |
| 027. | Illinois, University-Urbana/Champaign | 20 | 32 | 64 | .24 | 6.8 | .79 | .30 |
| | *Geology* | 55 | 59 | 69 | 48 | 52 | 52 | 56 |
| 028. | Indiana University-Bloomington | 23 | 25 | 16 | .16 | 5.9 | .81 | .38 |
| | *Geology* | 59 | 54 | 46 | 42 | 62 | 54 | 61 |
| 029. | Iowa State University-Ames | 18 | 6 | 9 | NA | NA | NA | NA |
| | *Earth Sciences* | 53 | 41 | 42 | | | | |
| 030. | Iowa, University of-Iowa City | 14 | 19 | 68 | .31 | 7.0 | .72 | .28 |
| | *Geology* | 48 | 50 | 71 | 54 | 50 | 46 | 54 |
| 031. | Johns Hopkins University | 11 | 27 | 41 | .44 | 5.5 | .89 | .44 |
| | *Earth and Planetary Sciences* | 44 | 55 | 58 | 64 | 66 | 60 | 66 |
| 032. | Kansas, University of | 18 | 17 | 18 | .14 | 7.3 | .82 | .18 |
| | *Geology* | 53 | 48 | 47 | 40 | 46 | 54 | 47 |
| 033. | Kentucky, University of | 9 | 5 | 7 | NA | NA | NA | NA |
| | *Geology* | 42 | 40 | 41 | | | | |
| 034. | Lehigh University | 9 | 7 | 4 | NA | NA | NA | NA |
| | *Geological Sciences* | 42 | 42 | 40 | | | | |
| 035. | Louisiana State University-Baton Rouge | 17 | 9 | 83 | .20 | 8.5 | NA | NA |
| | *Geology* | 52 | 43 | 78 | 45 | 33 | | |
| 036. | Massachusetts Institute of Technology | 25 | 87 | 107 | .21 | 6.0 | .77 | .37 |
| | *Earth and Planetary Sciences* | 62 | 97 | 90 | 46 | 61 | 50 | 61 |
| 037. | Massachusetts, University of-Amherst | 16 | 18 | 25 | .19 | 8.3 | .75 | .13 |
| | *Geology and Geography* | 51 | 49 | 50 | 44 | 35 | 49 | 43 |
| 038. | Miami, University of-Florida | 17 | 14 | 25 | NA | NA | NA | NA |
| | *Marine Geol & Geophys/ Marine & Atmos Chem* | 52 | 46 | 50 | | | | |
| 039. | Michigan State University-East Lansing | 14 | 11 | 9 | .29 | 7.0 | 1.00 | .29 |
| | *Geology* | 48 | 44 | 42 | 52 | 50 | 70 | 55 |
| 040. | Michigan, University of-Ann Arbor | 20 | 20 | 10 | .33 | 7.0 | .80 | .33 |
| | *Geological Sciences* | 55 | 51 | 43 | 55 | 50 | 53 | 58 |

\* indicates program was initiated since 1970.

NOTE: On the first line of data for every program, raw values for each measure are reported; on the second line values are reported in standardized form, with mean = 50 and standard deviation = 10. "NA" indicates that the value for a measure is not available.

TABLE 5.1  Program Measures (Raw and Standardized Values) in Geosciences

| Prog No. | Survey Results | | | | University Library | Research Support | | Published Articles | | Survey Ratings Standard Error | | | |
|---|---|---|---|---|---|---|---|---|---|---|---|---|---|
| | (08) | (09) | (10) | (11) | (12) | (13) | (14) | (15) | (16) | (08) | (09) | (10) | (11) |
| 021. | 2.2 | 1.4 | 1.0 | 0.7 | −0.4 | .30 | 2610 | 37 | | .10 | .09 | .10 | .07 |
| | 41 | 42 | 45 | 44 | 43 | 42 | 47 | 48 | 44 | | | | |
| 022. | 1.6 | 1.0 | 0.9 | 0.3 | NA | .20 | 738 | 10 | | .14 | .14 | .12 | .05 |
| | 33 | 34 | 40 | 33 | | 38 | 42 | 41 | 41 | | | | |
| 023. | 4.1 | 2.4 | 0.9 | 1.5 | 3.0 | .85 | 2324 | 96 | | .08 | .06 | .08 | .06 |
| | 65 | 64 | 42 | 68 | 75 | 68 | 46 | 63 | 70 | | | | |
| 024. | 2.9 | 1.6 | 1.3 | 0.9 | −0.1 | .52 | 9001 | 48 | | .11 | .08 | .08 | .07 |
| | 50 | 47 | 56 | 51 | 45 | 52 | 62 | 51 | 53 | | | | |
| 025. | 2.2 | 1.1 | 1.3 | 0.6 | −0.9 | .07 | 610 | 25 | | .11 | .08 | .08 | .06 |
| | 42 | 36 | 55 | 41 | 38 | 32 | 42 | 45 | 42 | | | | |
| 026. | 1.6 | 1.1 | 1.1 | 0.3 | NA | .08 | 846 | 3 | | .11 | .12 | .08 | .06 |
| | 34 | 35 | 48 | 35 | | 32 | 43 | 40 | 41 | | | | |
| 027. | 3.2 | 1.9 | 0.9 | 1.0 | 2.0 | .30 | 7160 | 53 | | .08 | .06 | .10 | .07 |
| | 53 | 54 | 42 | 54 | 65 | 42 | 57 | 52 | 51 | | | | |
| 028. | 3.2 | 2.0 | 1.3 | 0.9 | 0.9 | .44 | NA | 28 | | .09 | .07 | .07 | .08 |
| | 53 | 57 | 56 | 51 | 56 | 48 | | 46 | 44 | | | | |
| 029. | 2.2 | 1.2 | 1.2 | 0.4 | −0.5 | .33 | 1082 | 13 | | .10 | .13 | .10 | .06 |
| | 41 | 37 | 52 | 38 | 42 | 44 | 43 | 42 | 42 | | | | |
| 030. | 2.4 | 1.7 | 1.2 | 0.6 | 0.3 | .14 | NA | 64 | | .15 | .10 | .10 | .07 |
| | 44 | 48 | 51 | 41 | 49 | 35 | | 55 | 51 | | | | |
| 031. | 3.6 | 2.1 | 1.0 | 1.0 | −0.4 | .82 | 2656 | 21 | | .08 | .07 | .08 | .07 |
| | 58 | 59 | 43 | 53 | 43 | 66 | 47 | 44 | 46 | | | | |
| 032. | 2.9 | 1.8 | 1.0 | 0.8 | 0.1 | .33 | 2945 | 23 | | .08 | .06 | .06 | .07 |
| | 49 | 50 | 44 | 49 | 48 | 44 | 48 | 45 | 44 | | | | |
| 033. | 2.1 | 1.0 | 1.3 | 0.6 | −0.1 | NA | 638 | 5 | | .09 | .09 | .15 | .07 |
| | 40 | 33 | 55 | 41 | 46 | | 42 | 40 | 41 | | | | |
| 034. | 2.1 | 1.2 | 0.9 | 0.6 | NA | NA | NA | 8 | | .13 | .09 | .09 | .06 |
| | 40 | 38 | 41 | 44 | | | | 41 | 42 | | | | |
| 035. | 2.5 | 1.7 | 1.3 | 0.6 | −0.3 | .29 | 6501 | 25 | | .11 | .09 | .11 | .07 |
| | 44 | 48 | 54 | 44 | 44 | 42 | 56 | 45 | 44 | | | | |
| 036. | 4.8 | 2.7 | 1.5 | 1.6 | −0.3 | .92 | 8537 | 147 | | .05 | .05 | .07 | .06 |
| | 72 | 72 | 61 | 70 | 44 | 71 | 61 | 77 | 78 | | | | |
| 037. | 3.0 | 1.8 | 1.4 | 1.0 | −0.7 | .44 | 2734 | 38 | | .09 | .07 | .08 | .07 |
| | 51 | 52 | 58 | 53 | 40 | 49 | 47 | 49 | 50 | | | | |
| 038. | 3.1 | 1.9 | 1.2 | 0.8 | NA | .65 | 11765 | 41 | | .11 | .07 | .10 | .07 |
| | 52 | 52 | 53 | 48 | | 58 | 68 | 49 | 50 | | | | |
| 039. | 2.4 | 1.5 | 1.0 | 0.7 | 0.3 | .21 | 2535 | 18 | | .10 | .08 | .09 | .07 |
| | 44 | 45 | 45 | 46 | 50 | 38 | 47 | 43 | 43 | | | | |
| 040. | 3.5 | 2.1 | 1.2 | 1.2 | 1.8 | .75 | 12188 | 102 | | .09 | .05 | .08 | .07 |
| | 57 | 59 | 51 | 59 | 64 | 63 | 69 | 65 | 61 | | | | |

NOTE: On the first line of data for every program, raw values for each measure are reported; on the second line values are reported in standardized form, with mean = 50 and standard deviation = 10. "NA" indicates that the value for a measure is not available. Since the scale used to compute measure (16) is entirely arbitrary, only values in standardized form are reported for this measure.

TABLE 5.1  Program Measures (Raw and Standardized Values) in Geosciences

| Prog No. | University - Department/Academic Unit | Program Size | | | Characteristics of Program Graduates | | | |
|---|---|---|---|---|---|---|---|---|
| | | (01) | (02) | (03) | (04) | (05) | (06) | (07) |
| 041. | Minnesota, University of | 21 | 20 | 22 | .13 | 7.8 | .71 | .08 |
| | *Geology and Geophysics* | *57* | *51* | *49* | *40* | *40* | *45* | *40* |
| 042. | Missouri, University of-Columbia | 8 | 5 | 6 | NA | NA | NA | NA |
| | *Geology* | *41* | *40* | *41* | | | | |
| 043. | Missouri, University of-Rolla | 4 | 2 | 5 | NA | NA | NA | NA |
| | *eological Engineering* | *36* | *38* | *40* | | | | |
| 044. | Missouri, University of-Rolla | 8 | 9 | 4 | NA | NA | NA | NA |
| | *Geology/Geophysics* | *41* | | | | | | |
| 045. | Montana, University of-Missoula | 13 | 8 | 6 | .20 | 6.0 | .80 | .10 |
| | *Geology* | *47* | *42* | *41* | *45* | *61* | *53* | *41* |
| 046. | New Mexico Institute of Mining & Technology | 24 | 12 | 9 | .18 | 7.5 | .73 | .00 |
| | *Geoscience* | *60* | *45* | *42* | *44* | *44* | *47* | *33* |
| 047. | New Mexico, University of-Albuquerque | 14 | 16 | NA | NA | NA | NA | NA |
| | *Geology* | *48* | *48* | | | | | |
| 048. | North Carolina, University of-Chapel Hill | 15 | 19 | 7 | .11 | 6.5 | .84 | .21 |
| | *Geology* | *49* | *50* | *41* | *38* | *55* | *56* | *49* |
| 049. | North Dakota, University of-Grand Forks | 8 | 10 | 18 | NA | NA | NA | NA |
| | *Geology* | *41* | *44* | *47* | | | | |
| 050. | Northwestern University | 13 | 14 | 18 | .20 | 6.8 | .87 | .27 |
| | *Geological Sciences* | *47* | *46* | *47* | *45* | *52* | *59* | *53* |
| 051. | Ohio State University-Columbus | 4 | 14 | 27 | .07 | 7.3 | .50 | .14 |
| | *Geodetic Science* | *36* | *46* | *51* | *35* | *46* | *27* | *44* |
| 052. | Ohio State University-Columbus | 26 | 23 | 29 | .26 | 7.4 | .70 | .17 |
| | *Geology and Mineralogy* | *63* | *53* | *52* | *50* | *46* | *44* | *46* |
| 053. | Oklahoma, University of-Norman | 13 | 5 | 11 | NA | NA | NA | NA |
| | *Geology and Geophysics* | *47* | *40* | *43* | | | | |
| 054. | Oregon State University-Corvallis | 12 | 9 | 10 | .50 | 7.5 | .70 | .10 |
| | *Geology* | *46* | *43* | *43* | *68* | *44* | *44* | *41* |
| 055. | Oregon, University of-Eugene | 13 | 11 | 15 | .46 | 6.5 | .82 | .18 |
| | *Geology* | *47* | *44* | *45* | *65* | *55* | *54* | *47* |
| 056. | Pennsylvania State University | 17 | 8 | 33 | .17 | 6.6 | .77 | .26 |
| | *Geochemistry and Mineralogy* | *52* | *42* | *54* | *43* | *54* | *50* | *53* |
| 057. | Pennsylvania State University | 14 | 33 | 27 | .37 | 6.1 | .85 | .18 |
| | *Geology* | *48* | *60* | *51* | *58* | *60* | *57* | *47* |
| 058. | Pennsylvania State University | 7 | 5 | 15 | NA | NA | NA | NA |
| | *Geophysics* | *39* | *40* | *45* | | | | |
| 059. | Pittsburgh, University of | 12 | 17 | 12 | .13 | 8.8 | .63 | .25 |
| | *Geology and Planetary Science* | *46* | *48* | *44* | *40* | *30* | *38* | *52* |
| 060. | Princeton University | 17 | 28 | 40 | .32 | 5.9 | .91 | .23 |
| | *Geological and Geophysical Sciences* | *52* | *56* | *57* | *54* | *62* | *62* | *50* |

\* indicates program was initiated since 1970.


NOTE: On the first line of data for every program, raw values for each measure are reported; on the second line values are reported in standardized form, with mean = 50 and standard deviation = 10. "NA" indicates that the value for a measure is not available.

TABLE 5.1 Program Measures (Raw and Standardized Values) in Geosciences

| Prog No. | Survey Results (08) | (09) | (10) | (11) | University Library (12) | Research Support (13) | (14) | Published Articles (15) | (16) | Survey Ratings Standard Error (08) | (09) | (10) | (11) |
|---|---|---|---|---|---|---|---|---|---|---|---|---|---|
| 041. | 3.3 | 2.0 | 1.1 | 1.1 | 1.2 | .62 | 2617 | 65 | | .08 | .05 | .06 | .06 |
| | 54 | 55 | 48 | 56 | 58 | 57 | 47 | 55 | 53 | | | | |
| 042. | 2.2 | 1.4 | 0.8 | 0.5 | -0.2 | NA | NA | 19 | | .13 | .09 | .11 | .06 |
| | 41 | 41 | 39 | 40 | 45 | | | 44 | 43 | | | | |
| 043. | 1.3 | 0.7 | NA | 0.2 | NA | NA | NA | 10 | | .24 | .20 | NA | .05 |
| | 30 | 26 | | 31 | | | | 41 | 42 | | | | |
| 044. | 1.8 | 1.1 | 0.9 | 0.4 | NA | NA | NA | 10 | | .19 | .16 | .09 | .06 |
| | 36 | 35 | 41 | 36 | | | | 41 | 42 | | | | |
| 045. | 2.2 | 1.5 | 1.2 | 0.6 | NA | .23 | NA | 0 | | .11 | .11 | .11 | .07 |
| | 41 | 45 | 51 | 42 | | 39 | | 39 | 40 | | | | |
| 046. | 2.5 | 1.6 | 1.3 | 0.6 | NA | .21 | 1962 | 12 | | .11 | .10 | .09 | .07 |
| | 44 | 47 | 56 | 43 | | 38 | 45 | 42 | 42 | | | | |
| 047. | 2.7 | 1.7 | 1.3 | 0.8 | -1.0 | .14 | 892 | 19 | | .09 | .07 | .09 | .07 |
| | 47 | 48 | 56 | 48 | 37 | 35 | 43 | 44 | 43 | | | | |
| 048. | 2.9 | 1.9 | 1.4 | 0.9 | 1.0 | .47 | 3087 | 17 | | .10 | .05 | .08 | .07 |
| | 50 | 53 | 58 | 51 | 56 | 50 | 48 | 43 | 44 | | | | |
| 049. | 1.4 | 0.8 | 0.9 | 0.3 | NA | NA | 703 | 1 | | .15 | .14 | .15 | .06 |
| | 31 | 27 | 41 | 35 | | | 42 | 39 | 40 | | | | |
| 050. | 3.6 | 2.1 | 0.8 | 1.2 | 0.3 | .77 | NA | 32 | | .08 | .05 | .08 | .06 |
| | 58 | 57 | 36 | 58 | 50 | 64 | | 47 | 49 | | | | |
| 051. | 3.2 | 1.7 | NA | 0.2 | 0.9 | NA | 1320 | 54 | | .29 | .16 | NA | .05 |
| | 53 | 49 | | 32 | 55 | | 44 | 53 | 46 | | | | |
| 052. | 2.9 | 1.8 | 1.1 | 0.8 | 0.9 | .50 | 1320 | 54 | | .10 | .07 | .08 | .07 |
| | 50 | 51 | 50 | 49 | 55 | 51 | 44 | 53 | 46 | | | | |
| 053. | 2.6 | 1.5 | 0.8 | 0.8 | -0.6 | .23 | 3117 | 10 | | .11 | .08 | .08 | .07 |
| | 46 | 43 | 37 | 47 | 41 | 39 | 48 | 41 | 41 | | | | |
| 054. | 2.6 | 1.6 | 1.2 | 1.0 | NA | .50 | 9024 | 52 | | .09 | .07 | .07 | .07 |
| | 45 | 47 | 51 | 52 | | 51 | 62 | 52 | 52 | | | | |
| 055. | 3.0 | 1.9 | 1.1 | 1.0 | -0.9 | .62 | 617 | 14 | | .09 | .05 | .07 | .07 |
| | 50 | 53 | 47 | 53 | 38 | 57 | 42 | 42 | 43 | | | | |
| 056. | 3.9 | 2.2 | 1.2 | 1.0 | 0.7 | .59 | 6180 | 54 | | .09 | .08 | .09 | .08 |
| | 62 | 60 | 51 | 54 | 53 | 56 | 55 | 53 | 54 | | | | |
| 057. | 3.3 | 2.0 | 0.9 | 0.8 | 0.7 | .36 | 6180 | 54 | | .10 | .07 | .08 | .07 |
| | 54 | 55 | 42 | 49 | 53 | 45 | 55 | 53 | 54 | | | | |
| 058. | 3.0 | 1.8 | 1.2 | 0.5 | 0.7 | NA | 6180 | 54 | | .13 | .08 | .08 | .07 |
| | 51 | 50 | 52 | 41 | 53 | | 55 | 53 | 54 | | | | |
| 059. | 2.3 | 1.4 | 1.0 | 0.5 | 0.1 | .33 | 1044 | 24 | | .12 | .10 | .11 | .06 |
| | 42 | 42 | 44 | 40 | 47 | 44 | 43 | 45 | 44 | | | | |
| 060. | 4.0 | 2.3 | 1.1 | 1.5 | 0.9 | .71 | 2832 | 48 | | .07 | .06 | .07 | .05 |
| | 63 | 62 | 48 | 69 | 55 | 61 | 47 | 51 | 53 | | | | |

NOTE: On the first line of data for every program, raw values for each measure are reported; on the second line values are reported in standardized form, with mean = 50 and standard deviation = 10. "NA" indicates that the value for a measure is not available. Since the scale used to compute measure (16) is entirely arbitrary, only values in standardized form are reported for this measure.

TABLE 5.1  Program Measures (Raw and Standardized Values) in Geosciences

| Prog No. | University - Department/Academic Unit | Program Size (01) | (02) | (03) | Characteristics of Program Graduates (04) | (05) | (06) | (07) |
|---|---|---|---|---|---|---|---|---|
| 061. | Princeton University *Geophysical Fluid Dynamics* | 17 *52* | 9 *43* | 12 *44* | NA | NA | NA | NA |
| 062. | Purdue University-West Lafayette *Geosciences* | 22 *58* | 19 *50* | 15 *45* | .36 *58* | 7.8 *41* | .70 *44* | .00 *33* |
| 063. | Rensselaer Polytechnic Institute *Geology* | 7 *39* | 8 *42* | 8 *42* | .31 *54* | 6.3 *58* | 1.00 *70* | .00 *33* |
| 064. | Rice University *Geology* | 11 *44* | 33 *60* | 25 *50* | .27 *50* | 5.5 *67* | .76 *49* | .10 *41* |
| 065. | SUNY at Albany *Geological Sciences** | 8 *41* | 6 *41* | 29 *52* | NA | NA | NA | NA |
| 066. | SUNY at Binghamton *Geological Sciences* | 10 *43* | 12 *45* | 11 *43* | .36 *58* | 7.5 *44* | .70 *44* | .20 *48* |
| 067. | SUNY at Stony Brook *Earth and Space Sciences* | 25 *62* | 25 *54* | 28 *51* | .16 *42* | 7.0 *50* | .94 *65* | .33 *58* |
| 068. | Saint Louis University *Earth and Atmospheric Sciences* | 5 *37* | 10 *44* | 17 *46* | NA | NA | NA | NA |
| 069. | South Carolina, University of-Columbia *Geology* | 20 *55* | 45 *68* | 20 *48* | .08 *36* | 6.2 *58* | .87 *58* | .32 *57* |
| 070. | Southern California, University of *Geological Sciences* | 13 *47* | 19 *50* | 26 *50* | .18 *43* | 7.8 *40* | .92 *63* | .31 *56* |
| 071. | Southern Methodist University *Geological Sciences* | 10 *43* | 11 *44* | 14 *45* | .80 *91* | 6.8 *52* | NA | NA |
| 072. | Stanford University *Applied Earth Sciences** | 10 *43* | 26 *55* | 23 *49* | .26 *50* | 6.9 *51* | .87 *59* | .10 *41* |
| 073. | Stanford University *Geology* | 14 *48* | 54 *74* | 45 *60* | .23 *48* | 7.3 *47* | .83 *56* | .19 *48* |
| 074. | Stanford University *Geophysics* | 8 *41* | 33 *60* | 39 *57* | .41 *61* | 5.3 *69* | .82 *54* | .19 *47* |
| 075. | Syracuse University *Geology* | 9 *42* | 20 *51* | 11 *43* | .33 *55* | 8.0 *39* | .61 *37* | .00 *33* |
| 076. | Tennessee, University of-Knoxville *Geological Sciences* | 9 *42* | 9 *43* | 8 *42* | NA | NA | NA | NA |
| 077. | Texas A & M University *Geology* | 17 *52* | 10 *44* | 14 *45* | .18 *44* | 6.2 *59* | .82 *54* | .09 *40* |
| 078. | Texas A & M University *Geophysics* | 14 *48* | 11 *44* | 11 *43* | .30 *53* | 7.5 *44* | .60 *36* | .00 *33* |
| 079. | Texas, University of-Austin *Geological Sciences* | 37 *77* | 33 *60* | 50 *62* | .36 *58* | 7.6 *43* | .80 *53* | .20 *48* |
| 080. | Texas, University of-Dallas *Geosciences* | 12 *46* | 17 *48* | 35 *55* | .13 *40* | 6.7 *53* | .71 *45* | .43 *65* |

\* indicates program was initiated since 1970.


NOTE: On the first line of data for every program, raw values for each measure are reported; on the second line values are reported in standardized form, with mean = 50 and standard deviation = 10. "NA" indicates that the value for a measure is not available.

TABLE 5.1  Program Measures (Raw and Standardized Values) in Geosciences

| Prog No. | Survey Results (08) | (09) | (10) | (11) | University Library (12) | Research Support (13) | (14) | Published Articles (15) | (16) | Survey Ratings Standard Error (08) | (09) | (10) | (11) |
|---|---|---|---|---|---|---|---|---|---|---|---|---|---|
| 061. | 4.2 | 2.2 | 1.2 | 0.6 | 0.9 | .47 | 2832 | 48 | | .11 | .11 | .11 | .07 |
| | 65 | 61 | 51 | 43 | 55 | 50 | 47 | 51 | 53 | | | | |
| 062. | 2.8 | 1.7 | 1.5 | 0.8 | −0.5 | .36 | 776 | 29 | | .10 | .07 | .10 | .07 |
| | 49 | 49 | 64 | 47 | 42 | 45 | 42 | 46 | 48 | | | | |
| 063. | 2.4 | 1.4 | 1.0 | 0.7 | NA | NA | NA | 14 | | .11 | .08 | .07 | .07 |
| | 44 | 43 | 45 | 46 | | | | 42 | 43 | | | | |
| 064. | 2.6 | 1.6 | 0.5 | 1.0 | −1.4 | .27 | NA | 34 | | .08 | .08 | .08 | .07 |
| | 47 | 47 | 27 | 54 | 33 | 41 | | 48 | 48 | | | | |
| 065. | 3.5 | 1.8 | 1.2 | 1.2 | −1.0 | NA | 2528 | 29 | | .11 | .07 | .06 | .08 |
| | 57 | 52 | 50 | 60 | 37 | | 47 | 46 | 47 | | | | |
| 066. | 2.6 | 1.6 | 0.9 | 0.8 | NA | .50 | NA | 11 | | .09 | .08 | .07 | .06 |
| | 46 | 46 | 40 | 49 | | 51 | | 42 | 42 | | | | |
| 067. | 3.7 | 2.1 | 1.2 | 1.2 | −0.6 | .60 | 3096 | 40 | | .08 | .06 | .08 | .07 |
| | 59 | 58 | 52 | 58 | 41 | 56 | 48 | 49 | 50 | | | | |
| 068. | 2.1 | 1.5 | 1.0 | 0.4 | NA | NA | 672 | 17 | | .21 | .11 | .12 | .07 |
| | 39 | 45 | 43 | 38 | | | 42 | 43 | 43 | | | | |
| 069. | 2.9 | 1.7 | 1.7 | 0.9 | −0.4 | .50 | 1626 | 23 | | .11 | .09 | .07 | .08 |
| | 50 | 50 | 70 | 51 | 43 | 51 | 44 | 45 | 43 | | | | |
| 070. | 3.1 | 1.9 | 1.3 | 1.0 | 0.4 | .62 | 3284 | 23 | | .08 | .05 | .07 | .07 |
| | 52 | 54 | 56 | 55 | 50 | 57 | 48 | 45 | 44 | | | | |
| 071. | 2.5 | 1.6 | 1.2 | 0.8 | NA | .40 | NA | 5 | | .09 | .07 | .08 | .07 |
| | 45 | 46 | 51 | 49 | | 47 | | 40 | 41 | | | | |
| 072. | 3.4 | 2.1 | 0.8 | 0.9 | 2.0 | .40 | 2790 | 123 | | .12 | .09 | .09 | .07 |
| | 56 | 57 | 38 | 51 | 66 | 47 | 47 | 70 | 67 | | | | |
| 073. | 3.7 | 2.3 | 0.6 | 1.4 | 2.0 | .50 | 2790 | 123 | | .09 | .06 | .08 | .07 |
| | 60 | 62 | 29 | 63 | 66 | 51 | 47 | 70 | 67 | | | | |
| 074. | 4.2 | 2.5 | 1.4 | 1.1 | 2.0 | NA | 2790 | 123 | | .07 | .08 | .07 | .08 |
| | 66 | 66 | 59 | 57 | 66 | | 47 | 70 | 67 | | | | |
| 075. | 2.3 | 1.5 | 0.9 | 0.6 | −0.3 | NA | NA | 2 | | .10 | .10 | .08 | .07 |
| | 42 | 44 | 42 | 43 | 43 | | | 39 | 40 | | | | |
| 076. | 2.0 | 1.3 | 1.1 | 0.7 | −0.4 | NA | 689 | 15 | | .09 | .08 | .09 | .06 |
| | 39 | 40 | 48 | 44 | 43 | | 42 | 43 | 45 | | | | |
| 077. | 3.0 | 1.8 | 1.3 | 0.9 | −0.5 | .24 | 5745 | 54 | | .10 | .07 | .08 | .07 |
| | 50 | 50 | 56 | 50 | 42 | 39 | 54 | 53 | 50 | | | | |
| 078. | 3.1 | 1.8 | 1.5 | 0.8 | −0.5 | .29 | 5745 | 54 | | .11 | .07 | .10 | .07 |
| | 53 | 50 | 61 | 48 | 42 | 42 | 54 | 53 | 50 | | | | |
| 079. | 3.8 | 2.2 | 1.4 | 1.4 | 1.6 | .32 | 10789 | 62 | | .08 | .05 | .06 | .06 |
| | 60 | 60 | 60 | 64 | 62 | 43 | 66 | 55 | 50 | | | | |
| 080. | 2.6 | 1.7 | 1.2 | 0.8 | NA | .75 | 652 | NA | | .09 | .07 | .09 | .07 |
| | 46 | 49 | 52 | 49 | | 63 | 42 | | NA | | | | |

NOTE: On the first line of data for every program, raw values for each measure are reported;
on the second line values are reported in standardized form, with mean = 50 and
standard deviation = 10. "NA" indicates that the value for a measure is not available.
Since the scale used to compute measure (16) is entirely arbitrary, only values in
standardized form are reported for this measure.

TABLE 5.1  Program Measures (Raw and Standardized Values) in Geosciences

| Prog No. | University - Department/Academic Unit | Program Size | | | Characteristics of Program Graduates | | | |
|---|---|---|---|---|---|---|---|---|
| | | (01) | (02) | (03) | (04) | (05) | (06) | (07) |
| 081. | Tulsa, University of | 5 | 7 | 14 | NA | NA | NA | NA |
| | *Geosciences* | *37* | *42* | *45* | | | | |
| 082. | Utah, University of-Salt Lake City | 24 | 20 | 30 | .21 | 7.5 | .64 | .07 |
| | *Geology and Geophysics* | *60* | *51* | *52* | *46* | *44* | *39* | *39* |
| 083. | Virginia Polytechnic Institute & State Univ | 22 | 13 | 12 | NA | NA | NA | NA |
| | *Geological Sciences* | *58* | *46* | *44* | | | | |
| 084. | Virginia, University of | 26 | 17 | 25 | .18 | 7.8 | .46 | .09 |
| | *Environmental Sciences* | *63* | *48* | *50* | *44* | *41* | *23* | *40* |
| 085. | Washington State University-Pullman | 10 | 13 | 9 | NA | NA | NA | NA |
| | *Geology* | *43* | *46* | *42* | | | | |
| 086. | Washington University-Saint Louis | 12 | 4 | 18 | NA | NA | NA | NA |
| | *Earth and Planetary Sciences* | *46* | *40* | *47* | | | | |
| 087. | Washington, University of-Seattle | 17 | 41 | 16 | .29 | 6.2 | .73 | .40 |
| | *Geological Sciences* | *52* | *65* | *46* | *52* | *58* | *46* | *63* |
| 088. | West Virginia University | 16 | 12 | 5 | NA | NA | NA | NA |
| | *Geology and Geography* | *51* | *45* | *40* | | | | |
| 089. | Wisconsin, University of-Madison | 19 | 33 | 45 | .22 | 6.4 | .87 | .27 |
| | *Geology and Geophysics* | *54* | *60* | *60* | *47* | *57* | *59* | *53* |
| 090. | Wyoming, University of | 17 | 18 | 22 | .39 | 7.0 | .69 | .15 |
| | *Geology* | *52* | *49* | *49* | *59* | *50* | *44* | *45* |
| 091. | Yale University | 23 | 33 | 40 | .27 | 6.1 | .83 | .53 |
| | *Geology and Geophysics* | *59* | *60* | *57* | *50* | *60* | *56* | *73* |

* indicates program was initiated since 1970.

NOTE: On the first line of data for every program, raw values for each measure are reported; on the second line values are reported in standardized form, with mean = 50 and standard deviation = 10. "NA" indicates that the value for a measure is not available.

TABLE 5.1  Program Measures (Raw and Standardized Values) in Geosciences

| Prog No. | Survey Results (08) | (09) | (10) | (11) | University Library (12) | Research Support (13) | (14) | Published Articles (15) | (16) | Survey Ratings Standard Error (08) | (09) | (10) | (11) |
|---|---|---|---|---|---|---|---|---|---|---|---|---|---|
| 081. | 1.0 | 0.7 | 0.7 | 0.3 | NA | NA | NA | 9 | | .15 | .11 | .19 | .05 |
| | 26 | 25 | 33 | 34 | | | | 41 | 41 | | | | |
| 082. | 3.0 | 1.9 | 1.4 | 0.8 | -0.6 | .33 | 3853 | 29 | | .10 | .05 | .09 | .07 |
| | 51 | 53 | 60 | 49 | 41 | 44 | 50 | 46 | 46 | | | | |
| 083. | 3.7 | 2.1 | 1.6 | 1.3 | -0.0 | .59 | 2586 | 48 | | .08 | .06 | .07 | .07 |
| | 59 | 58 | 66 | 61 | 46 | 56 | 47 | 51 | 50 | | | | |
| 084. | 2.3 | 1.4 | 1.4 | 0.4 | 0.7 | .27 | 1394 | 13 | | .15 | .12 | .19 | .06 |
| | 43 | 42 | 59 | 36 | 54 | 41 | 44 | 42 | 45 | | | | |
| 085. | 2.1 | 1.3 | 1.0 | 0.5 | -0.3 | .10 | NA | 14 | | .12 | .10 | .07 | .07 |
| | 40 | 40 | 46 | 39 | 44 | 33 | | 42 | 43 | | | | |
| 086. | 2.7 | 1.8 | 1.7 | 0.6 | -0.4 | .50 | 830 | 19 | | .13 | .10 | .09 | .08 |
| | 47 | 50 | 70 | 43 | 43 | 51 | 43 | 44 | 44 | | | | |
| 087. | 3.3 | 2.0 | 1.2 | 1.2 | 1.5 | .47 | 13047 | 90 | | .08 | .05 | .07 | .07 |
| | 55 | 56 | 50 | 58 | 61 | 50 | 71 | 62 | 65 | | | | |
| 088. | 2.1 | 1.3 | 1.1 | 0.4 | NA | .06 | 644 | 9 | | .15 | .12 | .12 | .06 |
| | 40 | 40 | 49 | 36 | | 31 | 42 | 41 | 42 | | | | |
| 089. | 3.7 | 2.3 | 1.1 | 1.2 | 1.6 | .74 | 11910 | 86 | | .08 | .06 | .07 | .07 |
| | 60 | 62 | 48 | 60 | 62 | 63 | 68 | 61 | 55 | | | | |
| 090. | 2.8 | 1.8 | 1.3 | 0.8 | NA | .47 | 2474 | 21 | | .08 | .07 | .09 | .07 |
| | 48 | 50 | 55 | 48 | | 50 | 46 | 44 | 44 | | | | |
| 091. | 4.1 | 2.3 | 0.8 | 1.5 | 2.1 | .78 | 715 | 36 | | .08 | .07 | .07 | .06 |
| | 64 | 62 | 37 | 68 | 67 | 65 | 42 | 48 | 51 | | | | |

NOTE: On the first line of data for every program, raw values for each measure are reported; on the second line values are reported in standardized form, with mean = 50 and standard deviation = 10. "NA" indicates that the value for a measure is not available. Since the scale used to compute measure (16) is entirely arbitrary, only values in standardized form are reported for this measure.

TABLE 5.2  Summary Statistics Describing Each Program Measure—Geosciences

| Measure | Number of Programs Evaluated | Mean | Standard Deviation | DECILES 1 | 2 | 3 | 4 | 5 | 6 | 7 | 8 | 9 |
|---|---|---|---|---|---|---|---|---|---|---|---|---|
| **Program Size** | | | | | | | | | | | | |
| 01 Raw Value | 91 | 16 | 8 | 7 | 9 | 10 | 12 | 14 | 17 | 18 | 20 | 25 |
| Std Value | 91 | 50 | 10 | 39 | 42 | 43 | 46 | 48 | 52 | 53 | 55 | 62 |
| 02 Raw Value | 91 | 19 | 15 | 6 | 9 | 11 | 12 | 16 | 18 | 20 | 27 | 33 |
| Std Value | 91 | 50 | 10 | 41 | 43 | 44 | 45 | 48 | 49 | 51 | 55 | 60 |
| 03 Raw Value | 89 | 25 | 21 | 7 | 9 | 12 | 15 | 18 | 24 | 28 | 39 | 45 |
| Std Value | 89 | 50 | 10 | 41 | 42 | 44 | 45 | 47 | 50 | 51 | 57 | 60 |
| **Program Graduates** | | | | | | | | | | | | |
| 04 Raw Value | 66 | .26 | .13 | .12 | .16 | .18 | .21 | .24 | .28 | .31 | .36 | .39 |
| Std Value | 66 | 50 | 10 | 39 | 42 | 44 | 46 | 48 | 52 | 54 | 58 | 60 |
| 05 Raw Value | 66 | 7.0 | .9 | 8.3 | 7.7 | 7.5 | 7.3 | 6.9 | 6.7 | 6.5 | 6.2 | 5.9 |
| Std Value | 66 | 50 | 10 | 35 | 42 | 44 | 46 | 51 | 53 | 55 | 59 | 62 |
| 06 Raw Value | 63 | .77 | .12 | .60 | .68 | .70 | .74 | .79 | .81 | .82 | .86 | .92 |
| Std Value | 63 | 50 | 10 | 36 | 43 | 44 | 48 | 52 | 53 | 54 | 58 | 63 |
| 07 Raw Value | 63 | .22 | .13 | .00 | .10 | .14 | .18 | .21 | .27 | .30 | .34 | .39 |
| Std Value | 63 | 50 | 10 | 33 | 41 | 44 | 47 | 49 | 54 | 56 | 59 | 63 |
| **Survey Results** | | | | | | | | | | | | |
| 08 Raw Value | 91 | 2.9 | .8 | 2.0 | 2.2 | 2.5 | 2.7 | 2.9 | 3.1 | 3.3 | 3.7 | 4.1 |
| Std Value | 91 | 50 | 10 | 39 | 41 | 45 | 47 | 50 | 52 | 55 | 59 | 64 |
| 09 Raw Value | 91 | 1.8 | .4 | 1.1 | 1.4 | 1.5 | 1.7 | 1.8 | 1.9 | 2.0 | 2.1 | 2.3 |
| Std Value | 91 | 50 | 10 | 35 | 42 | 44 | 49 | 51 | 53 | 56 | 58 | 63 |
| 10 Raw Value | 89 | 1.1 | .3 | .8 | .9 | 1.0 | 1.1 | 1.2 | 1.2 | 1.3 | 1.4 | 1.5 |
| Std Value | 89 | 50 | 10 | 37 | 41 | 45 | 49 | 52 | 52 | 56 | 60 | 63 |
| 11 Raw Value | 91 | .9 | .4 | .4 | .5 | .6 | .8 | .8 | .9 | 1.0 | 1.2 | 1.4 |
| Std Value | 91 | 50 | 10 | 37 | 40 | 43 | 48 | 48 | 51 | 54 | 59 | 65 |
| **University Library** | | | | | | | | | | | | |
| 12 Raw Value | 69 | .4 | 1.0 | -1.0 | -.6 | -.4 | -.3 | .0 | .7 | .9 | 1.5 | 2.0 |
| Std Value | 69 | 50 | 10 | 37 | 41 | 43 | 44 | 47 | 53 | 55 | 61 | 66 |
| **Research Support** | | | | | | | | | | | | |
| 13 Raw Value | 72 | .47 | .22 | .20 | .27 | .32 | .38 | .47 | .50 | .60 | .66 | .75 |
| Std Value | 72 | 50 | 10 | 38 | 41 | 43 | 46 | 50 | 51 | 56 | 59 | 63 |
| 14 Raw Value | 73 | 3996 | 4279 | 677 | 840 | 1320 | 2250 | 2637 | 2873 | 3341 | 6180 | 9785 |
| Std Value | 73 | 50 | 10 | 42 | 43 | 44 | 46 | 47 | 47 | 48 | 55 | 64 |
| **Publication Records** | | | | | | | | | | | | |
| 15 Raw Value | 90 | 44 | 39 | 9 | 13 | 17 | 23 | 32 | 38 | 54 | 64 | 102 |
| Std Value | 90 | 50 | 10 | 41 | 42 | 43 | 45 | 47 | 49 | 53 | 55 | 65 |
| 16 Std Value | 90 | 50 | 10 | 41 | 42 | 43 | 44 | 46 | 49 | 51 | 54 | 67 |

NOTE: Standardized values reported in the preceding table have been computed from exact values of the mean and standard deviation and not the rounded values reported here.  Since the scale used to compute measure 16 is entirely arbitrary, only data in standardized form are reported for this measure.

TABLE 5.3  Intercorrelations Among Program Measures on 91 Programs in Geosciences

Measure

|  | 01 | 02 | 03 | 04 | 05 | 06 | 07 | 08 | 09 | 10 | 11 | 12 | 13 | 14 | 15 | 16 |
|---|---|---|---|---|---|---|---|---|---|---|---|---|---|---|---|---|
| **Program Size** | | | | | | | | | | | | | | | | |
| 01 |  | .42 | .40 | -.06 | -.08 | -.14 | .21 | .45 | .42 | .47 | .39 | .22 | .18 | .61 | .39 | .36 |
| 02 |  |  | .72 | -.01 | .29 | .05 | .36 | .64 | .67 | .06 | .67 | .43 | .40 | .25 | .73 | .74 |
| 03 |  |  |  | -.04 | .16 | .06 | .45 | .61 | .62 | .17 | .59 | .30 | .33 | .28 | .66 | .64 |
| **Program Graduates** | | | | | | | | | | | | | | | | |
| 04 |  |  |  |  | .24 | .00 | .04 | .08 | .10 | -.10 | .17 | -.01 | .09 | .22 | .03 | .07 |
| 05 |  |  |  |  |  | .51 | .49 | .50 | .51 | -.09 | .44 | .26 | .38 | -.05 | .27 | .31 |
| 06 |  |  |  |  |  |  | .30 | .24 | .29 | -.13 | .35 | .10 | .20 | -.04 | .01 | .00 |
| 07 |  |  |  |  |  |  |  | .58 | .58 | .05 | .59 | .31 | .61 | .06 | .33 | .39 |
| **Survey Results** | | | | | | | | | | | | | | | | |
| 08 |  |  |  |  |  |  |  |  | .97 | .29 | .87 | .58 | .72 | .27 | .75 | .77 |
| 09 |  |  |  |  |  |  |  |  |  | .29 | .87 | .58 | .72 | .25 | .73 | .75 |
| 10 |  |  |  |  |  |  |  |  |  |  | .19 | -.08 | .02 | .13 | .09 | .09 |
| 11 |  |  |  |  |  |  |  |  |  |  |  | .43 | .70 | .18 | .66 | .70 |
| **University Library** | | | | | | | | | | | | | | | | |
| 12 |  |  |  |  |  |  |  |  |  |  |  |  | .36 | .33 | .66 | .66 |
| **Research Support** | | | | | | | | | | | | | | | | |
| 13 |  |  |  |  |  |  |  |  |  |  |  |  |  | .20 | .45 | .51 |
| 14 |  |  |  |  |  |  |  |  |  |  |  |  |  |  | .42 | .35 |
| **Publication Records** | | | | | | | | | | | | | | | | |
| 15 |  |  |  |  |  |  |  |  |  |  |  |  |  |  |  | .97 |
| 16 |  |  |  |  |  |  |  |  |  |  |  |  |  |  |  |  |

NOTE: Since in computing correlation coefficients program data must be available for both of the measures being correlated, the actual number of programs on which each coefficient is based varies.

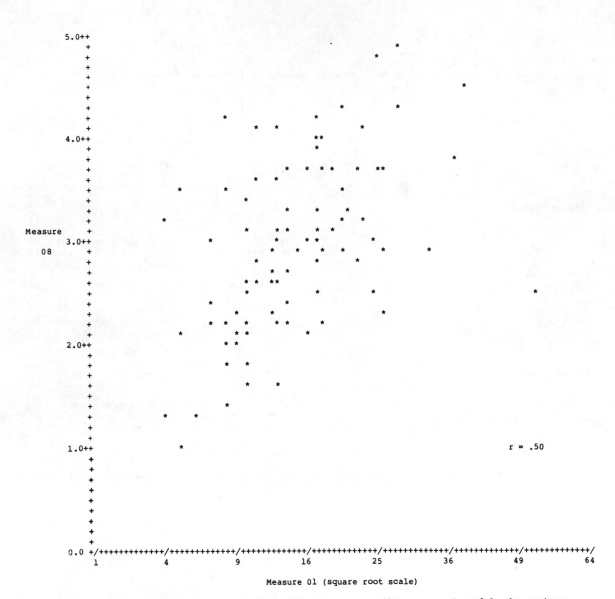

Measure 01 (square root scale)

FIGURE 5.1  Mean rating of scholarly quality of faculty (measure 08) versus number of faculty members
(measure 01)--91 programs in geosciences.

91

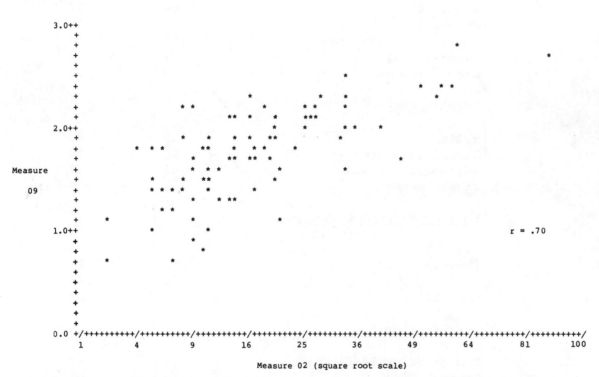

FIGURE 5.2  Mean rating of program effectiveness in educating research scholars/scientists (measure 09) versus number of graduates in last five years (measure 02)--91 programs in geosciences.

TABLE 5.4  Characteristics of Survey Participants in Geosciences

|  | Respondents | |
|---|---|---|
|  | N | % |
| **Field of Specialization** | | |
| Geochemistry | 18 | 10 |
| Geology | 66 | 37 |
| Geophysics | 29 | 16 |
| Mineralogy/Petrology | 31 | 18 |
| Paleontology | 13 | 7 |
| Other/Unknown | 20 | 11 |
| **Faculty Rank** | | |
| Professor | 115 | 65 |
| Associate Professor | 36 | 20 |
| Assistant Professor | 21 | 12 |
| Other/Unknown | 5 | 3 |
| **Year of Highest Degree** | | |
| Pre-1950 | 8 | 5 |
| 1950-59 | 51 | 29 |
| 1960-69 | 70 | 40 |
| Post-1969 | 48 | 27 |
| **Evaluator Selection** | | |
| Nominated by Institution | 154 | 87 |
| Other | 23 | 13 |
| **Survey Form** | | |
| With Faculty Names | 157 | 89 |
| Without Names | 20 | 11 |
| **Total Evaluators** | 177 | 100 |

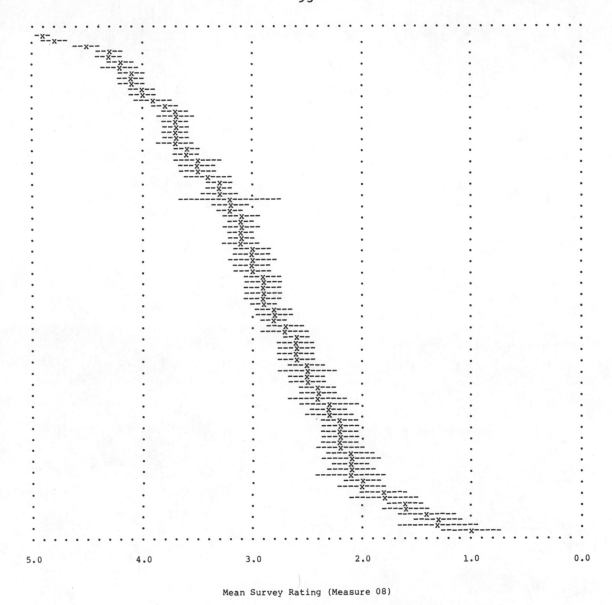

5.0            4.0            3.0            2.0            1.0            0.0

Mean Survey Rating (Measure 08)

FIGURE 5.3  Mean rating of scholarly quality of faculty in 91 programs in geosciences.

NOTE:  Programs are listed in sequence of mean rating, with the highest-rated program appearing at the top of the page.  The broken lines (---) indicate a confidence interval of ±1.5 standard errors around the reported mean (x) of each program.

# VI
# Mathematics Programs

In this chapter 115 research-doctorate programs in mathematics are assessed.  These programs, according to the information supplied by their universities, have accounted for 2,731 doctoral degrees awarded during the FY1976-80 period--approximately 92 percent of the aggregate number of mathematics doctorates earned from U.S. universities in this five-year span.[1]  On the average, 35 full-time and part-time students intending to earn doctorates were enrolled in a program in December 1980, with an average faculty size of 33 members.[2]  The 115 programs, listed in Table 6.1, represent 114 different universities--two programs are included from the University of Maryland (College Park).  All but six of the programs were initiated prior to 1970.  In addition to the 114 universities represented in this discipline, another 3 were initially identified as meeting the criteria[3] for inclusion in the assessment:

> Idaho State University
> Lehigh University
> University of Northern Colorado

Mathematics programs at these three institutions have not been included in the evaluations in this discipline, since in each case the study coordinator either indicated that the institution did not at that time have a research-doctorate program in mathematics or failed to provide the information requested by the committee.

Before examining the individual program results presented in Table 6.1, the reader is urged to refer to Chapter II, in which each of the 16 measures used in the assessment is discussed.  Summary

---

[1] Data from the NRC's Survey of Earned Doctorates indicate that 2,958 research doctorates in mathematics were awarded by U.S. universities between FY1976 and FY1980.

[2] See the reported means for measures 03 and 01 in Table 6.2.

[3] As mentioned in Chapter I, the primary criterion for inclusion was that a university had awarded at least 7 doctorates in mathematics during the FY1976-78 period.

statistics describing every measure are given in Table 6.2. For 10 of the measures, data are reported for at least 108 of the 115 mathematics programs. For measures 04-07, which pertain to characteristics of the program graduates, data are presented for 95 (or more) of the programs; the other 20 programs had too few graduates on which to base statistics.[4] For measure 12, a composite index of the size of a university library, data are available for 82 programs; for measure 14, total university expenditures for research in this discipline, data are available for 83 programs. With respect to the latter measure, it should be noted that reported data include expenditures for research in statistics as well as in mathematics. The programs not evaluated on measures 12 and 14 are typically smaller--in terms of faculty size and graduate student enrollment--than other mathematics programs. Were data on measures 12 and 14 available for all 91 programs, it is likely that the reported means for these measures would be appreciably lower (and that some of the correlations of these measures with others would be higher).

Intercorrelations among the 16 measures (Pearson product-moment coefficients) are given in Table 6.3. Of particular note are the high positive correlations of the measures of the numbers of doctoral graduates and students (02, 03) with measures of publication records (15, 16) and reputational survey ratings (08, 09, and 11). Figure 6.1 illustrates the relation between the mean rating of the scholarly quality of faculty (measure 08) and the number of faculty members (measure 01) for each of the 114 mathematics programs. Figure 6.2 plots the mean rating of program effectiveness (measure 09) against the total number of FY1976-80 program graduates (measure 02). Although in both figures there is a significant positive correlation between program size and reputational rating, it is quite apparent that some of the smaller programs received high mean ratings and some of the larger programs received low mean ratings.

Table 6.4 describes the 223 faculty members who participated in the evaluation of mathematics programs. These individuals constituted 64 percent of those asked to respond to the survey in this discipline and 6 percent of the faculty population in the 115 research-doctorate programs being evaluated.[5] More than one-third of the survey participants were in the specialty area of analysis/functional analysis, and almost two-thirds were full professors. More than two-thirds had earned their highest degree prior to 1970.

Several exceptions should be noted with regard to the survey evaluations in this discipline. In the initial survey mailing the list of faculty in the Brown University program included only applied mathematicians. At the request of the study coordinator at this

---

[4]As mentioned in Chapter II, data for measures 04-07 are not reported if they are based on the survey responses of fewer than 10 FY1975-79 program graduates.
[5]See Table 2.3 in Chapter II.

institution and a member of the study committee, the names of another 24 mathematics faculty members were added to the list, and revised survey forms that included the Brown program along with 11 other (randomly selected) mathematics programs were sent to 178 evaluators in this discipline.[6] The responses to the second survey were used to compute mean ratings for the Brown program. Another problem with the survey evaluations in mathematics involved the mislabeling of the location of an institution. In the program listing on the survey form, New Mexico State University at Las Cruces was identified as being located in Alamogordo, which has a junior college branch of the same institution. Since a large majority of faculty evaluators indicated that they were unfamiliar with this program and it is quite possible that some of them were misled by the inaccurate identification of this institution, the committee has decided not to report the survey results for this program. Two other instances of mislabeling were called to the attention of the committee. The program at the Courant Institute was identified as "New York University--Mathematics," and the Wesleyan University program in the Department of Mathematics was called "Physical Sciences." The committee has decided in both instances to report the survey results but cautions that the reputational ratings may have been influenced by the use of inaccurate program titles on the survey form.

To assist the reader in interpreting results of the survey evaluations, estimated standard errors have been computed for mean ratings of the scholarly quality of faculty in 114 mathematics programs (and are given in Table 6.1). For each program the mean rating and an associated "confidence interval" of 1.5 standard errors are illustrated in Figure 6.3 (listed in order of highest to lowest mean rating). In comparing two programs, if their confidence intervals do not overlap, one may conclude that there is a significant difference in their mean ratings at a .05 level of significance.[7] From this figure it is also apparent that one should have somewhat more confidence in the accuracy of the mean ratings of higher-rated programs than lower-rated programs. This generalization results primarily from the fact that evaluators are not as likely to be familiar with the less prestigious programs, and consequently the mean ratings of these programs are usually based on fewer survey responses.

---

[6] See Chapter IX for a comparison of the "resurvey" results with the original survey ratings for these 11 other mathematics programs.
[7] See pp. 29-31 for a discussion of the interpretation of mean ratings and associated confidence intervals.

TABLE 6.1  Program Measures (Raw and Standardized Values) in Mathematics

| Prog No. | University – Department/Academic Unit | Program Size | | | Characteristics of Program Graduates | | | |
|---|---|---|---|---|---|---|---|---|
| | | (01) | (02) | (03) | (04) | (05) | (06) | (07) |
| 001. | Adelphi University | 14 | 9 | 37 | .29 | 8.5 | .79 | .07 |
| | *Mathematics and Computer Science* | *39* | *43* | *51* | *52* | *34* | *61* | *38* |
| 002. | Alabama, University of–Tuscaloosa | 19 | 19 | 4 | .25 | 7.8 | .53 | .13 |
| | *Mathematics* | *42* | *48* | *40* | *50* | *40* | *45* | *42* |
| 003. | Arizona, University of–Tucson | 46 | 14 | 14 | .17 | 5.5 | .75 | .33 |
| | *Mathematics* | *58* | *46* | *43* | *44* | *59* | *58* | *56* |
| 004. | Auburn University | 34 | 13 | 26 | .15 | 7.0 | .69 | .31 |
| | *Mathematics* | *51* | *45* | *47* | *43* | *46* | *55* | *54* |
| 005. | Boston University | 17 | 9 | 25 | .08 | 7.5 | .73 | .27 |
| | *Mathematics* | *40* | *43* | *47* | *38* | *42* | *57* | *52* |
| 006. | Bowling Green State University | 25 | 11 | 13 | NA | NA | NA | NA |
| | *Mathematics and Statistics** | *45* | *44* | *43* | | | | |
| 007. | Brandeis University | 12 | 21 | 29 | .14 | 5.5 | .80 | .40 |
| | *Mathematics* | *37* | *49* | *48* | *42* | *59* | *61* | *60* |
| 008. | Brown University | 43 | 80 | 78 | .11 | 6.2 | .63 | .26 |
| | *Mathematics and Applied Mathematics* | *56* | *75* | *64* | *40* | *53* | *51* | *51* |
| 009. | CUNY–Graduate School | 26 | 33 | 57 | .06 | 7.0 | .42 | .10 |
| | *Mathematics* | *46* | *54* | *57* | *37* | *46* | *38* | *39* |
| 010. | California Institute of Technology | 20 | 13 | 17 | .25 | 4.2 | .70 | .45 |
| | *Mathematics* | *42* | *45* | *44* | *50* | *70* | *55* | *64* |
| 011. | California, University of–Berkeley | 72 | 160 | 209 | .30 | 5.7 | .72 | .45 |
| | *Mathematics* | *74* | *99* | *99* | *53* | *57* | *56* | *63* |
| 012. | California, University of–Davis | 32 | 16 | 12 | .09 | 6.1 | .40 | .10 |
| | *Mathematics* | *49* | *47* | *43* | *39* | *54* | *37* | *40* |
| 013. | California, University of–Los Angeles | 85 | 50 | 115 | .15 | 6.1 | .71 | .37 |
| | *Mathematics/Biomathematics* | *82* | *62* | *76* | *43* | *54* | *56* | *58* |
| 014. | California, University of–Riverside | 16 | 21 | 23 | .17 | 5.4 | .47 | .18 |
| | *Mathematics* | *40* | *49* | *46* | *44* | *60* | *42* | *45* |
| 015. | California, University of–San Diego | 42 | 31 | 41 | .14 | 5.8 | .71 | .49 |
| | *Mathematics* | *56* | *53* | *52* | *42* | *57* | *56* | *66* |
| 016. | California, University of–Santa Barbara | 35 | 26 | 56 | .00 | 5.0 | .43 | .21 |
| | *Mathematics* | *51* | *51* | *57* | *32* | *63* | *39* | *47* |
| 017. | Carnegie–Mellon University | 22 | 24 | 40 | .35 | 7.5 | .74 | .33 |
| | *Mathematics* | *43* | *50* | *52* | *57* | *42* | *58* | *56* |
| 018. | Case Western Reserve University | 23 | 16 | 10 | .25 | 6.2 | NA | NA |
| | *Mathematics and Statistics* | *44* | *47* | *42* | *50* | *53* | | |
| 019. | Chicago, University of | 30 | 60 | 77 | .30 | 4.8 | .91 | .63 |
| | *Mathematics* | *48* | *66* | *64* | *53* | *65* | *68* | *76* |
| 020. | Cincinnati, University of | 30 | 12 | 15 | .18 | 7.0 | .36 | .18 |
| | *Mathematics* | *48* | *45* | *44* | *45* | *46* | *35* | *45* |

\* indicates program was initiated since 1970.

NOTE: On the first line of data for every program, raw values for each measure are reported; on the second line values are reported in standardized form, with mean = 50 and standard deviation = 10. "NA" indicates that the value for a measure is not available.

TABLE 6.1  Program Measures (Raw and Standardized Values) in Mathematics

| Prog No. | Survey Results (08) | (09) | (10) | (11) | University Library (12) | Research Support (13) | (14) | Published Articles (15) | (16) | Survey Ratings Standard Error (08) | (09) | (10) | (11) |
|---|---|---|---|---|---|---|---|---|---|---|---|---|---|
| 001. | 0.9 | 0.7 | NA | 0.2 | NA | .00 | NA | 0 | | .20 | .17 | NA | .04 |
| | 33 | 35 | | 35 | | 31 | | 37 | 41 | | | | |
| 002. | 1.1 | 0.4 | 1.1 | 0.2 | -1.3 | .05 | NA | 12 | | .13 | .10 | .10 | .05 |
| | 34 | 30 | 45 | 36 | 36 | 34 | | 41 | 42 | | | | |
| 003. | 2.6 | 1.6 | 1.3 | 0.7 | 0.9 | .28 | 294 | 30 | | .12 | .11 | .10 | .07 |
| | 50 | 50 | 58 | 49 | 58 | 48 | 46 | 47 | 44 | | | | |
| 004. | 1.9 | 1.1 | 1.1 | 0.4 | NA | .18 | NA | 23 | | .15 | .13 | .09 | .06 |
| | 42 | 42 | 47 | 41 | | 42 | | 45 | 45 | | | | |
| 005. | 1.7 | 0.8 | 1.2 | 0.3 | -0.4 | .18 | 151 | 6 | | .13 | .13 | .15 | .05 |
| | 40 | 36 | 52 | 38 | 46 | 42 | 44 | 39 | 41 | | | | |
| 006. | 1.3 | 0.6 | NA | 0.3 | NA | .16 | NA | 15 | | .15 | .14 | NA | .06 |
| | 37 | 34 | | 39 | | 41 | | 42 | 44 | | | | |
| 007. | 3.8 | 2.1 | 0.8 | 1.1 | NA | .50 | 209 | 43 | | .07 | .06 | .08 | .08 |
| | 61 | 59 | 31 | 59 | | 61 | 45 | 51 | 53 | | | | |
| 008. | 4.1 | 2.4 | 1.3 | 1.4 | -1.1 | .58 | 1214 | 43 | | .06 | .06 | .06 | .06 |
| | 64 | 64 | 57 | 66 | 39 | 66 | 58 | 51 | 54 | | | | |
| 009. | 3.5 | 1.9 | 0.9 | 1.0 | NA | .54 | 199 | 98 | | .08 | .09 | .07 | .07 |
| | 59 | 56 | 38 | 56 | | 63 | 45 | 70 | 58 | | | | |
| 010. | 3.8 | 2.2 | 1.1 | 1.3 | NA | .45 | 212 | 47 | | .08 | .07 | .07 | .06 |
| | 62 | 62 | 46 | 63 | | 58 | 45 | 53 | 51 | | | | |
| 011. | 4.9 | 2.7 | 1.2 | 1.6 | 2.2 | .39 | 1292 | 141 | | .03 | .05 | .05 | .06 |
| | 72 | 71 | 50 | 71 | 71 | 54 | 59 | 84 | 91 | | | | |
| 012. | 2.3 | 1.4 | 1.1 | 0.4 | 0.6 | .16 | NA | 22 | | .15 | .14 | .13 | .06 |
| | 46 | 48 | 45 | 41 | 55 | 41 | | 44 | 44 | | | | |
| 013. | 4.0 | 2.3 | 1.3 | 1.3 | 2.0 | .44 | 472 | 80 | | .06 | .06 | .06 | .07 |
| | 63 | 64 | 59 | 64 | 68 | 57 | 48 | 64 | 65 | | | | |
| 014. | 2.3 | 1.6 | 1.0 | 0.5 | -1.0 | .25 | 259 | 19 | | .14 | .13 | .08 | .07 |
| | 47 | 50 | 44 | 44 | 40 | 46 | 45 | 43 | 43 | | | | |
| 015. | 3.4 | 2.0 | 1.4 | 1.1 | -0.0 | .48 | 424 | 56 | | .08 | .06 | .08 | .07 |
| | 57 | 58 | 66 | 60 | 49 | 59 | 48 | 56 | 61 | | | | |
| 016. | 2.7 | 1.7 | 1.0 | 0.7 | -0.1 | .31 | 352 | 57 | | .11 | .10 | .10 | .07 |
| | 50 | 52 | 44 | 49 | 48 | 50 | 47 | 56 | 53 | | | | |
| 017. | 3.0 | 1.9 | 1.0 | 0.6 | NA | .59 | 664 | 35 | | .14 | .11 | .09 | .07 |
| | 53 | 57 | 44 | 45 | | 66 | 51 | 49 | 45 | | | | |
| 018. | 2.6 | 1.5 | 1.1 | 0.8 | -1.3 | .17 | NA | 20 | | .10 | .10 | .10 | .07 |
| | 50 | 50 | 49 | 51 | 36 | 42 | | 44 | 47 | | | | |
| 019. | 4.8 | 2.7 | 1.0 | 1.5 | 0.9 | .60 | 1119 | 69 | | .04 | .05 | .06 | .05 |
| | 71 | 70 | 42 | 68 | 58 | 67 | 56 | 60 | 58 | | | | |
| 020. | 1.9 | 1.2 | 1.3 | 0.5 | -0.2 | .20 | NA | 28 | | .12 | .12 | .11 | .06 |
| | 42 | 44 | 58 | 43 | 47 | 43 | | 46 | 44 | | | | |

NOTE: On the first line of data for every program, raw values for each measure are reported;
on the second line values are reported in standardized form, with mean = 50 and
standard deviation = 10. "NA" indicates that the value for a measure is not available.
Since the scale used to compute measure (16) is entirely arbitrary, only values in
standardized form are reported for this measure.

TABLE 6.1  Program Measures (Raw and Standardized Values) in Mathematics

| Prog No. | University - Department/Academic Unit | Program Size | | | Characteristics of Program Graduates | | | |
|---|---|---|---|---|---|---|---|---|
| | | (01) | (02) | (03) | (04) | (05) | (06) | (07) |
| 021. | Claremont Graduate School | 15 | 6 | 11 | .70 | NA | .20 | .00 |
| | *Mathematics* | *39* | *42* | *42* | *81* | | *25* | *33* |
| 022. | Clarkson College of Technology | 17 | 4 | 16 | NA | NA | NA | NA |
| | *Mathematics and Computer Science* | *40* | *41* | *44* | | | | |
| 023. | Clemson University | 28 | 12 | 19 | NA | NA | NA | NA |
| | *Mathematical Sciences* | *47* | *45* | *45* | | | | |
| 024. | Colorado State University-Fort Collins | 30 | 11 | 23 | .10 | 5.5 | NA | NA |
| | *Mathematics* | *48* | *44* | *46* | *39* | *59* | | |
| 025. | Colorado, University of | 44 | 34 | 45 | .17 | 7.0 | .63 | .30 |
| | *Mathematics* | *57* | *55* | *53* | *44* | *46* | *51* | *53* |
| 026. | Columbia University | 19 | 30 | 46 | .18 | 6.3 | .56 | .37 |
| | *Mathematics* | *42* | *53* | *54* | *45* | *52* | *47* | *58* |
| 027. | Connecticut, University of-Storrs | 23 | 6 | 12 | NA | NA | NA | NA |
| | *Mathematics* | *44* | *42* | *43* | | | | |
| 028. | Cornell University-Ithaca | 38 | 44 | 39 | .27 | 5.3 | .67 | .35 |
| | *Mathematics* | *53* | *59* | *51* | *51* | *61* | *54* | *57* |
| 029. | Dartmouth College | 19 | 16 | 13 | .28 | 5.3 | .71 | .29 |
| | *Mathematics* | *42* | *47* | *43* | *52* | *61* | *56* | *53* |
| 030. | Delaware, University of-Newark | 32 | 16 | 25 | .17 | 9.5 | .75 | .17 |
| | *Mathematical Sciences* | *49* | *47* | *47* | *44* | *25* | *58* | *44* |
| 031. | Denver, University of | 16 | 8 | 8 | NA | NA | NA | NA |
| | *Mathematics and Computer Science* | *40* | *43* | *42* | | | | |
| 032. | Duke University | 21 | 16 | 21 | .57 | 5.8 | .50 | .21 |
| | *Mathematics* | *43* | *47* | *46* | *72* | *57* | *43* | *47* |
| 033. | Emory University | 17 | 13 | 8 | .50 | 6.5 | .50 | .08 |
| | *Mathematics* | *40* | *45* | *42* | *67* | *51* | *43* | *38* |
| 034. | Florida State University-Tallahassee | 28 | 14 | 16 | .25 | 5.9 | .57 | .07 |
| | *Mathematics* | *47* | *46* | *44* | *50* | *56* | *48* | *38* |
| 035. | Florida, University of-Gainesville | 34 | 18 | 41 | .13 | 6.0 | .38 | .13 |
| | *Mathematics* | *51* | *47* | *52* | *41* | *55* | *36* | *41* |
| 036. | Georgia Institute of Technology | 41 | 8 | 11 | .30 | 7.2 | .50 | .30 |
| | *Mathematics* | *55* | *43* | *42* | *53* | *45* | *43* | *53* |
| 037. | Georgia, University of-Athens | 23 | 12 | 10 | .40 | 7.5 | NA | NA |
| | *Mathematics* | *44* | *45* | *42* | *60* | *42* | | |
| 038. | Harvard University | 26 | 28 | 48 | .72 | 5.3 | .95 | .65 |
| | *Mathematics* | *46* | *52* | *54* | *83* | *61* | *71* | *78* |
| 039. | Houston, University of | 25 | 27 | 54 | .29 | 7.8 | .57 | .21 |
| | *Mathematics* | *45* | *51* | *56* | *53* | *39* | *48* | *47* |
| 040. | Illinois Institute of Technology | 6 | 11 | 17 | .42 | 10.0 | .67 | .08 |
| | *Mathematics* | *34* | *44* | *44* | *62* | *21* | *53* | *38* |

\* indicates program was initiated since 1970.

NOTE: On the first line of data for every program, raw values for each measure are reported; on the second line values are reported in standardized form, with mean = 50 and standard deviation = 10. "NA" indicates that the value for a measure is not available.

TABLE 6.1  Program Measures (Raw and Standardized Values) in Mathematics

| Prog No. | Survey Results (08) | (09) | (10) | (11) | University Library (12) | Research Support (13) | (14) | Published Articles (15) | (16) | Survey Ratings Standard Error (08) | (09) | (10) | (11) |
|---|---|---|---|---|---|---|---|---|---|---|---|---|---|
| 021. | 2.3 | 1.4 | 1.0 | 0.6 | NA | .40 | NA | 4 | | .13 | .12 | .11 | .07 |
| | *47* | *47* | *40* | *47* | | *55* | | *38* | *41* | | | | |
| 022. | 1.5 | 0.9 | 1.1 | 0.4 | NA | .41 | 152 | 13 | | .18 | .13 | .10 | .06 |
| | *39* | *38* | *47* | *40* | | *56* | *44* | *41* | *42* | | | | |
| 023. | 1.7 | 1.1 | 1.3 | 0.4 | NA | .07 | 162 | 16 | | .13 | .16 | .12 | .06 |
| | *40* | *43* | *57* | *41* | | *35* | *44* | *42* | *43* | | | | |
| 024. | 1.9 | 1.0 | 1.2 | 0.4 | -1.1 | .30 | 524 | 22 | | .15 | .09 | .12 | .06 |
| | *42* | *40* | *55* | *40* | *38* | *49* | *49* | *44* | *43* | | | | |
| 025. | 2.9 | 1.8 | 1.2 | 0.7 | -0.9 | .39 | 272 | 54 | | .09 | .08 | .08 | .07 |
| | *52* | *54* | *55* | *49* | *41* | *54* | *46* | *55* | *53* | | | | |
| 026. | 4.4 | 2.3 | 1.1 | 1.3 | 1.7 | .53 | 470 | 41 | | .06 | .07 | .06 | .07 |
| | *67* | *63* | *45* | *64* | *66* | *62* | *48* | *51* | *60* | | | | |
| 027. | 2.0 | 1.0 | 1.2 | 0.5 | -0.5 | .22 | 138 | 16 | | .11 | .10 | .09 | .06 |
| | *43* | *41* | *52* | *43* | *44* | *44* | *44* | *42* | *43* | | | | |
| 028. | 4.0 | 2.3 | 1.1 | 1.4 | 1.6 | .21 | 528 | 62 | | .05 | .06 | .06 | .07 |
| | *64* | *64* | *47* | *67* | *65* | *44* | *49* | *58* | *63* | | | | |
| 029. | 2.6 | 1.6 | 0.9 | 0.7 | -1.1 | .21 | NA | 9 | | .11 | .10 | .07 | .07 |
| | *50* | *50* | *39* | *49* | *38* | *44* | | *40* | *42* | | | | |
| 030. | 2.2 | 1.3 | 1.5 | 0.7 | NA | .22 | 186 | 17 | | .13 | .11 | .11 | .07 |
| | *45* | *46* | *71* | *48* | | *44* | *45* | *43* | *43* | | | | |
| 031. | 1.3 | 0.7 | 0.9 | 0.4 | NA | .00 | 171 | 3 | | .16 | .11 | .08 | .06 |
| | *37* | *35* | *35* | *40* | | *31* | *44* | *38* | *41* | | | | |
| 032. | 2.8 | 1.5 | 1.2 | 0.8 | 0.3 | .52 | 114 | 39 | | .09 | .08 | .11 | .07 |
| | *52* | *49* | *53* | *51* | *53* | *62* | *44* | *50* | *47* | | | | |
| 033. | 1.5 | 0.9 | 0.6 | 0.3 | -0.6 | .24 | NA | 14 | | .15 | .13 | .15 | .05 |
| | *38* | *39* | *19* | *38* | *43* | *45* | | *42* | *42* | | | | |
| 034. | 2.4 | 1.5 | 0.8 | 0.5 | -0.4 | .21 | 702 | 20 | | .11 | .12 | .09 | .06 |
| | *47* | *49* | *32* | *44* | *45* | *44* | *51* | *44* | *43* | | | | |
| 035. | 2.3 | 1.4 | 1.2 | 0.5 | 0.8 | .12 | 157 | 43 | | .12 | .10 | .12 | .06 |
| | *46* | *47* | *51* | *44* | *57* | *38* | *44* | *51* | *46* | | | | |
| 036. | 2.2 | 1.3 | 1.3 | 0.5 | NA | .10 | 596 | 39 | | .12 | .11 | .11 | .06 |
| | *45* | *46* | *58* | *43* | | *37* | *50* | *50* | *47* | | | | |
| 037. | 2.3 | 1.3 | 1.3 | 0.6 | 0.4 | .48 | 607 | 42 | | .10 | .09 | .11 | .07 |
| | *46* | *46* | *59* | *47* | *53* | *60* | *50* | *51* | *50* | | | | |
| 038. | 4.8 | 2.7 | 1.0 | 1.6 | 3.0 | .35 | 789 | 51 | | .05 | .05 | .05 | .06 |
| | *71* | *71* | *42* | *71* | *79* | *52* | *52* | *54* | *65* | | | | |
| 039. | 1.9 | 1.2 | 1.2 | 0.4 | -0.9 | .08 | 216 | 24 | | .13 | .11 | .10 | .06 |
| | *42* | *44* | *54* | *40* | *41* | *36* | *45* | *45* | *43* | | | | |
| 040. | 1.1 | 0.5 | NA | 0.3 | NA | NA | NA | 12 | | .15 | .15 | NA | .05 |
| | *34* | *32* | | *38* | | | | *41* | *42* | | | | |

NOTE: On the first line of data for every program, raw values for each measure are reported; on the second line values are reported in standardized form, with mean = 50 and standard deviation = 10. "NA" indicates that the value for a measure is not available. Since the scale used to compute measure (16) is entirely arbitrary, only values in standardized form are reported for this measure.

TABLE 6.1  Program Measures (Raw and Standardized Values) in Mathematics

| Prog No. | University - Department/Academic Unit | Program Size | | | Characteristics of Program Graduates | | | |
|---|---|---|---|---|---|---|---|---|
| | | (01) | (02) | (03) | (04) | (05) | (06) | (07) |
| 041. | Illinois, University of–Chicago Circle | 62 | 22 | 54 | .32 | 7.4 | .50 | .13 |
| | *Mathematics* | *68* | *49* | *56* | *55* | *43* | *43* | *41* |
| 042. | Illinois, University-Urbana/Champaign | 84 | 79 | 76 | .15 | 6.7 | .74 | .42 |
| | *Mathematics* | *81* | *75* | *63* | *43* | *49* | *58* | *62* |
| 043. | Indiana University-Bloomington | 33 | 21 | 47 | .23 | 6.4 | .77 | .43 |
| | *Mathematics* | *50* | *49* | *54* | *48* | *51* | *59* | *63* |
| 044. | Iowa State University-Ames | 42 | 9 | 22 | NA | NA | NA | NA |
| | *Mathematics* | *56* | *43* | *46* | | | | |
| 045. | Iowa, University of-Iowa City | 33 | 24 | 38 | .15 | 5.9 | .35 | .12 |
| | *Mathematics* | *50* | *50* | *51* | *43* | *56* | *34* | *41* |
| 046. | Johns Hopkins University | 11 | 16 | 21 | .07 | 5.1 | .62 | .08 |
| | *Mathematics* | *37* | *47* | *46* | *37* | *62* | *50* | *38* |
| 047. | Kansas, University of | 34 | 12 | 16 | .31 | 7.4 | .31 | .23 |
| | *Mathematics* | *51* | *45* | *44* | *54* | *43* | *32* | *49* |
| 048. | Kent State University | 24 | 18 | 21 | .21 | 6.4 | .87 | .27 |
| | *Mathematics\** | *45* | *47* | *46* | *47* | *52* | *65* | *51* |
| 049. | Kentucky, University of | 37 | 22 | 27 | .32 | 7.0 | .42 | .11 |
| | *Mathematics* | *52* | *49* | *48* | *54* | *46* | *39* | *40* |
| 050. | Louisiana State University-Baton Rouge | 31 | 14 | 15 | .35 | 6.4 | .47 | .24 |
| | *Mathematics* | *49* | *46* | *44* | *57* | *51* | *42* | *49* |
| 051. | Maryland, University of-College Park | 85 | 3 | 59 | NA | NA | NA | NA |
| | *Applied Mathematics\** | *82* | *41* | *58* | | | | |
| 052. | Maryland, University of-College Park | 80 | 31 | 65 | .23 | 7.1 | .54 | .14 |
| | *Mathematics* | *79* | *53* | *60* | *49* | *45* | *46* | *43* |
| 053. | Massachusetts Institute of Technology | 56 | 116 | 109 | .42 | 4.8 | .67 | .42 |
| | *Mathematics* | *64* | *91* | *74* | *62* | *65* | *53* | *62* |
| 054. | Massachusetts, University of-Amherst | 49 | 19 | 29 | .13 | 6.0 | .44 | .26 |
| | *Mathematics and Statistics* | *60* | *48* | *48* | *42* | *55* | *39* | *51* |
| 055. | Michigan State University-East Lansing | 50 | 25 | 55 | .26 | 5.8 | .65 | .32 |
| | *Mathematics* | *60* | *51* | *56* | *50* | *57* | *52* | *55* |
| 056. | Michigan, University of-Ann Arbor | 65 | 67 | 89 | .32 | 5.9 | .66 | .42 |
| | *Mathematics* | *69* | *69* | *67* | *55* | *56* | *53* | *62* |
| 057. | Minnesota, University of | 59 | 34 | 92 | .39 | 5.9 | .61 | .28 |
| | *Mathematics* | *66* | *55* | *68* | *60* | *56* | *50* | *52* |
| 058. | Missouri, University of-Columbia | 23 | 9 | 11 | .14 | 6.5 | .64 | .29 |
| | *Mathematics* | *44* | *43* | *42* | *42* | *51* | *52* | *52* |
| 059. | Missouri, University of-Rolla | 13 | 11 | 4 | NA | NA | NA | NA |
| | *Mathematics* | *38* | *44* | *40* | | | | |
| 060. | Montana, University of-Missoula | 19 | 11 | 12 | NA | NA | NA | NA |
| | *Mathematics* | *42* | *44* | *43* | | | | |

\* indicates program was initiated since 1970.

NOTE: On the first line of data for every program, raw values for each measure are reported; on the second line values are reported in standardized form, with mean = 50 and standard deviation = 10. "NA" indicates that the value for a measure is not available.

TABLE 6.1 Program Measures (Raw and Standardized Values) in Mathematics

| Prog No. | Survey Results (08) | (09) | (10) | (11) | University Library (12) | Research Support (13) | (14) | Published Articles (15) | (16) | Survey Ratings Standard Error (08) | (09) | (10) | (11) |
|---|---|---|---|---|---|---|---|---|---|---|---|---|---|
| 041. | 3.0 | 1.7 | 1.5 | 1.0 | NA | .32 | 883 | 49 | | .08 | .07 | .08 | .07 |
| | 54 | 52 | 71 | 56 | | 50 | 53 | 53 | 53 | | | | |
| 042. | 4.0 | 2.3 | 1.1 | 1.3 | 2.0 | .37 | 773 | 135 | | .06 | .06 | .06 | .06 |
| | 63 | 63 | 47 | 65 | 68 | 53 | 52 | 82 | 72 | | | | |
| 043. | 3.5 | 2.0 | 1.2 | 1.2 | 0.9 | .55 | 344 | 40 | | .07 | .04 | .09 | .06 |
| | 58 | 58 | 51 | 62 | 58 | 64 | 47 | 50 | 52 | | | | |
| 044. | 2.1 | 1.2 | 1.0 | 0.5 | -0.5 | .07 | 2140 | 36 | | .13 | .11 | .08 | .07 |
| | 45 | 44 | 39 | 43 | 44 | 35 | 69 | 49 | 44 | | | | |
| 045. | 2.2 | 1.5 | 1.1 | 0.5 | 0.3 | .36 | NA | 43 | | .12 | .11 | .07 | .07 |
| | 46 | 49 | 50 | 43 | 52 | 53 | | 51 | 50 | | | | |
| 046. | 3.4 | 1.7 | 0.8 | 0.8 | -0.4 | .64 | 184 | 35 | | .12 | .09 | .08 | .07 |
| | 57 | 53 | 29 | 51 | 45 | 69 | 44 | 49 | 56 | | | | |
| 047. | 2.3 | 1.3 | 0.9 | 0.5 | 0.1 | .24 | 109 | 44 | | .10 | .10 | .11 | .06 |
| | 46 | 46 | 39 | 44 | 50 | 45 | 44 | 52 | 49 | | | | |
| 048. | 1.8 | 1.1 | 1.2 | 0.5 | -1.8 | .25 | NA | 12 | | .14 | .12 | .11 | .06 |
| | 41 | 42 | 52 | 42 | 32 | 46 | | 41 | 42 | | | | |
| 049. | 2.8 | 1.8 | 1.6 | 0.8 | -0.1 | .49 | 175 | 72 | | .10 | .08 | .08 | .07 |
| | 52 | 55 | 72 | 52 | 48 | 60 | 44 | 61 | 57 | | | | |
| 050. | 2.7 | 1.6 | 1.1 | 0.7 | -0.3 | .29 | 709 | 31 | | .10 | .10 | .12 | .07 |
| | 50 | 51 | 45 | 48 | 46 | 48 | 51 | 47 | 45 | | | | |
| 051. | 3.5 | 1.9 | 1.2 | 0.7 | 0.2 | .44 | 382 | 93 | | .10 | .10 | .12 | .08 |
| | 58 | 56 | 52 | 47 | 51 | 57 | 47 | 68 | 70 | | | | |
| 052. | 3.5 | 2.0 | 1.3 | 1.2 | 0.2 | .33 | 382 | 93 | | .09 | .07 | .08 | .07 |
| | 58 | 58 | 60 | 61 | 51 | 51 | 47 | 68 | 70 | | | | |
| 053. | 4.9 | 2.7 | 1.1 | 1.6 | -0.3 | .41 | 1289 | 126 | | .03 | .06 | .05 | .05 |
| | 72 | 70 | 50 | 71 | 46 | 56 | 59 | 79 | 84 | | | | |
| 054. | 2.7 | 1.7 | 1.3 | 0.8 | -0.7 | .31 | 158 | 36 | | .10 | .08 | .08 | .06 |
| | 50 | 52 | 58 | 51 | 42 | 49 | 44 | 49 | 48 | | | | |
| 055. | 2.7 | 1.7 | 1.2 | 0.9 | 0.3 | .16 | 298 | 51 | | .09 | .07 | .07 | .06 |
| | 50 | 52 | 51 | 54 | 53 | 41 | 46 | 54 | 49 | | | | |
| 056. | 4.1 | 2.4 | 1.2 | 1.5 | 1.8 | .29 | 844 | 75 | | .07 | .07 | .06 | .06 |
| | 64 | 65 | 50 | 70 | 67 | 49 | 53 | 62 | 58 | | | | |
| 057. | 3.9 | 2.2 | 1.2 | 1.2 | 1.2 | .53 | 1207 | 73 | | .07 | .07 | .06 | .07 |
| | 62 | 62 | 54 | 62 | 61 | 62 | 58 | 62 | 60 | | | | |
| 058. | 1.7 | 0.9 | 1.0 | 0.5 | -0.2 | .04 | NA | 24 | | .11 | .11 | .12 | .06 |
| | 41 | 39 | 40 | 43 | 47 | 34 | | 45 | 45 | | | | |
| 059. | 1.1 | 0.9 | 0.9 | 0.3 | NA | .15 | NA | 22 | | .11 | .14 | .07 | .06 |
| | 35 | 38 | 36 | 39 | | 40 | | 44 | 43 | | | | |
| 060. | 1.2 | 0.6 | 1.1 | 0.3 | NA | .26 | NA | 0 | | .12 | .12 | .07 | .05 |
| | 35 | 34 | 48 | 38 | | 47 | | 37 | 41 | | | | |

NOTE: On the first line of data for every program, raw values for each measure are reported; on the second line values are reported in standardized form, with mean = 50 and standard deviation = 10. "NA" indicates that the value for a measure is not available. Since the scale used to compute measure (16) is entirely arbitrary, only values in standardized form are reported for this measure.

TABLE 6.1  Program Measures (Raw and Standardized Values) in Mathematics

| Prog No. | University - Department/Academic Unit | Program Size | | | Characteristics of Program Graduates | | | |
|---|---|---|---|---|---|---|---|---|
| | | (01) | (02) | (03) | (04) | (05) | (06) | (07) |
| 061. | Nebraska, University of-Lincoln | 34 | 16 | 16 | .31 | 9.3 | .75 | .50 |
| | *Mathematics and Statistics* | *51* | *47* | *44* | *54* | *27* | *58* | *67* |
| 062. | New Mexico State University-Las Cruces | 30 | 6 | 15 | .50 | 8.0 | .64 | .18 |
| | *Mathematical Sciences* | *48* | *42* | *44* | *67* | *38* | *52* | *45* |
| 063. | New Mexico, University of-Albuquerque | 29 | 11 | 25 | .30 | 7.5 | .90 | .20 |
| | *Mathematics and Statistics* | *48* | *44* | *47* | *53* | *42* | *67* | *47* |
| 064. | New York University | 37 | 70 | 91 | .25 | 6.5 | .77 | .29 |
| | *Mathematics (Courant Institute)* | *52* | *71* | *68* | *50* | *51* | *60* | *52* |
| 065. | North Carolina State University-Raleigh | 52 | 13 | 12 | .13 | 7.0 | .13 | .07 |
| | *Mathematics* | *62* | *45* | *43* | *42* | *46* | *21* | *37* |
| 066. | North Carolina, University of-Chapel Hill | 29 | 13 | 21 | .50 | 5.8 | .63 | .19 |
| | *Mathematics* | *48* | *45* | *46* | *67* | *56* | *51* | *46* |
| 067. | Northeastern University | 38 | 12 | 28 | NA | NA | NA | NA |
| | *Mathematics* | *53* | *45* | *48* | | | | |
| 068. | Northwestern University | 39 | 29 | 31 | .38 | 6.0 | .62 | .27 |
| | *Mathematics* | *54* | *52* | *49* | *59* | *55* | *50* | *51* |
| 069. | Notre Dame, University of | 19 | 28 | 36 | .19 | 5.8 | .62 | .43 |
| | *Mathematics* | *42* | *52* | *50* | *46* | *57* | *51* | *62* |
| 070. | Ohio State University-Columbus | 55 | 45 | 97 | .17 | 6.8 | .50 | .25 |
| | *Mathematics* | *63* | *60* | *70* | *44* | *48* | *43* | *50* |
| 071. | Ohio University-Athens | 25 | 8 | 39 | NA | NA | NA | NA |
| | *Mathematics\** | *45* | *43* | *51* | | | | |
| 072. | Oklahoma State University-Stillwater | 26 | 10 | 16 | NA | NA | NA | NA |
| | *Mathematics* | *46* | *44* | *44* | | | | |
| 073. | Oklahoma, University of-Norman | 26 | 17 | 10 | .21 | 8.2 | .29 | .14 |
| | *Mathematics* | *46* | *47* | *42* | *47* | *36* | *30* | *43* |
| 074. | Oregon State University-Corvallis | 32 | 19 | 15 | .11 | 9.8 | .50 | .22 |
| | *Mathematics* | *49* | *48* | *44* | *40* | *23* | *43* | *48* |
| 075. | Oregon, University of-Eugene | 30 | 22 | 38 | .35 | 7.3 | .60 | .30 |
| | *Mathematics* | *48* | *49* | *51* | *57* | *44* | *49* | *53* |
| 076. | Pennsylvania State University | 45 | 23 | 42 | .14 | 8.0 | .64 | .18 |
| | *Mathematics* | *57* | *50* | *52* | *42* | *38* | *52* | *45* |
| 077. | Pennsylvania, University of | 31 | 23 | 24 | .25 | 5.8 | .60 | .35 |
| | *Mathematics* | *49* | *50* | *47* | *50* | *56* | *49* | *57* |
| 078. | Pittsburgh, University of | 34 | 11 | 37 | .27 | 7.5 | .82 | .18 |
| | *Mathematics and Statistics* | *51* | *44* | *51* | *51* | *42* | *63* | *45* |
| 079. | Polytech Institute of New York | 13 | 17 | 20 | .17 | 8.6 | .46 | .04 |
| | *Mathematics* | *38* | *47* | *45* | *44* | *33* | *41* | *36* |
| 080. | Princeton University | 25 | 63 | 49 | .62 | 4.5 | .89 | .63 |
| | *Mathematics* | *45* | *68* | *55* | *76* | *68* | *67* | *76* |

\* indicates program was initiated since 1970.

NOTE: On the first line of data for every program, raw values for each measure are reported; on the second line values are reported in standardized form, with mean = 50 and standard deviation = 10. "NA" indicates that the value for a measure is not available.

TABLE 6.1  Program Measures (Raw and Standardized Values) in Mathematics

| Prog No. | Survey Results (08) | (09) | (10) | (11) | University Library (12) | Research Support (13) | (14) | Published Articles (15) | (16) | Survey Ratings Standard Error (08) | (09) | (10) | (11) |
|---|---|---|---|---|---|---|---|---|---|---|---|---|---|
| 061. | 1.8 | 1.0 | 1.1 | 0.4 | −0.5 | .18 | NA | 23 | | .11 | .10 | .08 | .05 |
| | 42 | 41 | 48 | 40 | 44 | 42 | | 45 | 43 | | | | |
| 062. | NA | NA | NA | NA | NA | .13 | 1049 | 16 | | NA | NA | NA | NA |
| | | | | | | 39 | 56 | 42 | 45 | | | | |
| 063. | 2.1 | 1.3 | 1.0 | 0.6 | −1.0 | .17 | 110 | 23 | | .15 | .12 | .09 | .07 |
| | 45 | 45 | 40 | 47 | 40 | 41 | 44 | 45 | 43 | | | | |
| 064. | 4.5 | 2.6 | 0.9 | 1.4 | 0.5 | .43 | 3788 | 67 | | .07 | .06 | .07 | .07 |
| | 69 | 68 | 34 | 66 | 54 | 57 | 90 | 59 | 70 | | | | |
| 065. | 2.1 | 1.2 | 1.3 | 0.4 | NA | .15 | 637 | 64 | | .11 | .12 | .10 | .05 |
| | 44 | 44 | 58 | 41 | | 40 | 50 | 58 | 46 | | | | |
| 066. | 3.0 | 1.7 | 1.3 | 1.0 | 1.0 | .48 | 4356 | 25 | | .08 | .08 | .09 | .07 |
| | 53 | 53 | 56 | 56 | 59 | 60 | 98 | 45 | 47 | | | | |
| 067. | 2.5 | 1.4 | 1.4 | 0.6 | NA | .26 | 114 | 15 | | .12 | .12 | .11 | .06 |
| | 48 | 48 | 64 | 45 | | 47 | 44 | 42 | 45 | | | | |
| 068. | 3.5 | 2.0 | 1.2 | 1.2 | 0.3 | .77 | 367 | 74 | | .06 | .05 | .06 | .07 |
| | 58 | 57 | 51 | 61 | 52 | 77 | 47 | 62 | 55 | | | | |
| 069. | 2.7 | 1.7 | 1.0 | 0.7 | −1.3 | .42 | 154 | 23 | | .11 | .08 | .10 | .06 |
| | 51 | 53 | 42 | 49 | 36 | 56 | 44 | 45 | 44 | | | | |
| 070. | 3.0 | 1.8 | 1.2 | 1.0 | 0.9 | .36 | 662 | 76 | | .09 | .08 | .10 | .07 |
| | 53 | 55 | 53 | 56 | 58 | 53 | 51 | 63 | 53 | | | | |
| 071. | 1.3 | 0.6 | 1.1 | 0.3 | NA | .00 | 152 | 15 | | .14 | .12 | .05 | .06 |
| | 37 | 33 | 45 | 39 | | 31 | 44 | 42 | 42 | | | | |
| 072. | 1.7 | 1.0 | 1.2 | 0.5 | −1.9 | .23 | NA | 12 | | .13 | .13 | .10 | .07 |
| | 40 | 41 | 50 | 43 | 30 | 45 | | 41 | 44 | | | | |
| 073. | 2.0 | 1.2 | 1.2 | 0.5 | −0.6 | .39 | NA | 28 | | .14 | .12 | .12 | .06 |
| | 44 | 44 | 52 | 42 | 44 | 54 | | 46 | 46 | | | | |
| 074. | 2.4 | 1.5 | 1.2 | 0.6 | NA | .31 | 345 | 15 | | .10 | .11 | .10 | .07 |
| | 47 | 48 | 54 | 46 | | 50 | 47 | 42 | 43 | | | | |
| 075. | 2.9 | 1.8 | 1.1 | 0.7 | −0.9 | .27 | 125 | 29 | | .10 | .11 | .06 | .07 |
| | 53 | 54 | 48 | 50 | 40 | 47 | 44 | 47 | 45 | | | | |
| 076. | 3.0 | 1.8 | 1.4 | 1.0 | 0.7 | .42 | 254 | 82 | | .09 | .07 | .07 | .08 |
| | 53 | 55 | 64 | 56 | 56 | 56 | 45 | 65 | 57 | | | | |
| 077. | 3.7 | 2.1 | 0.9 | 1.1 | 0.7 | .42 | 480 | 30 | | .07 | .06 | .06 | .07 |
| | 60 | 59 | 39 | 59 | 56 | 56 | 48 | 47 | 53 | | | | |
| 078. | 2.5 | 1.7 | 1.2 | 0.6 | 0.1 | .29 | 1113 | 34 | | .09 | .08 | .08 | .06 |
| | 48 | 52 | 52 | 45 | 50 | 49 | 56 | 48 | 46 | | | | |
| 079. | 2.1 | 1.2 | NA | 0.4 | NA | .15 | NA | 25 | | .16 | .20 | NA | .05 |
| | 44 | 44 | | 40 | | 40 | | 45 | 44 | | | | |
| 080. | 4.9 | 2.8 | 1.1 | 1.6 | 0.9 | .60 | 1389 | 91 | | .03 | .04 | .03 | .05 |
| | 73 | 73 | 45 | 73 | 58 | 67 | 60 | 68 | 78 | | | | |

NOTE: On the first line of data for every program, raw values for each measure are reported; on the second line values are reported in standardized form, with mean = 50 and standard deviation = 10. "NA" indicates that the value for a measure is not available. Since the scale used to compute measure (16) is entirely arbitrary, only values in standardized form are reported for this measure.

TABLE 6.1 Program Measures (Raw and Standardized Values) in Mathematics

| Prog No. | University - Department/Academic Unit | Program Size | | | Characteristics of Program Graduates | | | |
|---|---|---|---|---|---|---|---|---|
| | | (01) | (02) | (03) | (04) | (05) | (06) | (07) |
| 081. | Purdue University-West Lafayette | 57 | 28 | 30 | .19 | 6.4 | .56 | .28 |
| | *Mathematics* | *65* | *52* | *49* | *46* | *52* | *47* | *52* |
| 082. | Rensselaer Polytechnic Institute | 29 | 18 | 20 | .29 | 5.3 | .88 | .50 |
| | *Mathematical Sciences* | *48* | *47* | *45* | *53* | *61* | *66* | *67* |
| 083. | Rice University | 14 | 9 | 11 | .25 | 4.9 | .42 | .17 |
| | *Mathematics* | *39* | *43* | *42* | *50* | *64* | *38* | *44* |
| 084. | Rochester, University of | 26 | 23 | 26 | .29 | 6.9 | .67 | .05 |
| | *Mathematics* | *46* | *50* | *47* | *52* | *47* | *53* | *36* |
| 085. | Rutgers, The State University-New Brunswick | 65 | 57 | 85 | .15 | 5.4 | .55 | .33 |
| | *Mathematics* | *69* | *65* | *66* | *43* | *60* | *46* | *56* |
| 086. | SUNY at Albany | 28 | 6 | 47 | NA | NA | NA | NA |
| | *Mathematics and Statistics* | *47* | *42* | *54* | | | | |
| 087. | SUNY at Binghamton | 17 | 9 | 28 | .20 | 7.5 | .80 | .20 |
| | *Mathematical Sciences* | *40* | *43* | *48* | *46* | *42* | *61* | *47* |
| 088. | SUNY at Buffalo | 36 | 29 | 45 | .17 | 6.9 | .59 | .31 |
| | *Mathematics* | *52* | *52* | *53* | *44* | *47* | *49* | *54* |
| 089. | SUNY at Stony Brook | 31 | 35 | 60 | .18 | 5.5 | .61 | .39 |
| | *Mathematics* | *49* | *55* | *58* | *45* | *59* | *50* | *59* |
| 090. | Saint Louis University | 7 | 12 | 29 | .00 | 6.5 | .73 | .00 |
| | *Mathematical Sciences* | *34* | *45* | *48* | *32* | *51* | *57* | *33* |
| 091. | South Carolina, University of-Columbia | 28 | 13 | 13 | .14 | 6.3 | .57 | .07 |
| | *Mathematics and Statistics* | *47* | *45* | *43* | *42* | *53* | *48* | *38* |
| 092. | South Florida, University of-Tampa | 26 | 16 | 11 | .00 | 7.3 | .73 | .27 |
| | *Mathematics** | *46* | *47* | *42* | *32* | *44* | *57* | *52* |
| 093. | Southern California, University of | 29 | 12 | 12 | NA | NA | NA | NA |
| | *Mathematics* | *48* | *45* | *43* | | | | |
| 094. | Southern Illinois University-Carbondale | 31 | 9 | 10 | NA | NA | NA | NA |
| | *Mathematics* | *49* | *43* | *42* | | | | |
| 095. | Stanford University | 26 | 48 | 51 | .49 | 5.4 | .82 | .57 |
| | *Mathematics* | *46* | *61* | *55* | *67* | *60* | *62* | *72* |
| 096. | Stevens Institute of Technology | 12 | 19 | 25 | .28 | 10.2 | .75 | .19 |
| | *Pure and Applied Mathematics* | *37* | *48* | *47* | *52* | *19* | *58* | *46* |
| 097. | Syracuse University | 28 | 13 | 17 | .17 | 6.5 | 1.00 | .33 |
| | *Mathematics* | *47* | *45* | *44* | *44* | *51* | *73* | *56* |
| 098. | Temple University | 35 | 24 | 15 | .09 | 7.4 | .38 | .19 |
| | *Mathematics* | *51* | *50* | *44* | *39* | *43* | *36* | *46* |
| 099. | Tennessee, University of-Knoxville | 28 | 14 | 34 | .25 | 7.3 | .75 | .08 |
| | *Mathematics* | *47* | *46* | *50* | *50* | *44* | *58* | *38* |
| 100. | Texas Tech University-Lubbock | 32 | 14 | 11 | .07 | 7.5 | .64 | .07 |
| | *Mathematics* | *49* | *46* | *42* | *37* | *42* | *52* | *38* |

* indicates program was initiated since 1970.

NOTE: On the first line of data for every program, raw values for each measure are reported; on the second line values are reported in standardized form, with mean = 50 and standard deviation = 10. "NA" indicates that the value for a measure is not available.

TABLE 6.1  Program Measures (Raw and Standardized Values) in Mathematics

| Prog No. | Survey Results (08) | (09) | (10) | (11) | University Library (12) | Research Support (13) | (14) | Published Articles (15) | (16) | Survey Ratings Standard Error (08) | (09) | (10) | (11) |
|---|---|---|---|---|---|---|---|---|---|---|---|---|---|
| 081. | 3.4 | 2.1 | 1.1 | 1.2 | -0.5 | .67 | 167 | 68 | | .06 | .05 | .05 | .06 |
| | 58 | 59 | 49 | 61 | 44 | 71 | 44 | 60 | 67 | | | | |
| 082. | 2.7 | 1.7 | 1.1 | 0.6 | NA | .48 | 355 | 17 | | .16 | .12 | .10 | .07 |
| | 51 | 52 | 48 | 45 | | 60 | 47 | 43 | 43 | | | | |
| 083. | 3.4 | 2.0 | 1.1 | 1.0 | -1.4 | .57 | 273 | 31 | | .10 | .07 | .08 | .07 |
| | 57 | 59 | 48 | 56 | 35 | 65 | 46 | 47 | 51 | | | | |
| 084. | 2.7 | 1.7 | 1.1 | 0.8 | -0.6 | .42 | 194 | 11 | | .09 | .07 | .07 | .06 |
| | 51 | 53 | 45 | 51 | 43 | 56 | 45 | 41 | 42 | | | | |
| 085. | 3.6 | 2.0 | 1.4 | 1.3 | 0.8 | .39 | 1050 | 84 | | .07 | .05 | .08 | .06 |
| | 59 | 59 | 64 | 63 | 57 | 54 | 56 | 65 | 67 | | | | |
| 086. | 2.4 | 1.5 | 1.2 | 0.6 | -1.0 | .39 | 152 | 20 | | .13 | .11 | .13 | .07 |
| | 48 | 50 | 51 | 47 | 40 | 55 | 44 | 44 | 46 | | | | |
| 087. | 1.7 | 1.1 | 1.1 | 0.5 | NA | .41 | NA | 6 | | .13 | .14 | .14 | .07 |
| | 41 | 41 | 47 | 44 | | 56 | | 39 | 41 | | | | |
| 088. | 2.7 | 1.7 | 1.1 | 0.9 | 0.3 | .39 | 356 | 50 | | .08 | .10 | .10 | .07 |
| | 51 | 52 | 46 | 53 | 52 | 54 | 47 | 54 | 48 | | | | |
| 089. | 3.7 | 2.0 | 1.3 | 1.1 | -0.6 | .39 | 725 | 40 | | .08 | .07 | .08 | .07 |
| | 61 | 59 | 59 | 60 | 43 | 54 | 51 | 50 | 59 | | | | |
| 090. | 0.7 | 0.5 | NA | 0.2 | NA | NA | NA | 3 | | .17 | .18 | NA | .05 |
| | 31 | 31 | | 36 | | | | 38 | 41 | | | | |
| 091. | 1.8 | 0.9 | 1.4 | 0.4 | -0.4 | .25 | 211 | 23 | | .13 | .09 | .09 | .06 |
| | 41 | 39 | 61 | 41 | 46 | 46 | 45 | 45 | 45 | | | | |
| 092. | 1.5 | 0.9 | 1.3 | 0.5 | NA | .08 | 103 | 12 | | .14 | .14 | .09 | .07 |
| | 38 | 39 | 57 | 43 | | 36 | 43 | 41 | 42 | | | | |
| 093. | 2.8 | 1.6 | 1.1 | 0.9 | 0.4 | .35 | 227 | 47 | | .10 | .10 | .09 | .07 |
| | 52 | 51 | 46 | 54 | 53 | 52 | 45 | 53 | 47 | | | | |
| 094. | 1.5 | 0.8 | 1.2 | 0.5 | -0.2 | .10 | NA | 30 | | .11 | .10 | .11 | .06 |
| | 39 | 36 | 50 | 42 | 47 | 37 | | 47 | 43 | | | | |
| 095. | 4.6 | 2.6 | 1.0 | 1.4 | 2.0 | .54 | 2697 | 64 | | .05 | .06 | .07 | .06 |
| | 69 | 69 | 44 | 68 | 69 | 63 | 77 | 58 | 69 | | | | |
| 096. | 1.2 | 0.8 | NA | 0.2 | NA | .33 | NA | 4 | | .14 | .16 | NA | .04 |
| | 35 | 38 | | 36 | | 51 | | 38 | 41 | | | | |
| 097. | 2.3 | 1.4 | 0.9 | 0.6 | -0.3 | .14 | NA | 22 | | .10 | .13 | .06 | .06 |
| | 47 | 47 | 34 | 46 | 46 | 40 | | 44 | 44 | | | | |
| 098. | 2.4 | 1.3 | 1.2 | 0.6 | -0.4 | .14 | NA | 35 | | .11 | .11 | .11 | .06 |
| | 47 | 45 | 53 | 45 | 45 | 40 | | 49 | 46 | | | | |
| 099. | 2.1 | 1.4 | 1.3 | 0.4 | -0.4 | .25 | NA | 50 | | .14 | .14 | .12 | .07 |
| | 44 | 47 | 60 | 42 | 45 | 46 | | 54 | 48 | | | | |
| 100. | 1.7 | 1.0 | 1.4 | 0.5 | NA | .16 | 122 | 23 | | .13 | .13 | .11 | .06 |
| | 40 | 40 | 64 | 42 | | 41 | 44 | 45 | 43 | | | | |

NOTE: On the first line of data for every program, raw values for each measure are reported; on the second line values are reported in standardized form, with mean = 50 and standard deviation = 10. "NA" indicates that the value for a measure is not available. Since the scale used to compute measure (16) is entirely arbitrary, only values in standardized form are reported for this measure.

TABLE 6.1  Program Measures (Raw and Standardized Values) in Mathematics

| Prog No. | University – Department/Academic Unit | Program Size | | | Characteristics of Program Graduates | | | |
|---|---|---|---|---|---|---|---|---|
| | | (01) | (02) | (03) | (04) | (05) | (06) | (07) |
| 101. | Texas, University of–Arlington | 23 | 16 | 23 | NA | 5.5 | NA | NA |
| | *Mathematics** | *44* | *47* | *46* | | *59* | | |
| 102. | Texas, University of–Austin | 57 | 20 | 42 | .16 | 7.3 | .52 | .13 |
| | *Mathematics* | *65* | *48* | *52* | *44* | *44* | *45* | *42* |
| 103. | Tulane University | 27 | 17 | 30 | .33 | 6.0 | .60 | .25 |
| | *Mathematics* | *46* | *47* | *49* | *56* | *55* | *49* | *50* |
| 104. | Utah, University of–Salt Lake City | 53 | 24 | 21 | .24 | 6.7 | .79 | .54 |
| | *Mathematics* | *62* | *50* | *46* | *49* | *49* | *61* | *70* |
| 105. | Vanderbilt University | 19 | 16 | 16 | .26 | 6.4 | .72 | .28 |
| | *Mathematics* | *42* | *47* | *44* | *51* | *51* | *57* | *52* |
| 106. | Virginia Polytechnic Institute & State Univ | 36 | 9 | 25 | .20 | NA | .80 | .20 |
| | *Mathematics* | *52* | *43* | *47* | *46* | | *61* | *47* |
| 107. | Virginia, University of | 26 | 16 | 31 | .50 | 5.8 | .47 | .16 |
| | *Mathematics* | *46* | *47* | *49* | *67* | *57* | *42* | *44* |
| 108. | Washington University–Saint Louis | 21 | 14 | 32 | .32 | 6.8 | .47 | .11 |
| | *Mathematics* | *43* | *46* | *49* | *54* | *49* | *42* | *40* |
| 109. | Washington, University of–Seattle | 57 | 31 | 27 | .23 | 6.3 | .52 | .20 |
| | *Mathematics* | *65* | *53* | *48* | *49* | *53* | *45* | *47* |
| 110. | Wayne State University | 46 | 12 | 16 | .29 | 7.3 | .50 | .15 |
| | *Mathematics* | *58* | *45* | *44* | *52* | *44* | *43* | *43* |
| 111. | Wesleyan University | 14 | 12 | 7 | .17 | 6.0 | .39 | .08 |
| | *Mathematics* | *39* | *45* | *41* | *44* | *55* | *36* | *38* |
| 112. | Western Michigan University | 32 | 9 | 7 | NA | NA | NA | NA |
| | *Mathematics* | *49* | *43* | *41* | | | | |
| 113. | Wisconsin, University of–Madison | 61 | 76 | 174 | .23 | 6.3 | .67 | .33 |
| | *Mathematics* | *67* | *73* | *94* | *48* | *52* | *54* | *55* |
| 114. | Wisconsin, University of–Milwaukee | 21 | 17 | 9 | .20 | 7.0 | .45 | .10 |
| | *Mathematical Sciences* | *43* | *47* | *42* | *46* | *46* | *40* | *40* |
| 115. | Yale University | 22 | 43 | 39 | .65 | 4.9 | .42 | .27 |
| | *Mathematics* | *43* | *59* | *51* | *78* | *64* | *38* | *51* |

* indicates program was initiated since 1970.

NOTE: On the first line of data for every program, raw values for each measure are reported; on the second line values are reported in standardized form, with mean = 50 and standard deviation = 10. "NA" indicates that the value for a measure is not available.

TABLE 6.1  Program Measures (Raw and Standardized Values) in Mathematics

| Prog No. | Survey Results (08) | (09) | (10) | (11) | University Library (12) | Research Support (13) | (14) | Published Articles (15) | (16) | Survey Ratings Standard Error (08) | (09) | (10) | (11) |
|---|---|---|---|---|---|---|---|---|---|---|---|---|---|
| 101. | 1.7 | 0.9 | 1.4 | 0.5 | NA | .04 | NA | 0 | | .14 | .11 | .10 | .07 |
| | 40 | 39 | 65 | 43 | | 34 | | 37 | 41 | | | | |
| 102. | 3.3 | 1.8 | 1.6 | 1.1 | 1.6 | .58 | 390 | 69 | | .09 | .07 | .07 | .06 |
| | 56 | 55 | 74 | 58 | 65 | 66 | 47 | 60 | 53 | | | | |
| 103. | 2.7 | 1.7 | 1.1 | 0.8 | -1.0 | .56 | 172 | 20 | | .09 | .09 | .08 | .07 |
| | 50 | 53 | 45 | 52 | 39 | 64 | 44 | 44 | 44 | | | | |
| 104. | 3.2 | 1.7 | 1.7 | 1.1 | -0.6 | .28 | 353 | 39 | | .10 | .10 | .07 | .07 |
| | 55 | 53 | 80 | 60 | 43 | 48 | 47 | 50 | 48 | | | | |
| 105. | 2.0 | 1.2 | 1.1 | 0.5 | -0.7 | .11 | NA | 24 | | .14 | .12 | .08 | .07 |
| | 44 | 44 | 50 | 43 | 42 | 38 | | 45 | 43 | | | | |
| 106. | 2.3 | 1.4 | 1.4 | 0.6 | -0.0 | .22 | 490 | 51 | | .14 | .13 | .10 | .07 |
| | 47 | 47 | 64 | 45 | 49 | 44 | 48 | 54 | 49 | | | | |
| 107. | 3.0 | 1.8 | 0.8 | 0.8 | 0.7 | .42 | 148 | 27 | | .10 | .11 | .08 | .07 |
| | 53 | 55 | 30 | 52 | 56 | 56 | 44 | 46 | 45 | | | | |
| 108. | 3.1 | 1.8 | 1.1 | 0.9 | -0.4 | .48 | 210 | 22 | | .09 | .09 | .10 | .07 |
| | 55 | 55 | 45 | 52 | 45 | 59 | 45 | 44 | 45 | | | | |
| 109. | 3.6 | 2.0 | 1.3 | 1.2 | 1.5 | .46 | 808 | 42 | | .07 | .07 | .07 | .07 |
| | 59 | 59 | 59 | 61 | 63 | 58 | 52 | 51 | 57 | | | | |
| 110. | 2.2 | 1.2 | 1.2 | 0.6 | -0.4 | .24 | 141 | 34 | | .10 | .12 | .12 | .06 |
| | 45 | 44 | 55 | 46 | 46 | 45 | 44 | 48 | 47 | | | | |
| 111. | 1.9 | 1.1 | 1.1 | 0.5 | NA | .29 | NA | 9 | | .15 | .13 | .08 | .07 |
| | 42 | 43 | 50 | 42 | | 48 | | 40 | 42 | | | | |
| 112. | 1.1 | 0.6 | 1.1 | 0.3 | NA | .06 | NA | 11 | | .14 | .13 | .12 | .05 |
| | 35 | 33 | 47 | 38 | | 35 | | 41 | 42 | | | | |
| 113. | 4.2 | 2.4 | 1.1 | 1.6 | 1.6 | .48 | 3582 | 140 | | .07 | .06 | .07 | .06 |
| | 65 | 66 | 45 | 72 | 64 | 59 | 88 | 84 | 75 | | | | |
| 114. | 1.6 | 1.0 | 1.0 | 0.4 | NA | .10 | NA | NA | | .14 | .13 | .12 | .06 |
| | 40 | 41 | 42 | 40 | | 37 | | | NA | | | | |
| 115. | 4.5 | 2.5 | 1.1 | 1.4 | 2.1 | .50 | 558 | 58 | | .06 | .07 | .06 | .06 |
| | 69 | 67 | 49 | 68 | 70 | 61 | 49 | 56 | 56 | | | | |

NOTE: On the first line of data for every program, raw values for each measure are reported; on the second line values are reported in standardized form, with mean = 50 and standard deviation = 10. "NA" indicates that the value for a measure is not available. Since the scale used to compute measure (16) is entirely arbitrary, only values in standardized form are reported for this measure.

TABLE 6.2 Summary Statistics Describing Each Program Measure--Mathematics

| Measure | Number of Programs Evaluated | Mean | Standard Deviation | DECILES | | | | | | | | |
|---|---|---|---|---|---|---|---|---|---|---|---|---|
| | | | | 1 | 2 | 3 | 4 | 5 | 6 | 7 | 8 | 9 |
| **Program Size** | | | | | | | | | | | | |
| 01 Raw Value | 115 | 33 | 17 | 16 | 19 | 24 | 26 | 29 | 32 | 35 | 43 | 57 |
| Std Value | 115 | 50 | 10 | 40 | 42 | 45 | 46 | 48 | 49 | 51 | 56 | 65 |
| 02 Raw Value | 115 | 24 | 22 | 9 | 11 | 12 | 14 | 16 | 19 | 24 | 29 | 47 |
| Std Value | 115 | 50 | 10 | 43 | 44 | 45 | 46 | 47 | 48 | 50 | 52 | 60 |
| 03 Raw Value | 115 | 35 | 31 | 11 | 13 | 16 | 21 | 25 | 30 | 39 | 47 | 71 |
| Std Value | 115 | 50 | 10 | 42 | 43 | 44 | 46 | 47 | 49 | 51 | 54 | 62 |
| **Program Graduates** | | | | | | | | | | | | |
| 04 Raw Value | 98 | .25 | .14 | .10 | .14 | .17 | .19 | .23 | .26 | .29 | .32 | .43 |
| Std Value | 98 | 50 | 10 | 39 | 42 | 44 | 46 | 49 | 51 | 53 | 55 | 63 |
| 05 Raw Value | 97 | 6.6 | 1.2 | 8.0 | 7.5 | 7.1 | 6.8 | 6.4 | 6.2 | 5.9 | 5.6 | 5.3 |
| Std Value | 97 | 50 | 10 | 38 | 42 | 46 | 48 | 51 | 53 | 56 | 58 | 61 |
| 06 Raw Value | 95 | .61 | .17 | .39 | .47 | .50 | .57 | .62 | .66 | .71 | .75 | .80 |
| Std Value | 95 | 50 | 10 | 37 | 42 | 44 | 48 | 51 | 53 | 56 | 58 | 61 |
| 07 Raw Value | 95 | .25 | .15 | .07 | .11 | .16 | .19 | .23 | .27 | .30 | .35 | .44 |
| Std Value | 95 | 50 | 10 | 38 | 41 | 44 | 46 | 49 | 51 | 53 | 57 | 63 |
| **Survey Results** | | | | | | | | | | | | |
| 08 Raw Value | 114 | 2.7 | 1.0 | 1.4 | 1.7 | 2.0 | 2.3 | 2.5 | 2.7 | 3.0 | 3.5 | 4.1 |
| Std Value | 114 | 50 | 10 | 38 | 41 | 44 | 47 | 49 | 50 | 53 | 58 | 64 |
| 09 Raw Value | 114 | 1.6 | .6 | .8 | 1.0 | 1.2 | 1.4 | 1.5 | 1.7 | 1.8 | 2.0 | 2.3 |
| Std Value | 114 | 50 | 10 | 37 | 40 | 44 | 47 | 49 | 53 | 54 | 58 | 63 |
| 10 Raw Value | 108 | 1.2 | .2 | .9 | 1.0 | 1.1 | 1.1 | 1.1 | 1.2 | 1.2 | 1.3 | 1.4 |
| Std Value | 108 | 50 | 10 | 36 | 42 | 47 | 47 | 47 | 53 | 53 | 58 | 64 |
| 11 Raw Value | 114 | .8 | .4 | .3 | .4 | .5 | .5 | .6 | .8 | 1.0 | 1.2 | 1.4 |
| Std Value | 114 | 50 | 10 | 38 | 41 | 43 | 43 | 46 | 51 | 56 | 61 | 66 |
| **University Library** | | | | | | | | | | | | |
| 12 Raw Value | 82 | .1 | 1.0 | -1.1 | -.9 | -.5 | -.4 | -.2 | .2 | .5 | .9 | 1.6 |
| Std Value | 82 | 50 | 10 | 38 | 40 | 44 | 45 | 47 | 51 | 54 | 58 | 65 |
| **Research Support** | | | | | | | | | | | | |
| 13 Raw Value | 113 | .32 | .17 | .09 | .16 | .21 | .25 | .30 | .38 | .42 | .48 | .54 |
| Std Value | 113 | 50 | 10 | 36 | 41 | 44 | 46 | 49 | 54 | 56 | 59 | 63 |
| 14 Raw Value | 83 | 616 | 783 | 139 | 158 | 193 | 255 | 353 | 461 | 610 | 797 | 1212 |
| Std Value | 83 | 50 | 10 | 44 | 44 | 45 | 45 | 47 | 48 | 50 | 52 | 58 |
| **Publication Records** | | | | | | | | | | | | |
| 15 Raw Value | 114 | 39 | 30 | 10 | 15 | 20 | 24 | 30 | 39 | 46 | 59 | 76 |
| Std Value | 114 | 50 | 10 | 40 | 42 | 44 | 45 | 47 | 50 | 52 | 57 | 63 |
| 16 Std Value | 114 | 50 | 10 | 42 | 43 | 43 | 45 | 46 | 48 | 52 | 57 | 65 |

NOTE: Standardized values reported in the preceding table have been computed from exact values of the mean and standard deviation and not the rounded values reported here. Since the scale used to compute measure 16 is entirely arbitrary, only data in standardized form are reported for this measure.

TABLE 6.3  Intercorrelations Among Program Measures on 115 Programs in Mathematics

Measure

| | 01 | 02 | 03 | 04 | 05 | 06 | 07 | 08 | 09 | 10 | 11 | 12 | 13 | 14 | 15 | 16 |
|---|---|---|---|---|---|---|---|---|---|---|---|---|---|---|---|---|
| **Program Size** | | | | | | | | | | | | | | | | |
| 01 | | .50 | .61 | -.13 | .10 | -.02 | .25 | .48 | .47 | .43 | .51 | .44 | .18 | .18 | .73 | .63 |
| 02 | | | .85 | .08 | .31 | .18 | .46 | .70 | .68 | .01 | .72 | .45 | .35 | .41 | .75 | .81 |
| 03 | | | | .02 | .23 | .20 | .37 | .64 | .64 | .07 | .67 | .50 | .35 | .44 | .78 | .78 |
| **Program Graduates** | | | | | | | | | | | | | | | | |
| 04 | | | | | .14 | .07 | .21 | .30 | .30 | -.22 | .28 | .31 | .23 | .29 | .04 | .15 |
| 05 | | | | | | .05 | .41 | .57 | .56 | -.28 | .53 | .24 | .36 | .17 | .30 | .40 |
| 06 | | | | | | | .57 | .19 | .19 | -.05 | .23 | .14 | .07 | .23 | .03 | .16 |
| 07 | | | | | | | | .63 | .61 | -.03 | .62 | .39 | .28 | .22 | .38 | .50 |
| **Survey Results** | | | | | | | | | | | | | | | | |
| 08 | | | | | | | | | .98 | -.01 | .96 | .65 | .70 | .42 | .75 | .83 |
| 09 | | | | | | | | | | -.01 | .94 | .63 | .71 | .42 | .74 | .80 |
| 10 | | | | | | | | | | | .06 | .05 | .00 | -.12 | .13 | .05 |
| 11 | | | | | | | | | | | | .66 | .66 | .43 | .75 | .83 |
| **University Library** | | | | | | | | | | | | | | | | |
| 12 | | | | | | | | | | | | | .32 | .33 | .60 | .59 |
| **Research Support** | | | | | | | | | | | | | | | | |
| 13 | | | | | | | | | | | | | | .18 | .46 | .51 |
| 14 | | | | | | | | | | | | | | | .35 | .42 |
| **Publication Records** | | | | | | | | | | | | | | | | |
| 15 | | | | | | | | | | | | | | | | .90 |
| 16 | | | | | | | | | | | | | | | | |

NOTE: Since in computing correlation coefficients program data must be available for both of the measures being correlated, the actual number of programs on which each coefficient is based varies.

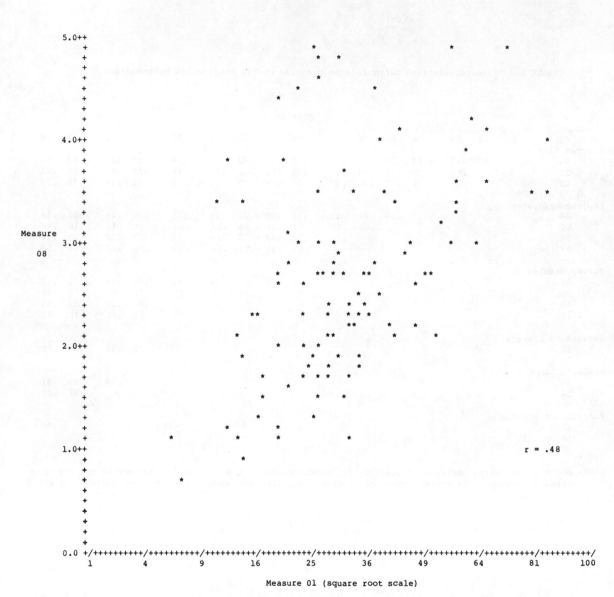

Measure 01 (square root scale)

FIGURE 6.1 Mean rating of scholarly quality of faculty (measure 08) versus number of faculty members (measure 01)--114 programs in mathematics.

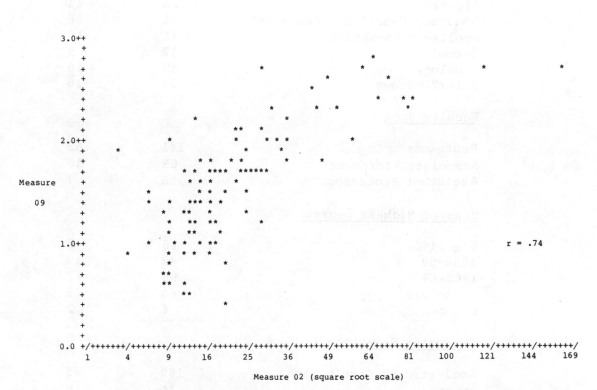

FIGURE 6.2  Mean rating of program effectiveness in educating research scholars/scientists (measure 09) versus number of graduates in last five years (measure 02)--114 programs in mathematics.

TABLE 6.4  Characteristics of Survey Participants in Mathematics

|  | Respondents | |
|---|---|---|
|  | N | % |
| **Field of Specialization** | | |
| Algebra | 25 | 11 |
| Analysis/Functional Analysis | 81 | 36 |
| Applied Mathematics | 31 | 14 |
| Geometry | 12 | 5 |
| Topology | 35 | 16 |
| Other/Unknown | 39 | 18 |
| **Faculty Rank** | | |
| Professor | 141 | 63 |
| Associate Professor | 66 | 30 |
| Assistant Professor | 16 | 7 |
| **Year of Highest Degree** | | |
| Pre-1950 | 16 | 7 |
| 1950-59 | 42 | 19 |
| 1960-69 | 96 | 43 |
| Post-1969 | 65 | 29 |
| Unknown | 4 | 2 |
| **Evaluator Selection** | | |
| Nominated by Institution | 189 | 85 |
| Other | 34 | 15 |
| **Survey Form** | | |
| With Faculty Names | 198 | 89 |
| Without Names | 25 | 11 |
| **Total Evaluators** | 223 | 100 |

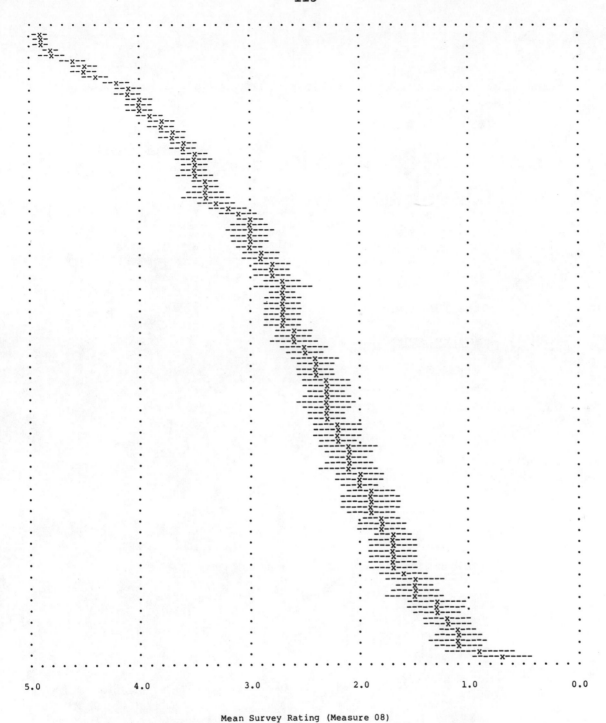

Mean Survey Rating (Measure 08)

FIGURE 6.3  Mean rating of scholarly quality of faculty in 114 programs in mathematics.

NOTE:  Programs are listed in sequence of mean rating, with the highest-rated program appearing at the top of the page.  The broken lines (---) indicate a confidence interval of ±1.5 standard errors around the reported mean (x) of each program.

# VII
# Physics Programs

In this chapter 123 research-doctorate programs in physics are assessed. These programs, according to the information supplied by their universities, have accounted for 4,271 doctoral degrees awarded during the FY1976-80 period--approximately 87 percent of the aggregate number of physics doctorates earned from U.S. universities in this five-year span.[1] On the average, 56 full-time and part-time students intending to earn doctorates were enrolled in a program in December 1980, with an average faculty size of 28 members.[2] The 123 programs, listed in Table 7.1, represent 122 different universities--only Stanford University has two physics programs included in the assessment. All but two of the programs were initiated prior to 1970. In addition to the 122 universities represented in this discipline, only one other institution--Purdue University--was initially identified as meeting the criteria[3] for inclusion in the assessment. Since no information was received (in response to the committee's request) on a physics program at this institution, it has not been included in the evaluations in this discipline.

Before examining the individual program results presented in Table 7.1, the reader is urged to refer to Chapter II, in which each of the 16 measures used in the assessment is discussed. Summary statistics describing every measure are given in Table 7.2. For all but two of the measures, data are reported for at least 109 of the 123 physics programs. For measure 12, a composite index of the size of a university library, data are available for 83 programs; for measure 14, total university expenditures for research in this

---

[1] Data from the NRC's Survey of Earned Doctorates indicate that 4,889 research doctorates in physics were awarded by U.S. universities between FY1976 and FY1980.

[2] See the reported means for measures 03 and 01 in Table 7.2.

[3] As mentioned in Chapter I, the primary criterion for inclusion was that a university had awarded at least 10 doctorates in physics during the FY1976-78 period.

discipline, data are available for 88 programs. The programs not evaluated on measures 12 and 14 are typically smaller—in terms of faculty size and graduate student enrollment—than other physics programs. Were data on measures 12 and 14 available for all 123 programs, it is likely that the reported means for these two measures would be appreciably lower (and that some of the correlations of these measures with others would be higher).

Intercorrelations among the 16 measures (Pearson product-moment coefficients) are given in Table 7.3. Of particular note are the high positive correlations of the measures of program size (01-03) with measures of publication records (15, 16), research expenditures (14), and reputational survey ratings (08, 09, and 11). Figure 7.1 illustrates the relation between the mean rating of the scholarly quality of faculty (measure 08) and the number of faculty members (measure 01) for each of the 121 physics programs. Figure 7.2 plots the mean rating of program effectiveness (measure 09) against the total number of FY1976-80 program graduates (measure 02). Although in both figures there is a significant positive correlation between program size and reputational rating, it is quite apparent that some of the smaller programs received high mean ratings and some of the larger programs received low mean ratings.

Table 7.4 describes the 211 faculty members who participated in the evaluation of physics programs. These individuals constituted 57 percent of those asked to respond to the survey in this discipline and 6 percent of the faculty population in the 123 research-doctorate programs being evaluated.[4] A majority of the survey participants specialized in elementary particles or solid state physics, and more than two-thirds held the rank of full professor. Approximately 85 percent had earned their highest degree prior to 1970.

One exception should be noted with regard to the survey evaluations in this discipline. In the program listing on the survey form, New Mexico State University at Las Cruces was identified as being located in Alamogordo, which has a junior college branch of the same institution. Since a large majority of faculty evaluators indicated that they were unfamiliar with this program and it is quite possible that some of them were misled by the inaccurate identification of this institution, the committee has decided not to report the survey results for this program.

To assist the reader in interpreting results of the survey evaluations, estimated standard errors have been computed for mean ratings of the scholarly quality of faculty in 121 physics programs (and are given in Table 7.1). For each program the mean rating and an associated "confidence interval" of 1.5 standard errors are illustrated in Figure 7.3 (listed in order of highest to lowest mean rating). In comparing two programs, if their confidence intervals do not overlap, one may safely conclude that there is a significant

---

[4] See Table 2.3 in Chapter II.

difference in their mean ratings at a .05 level of significance.[5] From this figure it is also apparent that one should have somewhat more confidence in the accuracy of the mean ratings of higher-rated programs than lower-rated programs. This generalization results primarily from the fact that evaluators are not as likely to be familiar with the less prestigious programs, and consequently the mean ratings of these programs are usually based on fewer survey responses.

---

[5] See pp. 29-31 for a discussion of the interpretation of mean ratings and associated confidence intervals.

TABLE 7.1  Program Measures (Raw and Standardized Values) in Physics

| Prog No. | University – Department/Academic Unit | Program Size (01) | (02) | (03) | Characteristics of Program Graduates (04) | (05) | (06) | (07) |
|---|---|---|---|---|---|---|---|---|
| 001. | American University | 6 | 17 | 13 | .33 | 9.8 | .71 | .00 |
| | *Physics* | *35* | *45* | *42* | *57* | *25* | *55* | *28* |
| 002. | Arizona State University-Tempe | 18 | 12 | 22 | .31 | 7.5 | .69 | .06 |
| | *Physics* | *43* | *43* | *44* | *55* | *46* | *53* | *33* |
| 003. | Arizona, University of-Tucson | 35 | 25 | 43 | .21 | 9.0 | .58 | .29 |
| | *Physics* | *55* | *47* | *48* | *45* | *32* | *44* | *53* |
| 004. | Auburn University | 13 | 9 | 27 | NA | NA | NA | NA |
| | *Physics* | *40* | *42* | *45* | | | | |
| 005. | Boston College | 10 | 15 | 23 | .05 | 6.5 | .47 | .26 |
| | *Physics* | *38* | *44* | *44* | *32* | *56* | *35* | *50* |
| 006. | Boston University | 23 | 10 | 31 | .21 | 8.5 | .31 | .15 |
| | *Physics* | *47* | *42* | *45* | *46* | *36* | *21* | *41* |
| 007. | Brandeis University | 20 | 30 | 40 | .42 | 7.0 | .70 | .43 |
| | *Physics* | *45* | *49* | *47* | *64* | *51* | *54* | *65* |
| 008. | Brown University | 31 | 61 | 83 | .20 | 7.1 | .88 | .37 |
| | *Physics* | *52* | *58* | *55* | *45* | *49* | *69* | *60* |
| 009. | CUNY-Graduate School | 51 | 45 | 95 | .18 | 8.1 | .58 | .23 |
| | *Physics* | *66* | *53* | *57* | *43* | *41* | *44* | *48* |
| 010. | California Institute of Technology | 37 | 69 | 112 | .42 | 6.2 | .69 | .31 |
| | *Physics, Mathematics, and Astronomy* | *56* | *61* | *61* | *64* | *59* | *53* | *54* |
| 011. | California, University of-Santa Cruz | 12 | 15 | 20 | .20 | NA | .70 | .10 |
| | *Physics** | *39* | *44* | *43* | *45* | | *54* | *36* |
| 012. | California, University of-Berkeley | 72 | 179 | 265 | .29 | 6.5 | .77 | .36 |
| | *Physics* | *80* | *94* | *89* | *52* | *56* | *60* | *59* |
| 013. | California, University of-Davis | 22 | 19 | 38 | .35 | 7.6 | .70 | .35 |
| | *Physics* | *46* | *45* | *47* | *58* | *45* | *54* | *58* |
| 014. | California, University of-Irvine | 23 | 41 | 52 | .14 | 6.1 | .58 | .19 |
| | *Physics* | *47* | *52* | *49* | *39* | *59* | *44* | *44* |
| 015. | California, University of-Los Angeles | 46 | 67 | 183 | .28 | 7.2 | .74 | .36 |
| | *Physics* | *63* | *60* | *74* | *52* | *49* | *57* | *59* |
| 016. | California, University of-Riverside | 20 | 29 | 38 | .14 | 7.3 | .64 | .29 |
| | *Physics* | *45* | *48* | *47* | *40* | *48* | *49* | *52* |
| 017. | California, University of-San Diego | 38 | 60 | 109 | .10 | 7.1 | .65 | .41 |
| | *Physics* | *57* | *58* | *60* | *36* | *50* | *50* | *63* |
| 018. | California, University of-Santa Barbara | 23 | 37 | 50 | .36 | 6.3 | .70 | .33 |
| | *Physics* | *47* | *51* | *49* | *58* | *58* | *54* | *56* |
| 019. | Carnegie-Mellon University | 30 | 28 | 57 | .36 | 6.5 | .76 | .35 |
| | *Physics* | *52* | *48* | *50* | *59* | *56* | *59* | *57* |
| 020. | Case Western Reserve University | 25 | 26 | 32 | .37 | 5.9 | .76 | .28 |
| | *Physics* | *48* | *47* | *46* | *60* | *62* | *59* | *52* |

* indicates program was initiated since 1970.

NOTE: On the first line of data for every program, raw values for each measure are reported; on the second line values are reported in standardized form, with mean = 50 and standard deviation = 10. "NA" indicates that the value for a measure is not available.

TABLE 7.1  Program Measures (Raw and Standardized Values) in Physics

| Prog No. | Survey Results | | | | University Library | Research Support | | Published Articles | | Survey Ratings Standard Error | | | |
|---|---|---|---|---|---|---|---|---|---|---|---|---|---|
| | (08) | (09) | (10) | (11) | (12) | (13) | (14) | (15) | (16) | (08) | (09) | (10) | (11) |
| 001. | 1.3 | 1.0 | 0.9 | 0.4 | NA | NA | NA | 9 | | .14 | .14 | .07 | .06 |
| | 37 | 36 | 42 | 41 | | | | | 40 | 41 | | | |
| 002. | 1.7 | 1.2 | 1.2 | 0.4 | -0.3 | .33 | NA | 39 | | .12 | .13 | .10 | .06 |
| | 40 | 41 | 54 | 42 | 46 | 48 | | 43 | 43 | | | | |
| 003. | 3.0 | 1.8 | 1.4 | 0.9 | 0.9 | .46 | 6187 | 137 | | .08 | .08 | .08 | .07 |
| | 54 | 53 | 61 | 55 | 58 | 56 | 57 | 53 | 50 | | | | |
| 004. | 1.2 | 0.9 | NA | 0.2 | NA | .08 | NA | 14 | | .14 | .16 | NA | .04 |
| | 35 | 34 | | 37 | | 32 | | 41 | 41 | | | | |
| 005. | 1.2 | 0.6 | NA | 0.1 | NA | .50 | 609 | 15 | | .14 | .12 | NA | .04 |
| | 35 | 29 | | 36 | | 59 | 45 | 41 | 41 | | | | |
| 006. | 2.3 | 1.5 | 1.2 | 0.5 | -0.4 | .44 | NA | 28 | | .11 | .11 | .10 | .07 |
| | 47 | 46 | 55 | 45 | 45 | 55 | | 42 | 43 | | | | |
| 007. | 3.2 | 1.9 | 0.9 | 0.9 | NA | .65 | 667 | 54 | | .08 | .05 | .07 | .07 |
| | 55 | 55 | 38 | 54 | | 69 | 45 | 45 | 45 | | | | |
| 008. | 3.5 | 2.0 | 1.2 | 1.0 | -1.1 | .39 | 1491 | 136 | | .09 | .06 | .09 | .08 |
| | 58 | 56 | 52 | 55 | 39 | 52 | 47 | 53 | 52 | | | | |
| 009. | 3.2 | 1.8 | 1.2 | 0.8 | NA | .22 | NA | 187 | | .10 | .09 | .11 | .08 |
| | 55 | 53 | 53 | 53 | | 41 | | 58 | 58 | | | | |
| 010. | 4.9 | 2.7 | 1.0 | 1.6 | NA | .35 | 7760 | 370 | | .03 | .06 | .05 | .05 |
| | 73 | 71 | 43 | 71 | | 49 | 61 | 76 | 71 | | | | |
| 011. | 2.4 | 1.6 | 1.2 | 0.7 | NA | .42 | 394 | 19 | | .12 | .09 | .10 | .07 |
| | 47 | 48 | 54 | 49 | | 54 | 44 | 41 | 42 | | | | |
| 012. | 4.9 | 2.7 | 0.8 | 1.6 | 2.2 | .35 | 1727 | 323 | | .04 | .06 | .06 | .05 |
| | 72 | 70 | 37 | 72 | 70 | 49 | 47 | 71 | 69 | | | | |
| 013. | 2.5 | 1.6 | 1.1 | 0.7 | 0.6 | .09 | 763 | 57 | | .10 | .08 | .08 | .07 |
| | 48 | 49 | 51 | 49 | 55 | 32 | 45 | 45 | 45 | | | | |
| 014. | 3.0 | 1.9 | 1.1 | 1.1 | NA | .74 | 2386 | 115 | | .08 | .05 | .05 | .06 |
| | 54 | 54 | 51 | 58 | | 75 | 49 | 51 | 51 | | | | |
| 015. | 3.8 | 2.2 | 1.2 | 1.3 | 2.0 | .44 | 5627 | 269 | | .07 | .05 | .05 | .06 |
| | 62 | 60 | 52 | 64 | 68 | 55 | 56 | 66 | 67 | | | | |
| 016. | 2.4 | 1.6 | 1.1 | 0.6 | -1.0 | .45 | 868 | 38 | | .10 | .09 | .09 | .07 |
| | 47 | 48 | 48 | 47 | 39 | 56 | 45 | 43 | 44 | | | | |
| 017. | 4.1 | 2.2 | 1.0 | 1.3 | -0.0 | .40 | 11341 | 136 | | .08 | .06 | .07 | .06 |
| | 65 | 61 | 46 | 65 | 49 | 52 | 69 | 53 | 52 | | | | |
| 018. | 3.8 | 2.1 | 1.7 | 1.2 | -0.1 | .61 | 1619 | 115 | | .08 | .05 | .06 | .07 |
| | 61 | 58 | 76 | 60 | 48 | 66 | 47 | 51 | 51 | | | | |
| 019. | 3.5 | 2.0 | 0.9 | 1.0 | NA | .40 | 1992 | 128 | | .09 | .06 | .08 | .07 |
| | 59 | 57 | 42 | 55 | | 53 | 48 | 52 | 51 | | | | |
| 020. | 2.9 | 1.8 | 0.9 | 0.9 | -1.3 | .56 | 1231 | 79 | | .09 | .07 | .07 | .07 |
| | 53 | 53 | 38 | 55 | 36 | 63 | 46 | 47 | 48 | | | | |

NOTE: On the first line of data for every program, raw values for each measure are reported; on the second line values are reported in standardized form, with mean = 50 and standard deviation = 10. "NA" indicates that the value for a measure is not available. Since the scale used to compute measure (16) is entirely arbitrary, only values in standardized form are reported for this measure.

TABLE 7.1 Program Measures (Raw and Standardized Values) in Physics

| Prog No. | University - Department/Academic Unit | Program Size | | | Characteristics of Program Graduates | | | |
|---|---|---|---|---|---|---|---|---|
| | | (01) | (02) | (03) | (04) | (05) | (06) | (07) |
| 021. | Catholic University of America | 12 | 27 | 42 | .09 | 8.3 | .71 | .06 |
| | *Physics* | *39* | *48* | *47* | *35* | *38* | *54* | *33* |
| 022. | Chicago, University of | 38 | 88 | 115 | .27 | 6.8 | .85 | .39 |
| | *Physics* | *57* | *66* | *61* | *51* | *53* | *66* | *62* |
| 023. | Cincinnati, University of | 24 | 23 | 38 | .19 | 8.0 | .47 | .11 |
| | *Physics* | *48* | *46* | *47* | *44* | *41* | *35* | *37* |
| 024. | Clarkson College of Technology | 11 | 8 | 11 | .20 | 6.5 | .70 | .20 |
| | *Physics* | *39* | *42* | *42* | *45* | *56* | *54* | *45* |
| 025. | Clemson University | 23 | 18 | 5 | .26 | 6.3 | .50 | .09 |
| | *Physics and Astronomy* | *47* | *45* | *41* | *50* | *57* | *37* | *36* |
| 026. | Colorado State University-Fort Collins | 21 | 14 | 26 | .15 | 9.3 | .69 | .31 |
| | *Physics* | *45* | *44* | *44* | *41* | *29* | *53* | *54* |
| 027. | Colorado, University of | 37 | 72 | 89 | .29 | 6.8 | .63 | .20 |
| | *Physics* | *56* | *61* | *56* | *53* | *52* | *48* | *45* |
| 028. | Columbia University | 33 | 81 | 120 | .16 | 7.6 | .77 | .36 |
| | *Physics* | *54* | *64* | *62* | *41* | *45* | *60* | *59* |
| 029. | Connecticut, University of-Storrs | 31 | 26 | 31 | .26 | 8.1 | .44 | .19 |
| | *Physics* | *52* | *47* | *45* | *50* | *40* | *32* | *44* |
| 030. | Cornell University-Ithaca | 45 | 89 | 166 | .27 | 6.5 | .87 | .36 |
| | *Physics* | *62* | *67* | *71* | *51* | *55* | *68* | *59* |
| 031. | Dartmouth College | 15 | 21 | 20 | .17 | 5.5 | .79 | .47 |
| | *Physics and Astronomy* | *41* | *46* | *43* | *42* | *65* | *61* | *68* |
| 032. | Delaware, University of-Newark | 24 | 20 | 27 | .39 | 8.9 | .74 | .44 |
| | *Physics* | *48* | *45* | *45* | *62* | *33* | *57* | *65* |
| 033. | Denver, University of | 17 | 10 | 17 | .27 | 6.3 | .55 | .18 |
| | *Physics* | *43* | *42* | *43* | *51* | *57* | *41* | *43* |
| 034. | Drexel University | 25 | 13 | 44 | .25 | 6.5 | .57 | .00 |
| | *Physics and Atmospheric Science* | *48* | *43* | *48* | *49* | *56* | *43* | *28* |
| 035. | Duke University | 19 | 27 | 46 | .33 | 5.9 | .64 | .46 |
| | *Physics* | *44* | *48* | *48* | *57* | *62* | *48* | *67* |
| 036. | Florida State University-Tallahassee | 27 | 37 | 35 | .21 | 6.4 | .73 | .33 |
| | *Physics* | *50* | *51* | *46* | *46* | *56* | *56* | *56* |
| 037. | Florida, University of-Gainesville | 25 | 21 | 51 | .35 | 6.3 | .63 | .08 |
| | *Physics* | *48* | *46* | *49* | *58* | *58* | *47* | *35* |
| 038. | Georgetown University | 10 | 18 | 14 | .35 | 8.0 | .75 | .19 |
| | *Physics* | *38* | *45* | *42* | *58* | *41* | *58* | *44* |
| 039. | Georgia Institute of Technology | 23 | 20 | 37 | .42 | 6.5 | .65 | .26 |
| | *Physics* | *47* | *45* | *47* | *64* | *56* | *50* | *50* |
| 040. | Harvard University | 28 | 84 | 93 | .35 | 5.6 | .88 | .41 |
| | *Physics* | *50* | *65* | *57* | *58* | *64* | *69* | *63* |

* indicates program was initiated since 1970.

NOTE: On the first line of data for every program, raw values for each measure are reported; on the second line values are reported in standardized form, with mean = 50 and standard deviation = 10. "NA" indicates that the value for a measure is not available.

TABLE 7.1  Program Measures (Raw and Standardized Values) in Physics

| Prog No. | Survey Results | | | | University Library | Research Support | | Published Articles | | Survey Ratings Standard Error | | | |
|---|---|---|---|---|---|---|---|---|---|---|---|---|---|
| | (08) | (09) | (10) | (11) | (12) | (13) | (14) | (15) | (16) | (08) | (09) | (10) | (11) |
| 021. | 2.0 | 1.5 | 0.9 | 0.5 | NA | .58 | 889 | 38 | | .12 | .11 | .10 | .07 |
| | 43 | 46 | 39 | 44 | | 65 | 45 | 43 | 43 | | | | |
| 022. | 4.6 | 2.6 | 0.9 | 1.5 | 0.9 | .50 | 7819 | 246 | | .07 | .06 | .08 | .06 |
| | 70 | 69 | 41 | 67 | 57 | 59 | 61 | 64 | 69 | | | | |
| 023. | 2.0 | 1.5 | 1.0 | 0.5 | -0.2 | .25 | NA | 47 | | .12 | .10 | .06 | .06 |
| | 43 | 46 | 46 | 45 | 47 | 43 | | 44 | 44 | | | | |
| 024. | 1.0 | 0.8 | 0.9 | 0.3 | NA | .55 | NA | 55 | | .17 | .12 | .08 | .06 |
| | 34 | 32 | 39 | 39 | | 62 | | 45 | 44 | | | | |
| 025. | 1.8 | 1.2 | NA | 0.2 | NA | .13 | NA | 42 | | .15 | .17 | NA | .05 |
| | 41 | 40 | | 38 | | 35 | | 44 | 43 | | | | |
| 026. | 1.7 | 1.5 | 1.2 | 0.3 | -1.1 | .19 | 414 | 74 | | .18 | .11 | .10 | .05 |
| | 40 | 46 | 53 | 39 | 38 | 39 | 44 | 47 | 45 | | | | |
| 027. | 3.1 | 1.9 | 1.0 | 1.0 | -0.9 | .22 | 3044 | 195 | | .08 | .06 | .08 | .07 |
| | 54 | 55 | 44 | 57 | 41 | 41 | 50 | 59 | 58 | | | | |
| 028. | 4.5 | 2.3 | 0.6 | 1.5 | 1.7 | .33 | 1701 | 172 | | .06 | .06 | .07 | .07 |
| | 69 | 63 | 29 | 67 | 66 | 48 | 47 | 56 | 58 | | | | |
| 029. | 2.3 | 1.6 | 1.1 | 0.7 | -0.5 | .19 | 1284 | 59 | | .08 | .09 | .06 | .07 |
| | 47 | 48 | 50 | 50 | 44 | 39 | 46 | 45 | 45 | | | | |
| 030. | 4.7 | 2.8 | 1.3 | 1.5 | 1.6 | .18 | 14914 | 356 | | .05 | .05 | .06 | .06 |
| | 70 | 74 | 58 | 69 | 65 | 38 | 78 | 74 | 76 | | | | |
| 031. | 2.2 | 1.9 | 1.1 | 0.4 | -1.1 | .20 | NA | 23 | | .21 | .18 | .08 | .06 |
| | 46 | 55 | 50 | 41 | 38 | 40 | | 42 | 42 | | | | |
| 032. | 2.0 | 1.3 | 1.4 | 0.4 | NA | .29 | 387 | 67 | | .11 | .10 | .09 | .07 |
| | 44 | 43 | 60 | 43 | | 46 | 44 | 46 | 46 | | | | |
| 033. | 1.6 | 0.8 | NA | 0.1 | NA | .47 | 1372 | 28 | | .17 | .14 | NA | .04 |
| | 40 | 32 | | 36 | | 57 | 46 | 42 | 42 | | | | |
| 034. | 1.8 | 1.3 | 1.1 | 0.5 | NA | .16 | 555 | 48 | | .11 | .11 | .10 | .07 |
| | 41 | 42 | 49 | 45 | | 37 | 45 | 44 | 44 | | | | |
| 035. | 3.0 | 1.9 | 1.0 | 0.9 | 0.3 | .26 | 1405 | 53 | | .10 | .07 | .06 | .08 |
| | 53 | 54 | 45 | 53 | 52 | 44 | 46 | 45 | 45 | | | | |
| 036. | 3.0 | 1.8 | 1.3 | 0.8 | -0.4 | .44 | 2081 | 89 | | .10 | .07 | .08 | .08 |
| | 53 | 52 | 59 | 53 | 45 | 55 | 48 | 48 | 49 | | | | |
| 037. | 2.6 | 1.7 | 1.5 | 0.7 | 0.8 | .40 | 1126 | 85 | | .13 | .09 | .10 | .08 |
| | 49 | 51 | 69 | 50 | 57 | 53 | 46 | 48 | 47 | | | | |
| 038. | 1.0 | NA | NA | 0.1 | -0.6 | .30 | NA | 13 | | .16 | NA | NA | .04 |
| | 34 | | | 36 | 43 | 46 | | 41 | 41 | | | | |
| 039. | 2.3 | 1.5 | 1.0 | 0.5 | NA | .52 | 1781 | 82 | | .10 | .09 | .09 | .07 |
| | 47 | 47 | 45 | 45 | | 61 | 47 | 48 | 47 | | | | |
| 040. | 4.9 | 2.8 | 1.0 | 1.7 | 3.0 | .68 | 5602 | 337 | | .03 | .05 | .06 | .06 |
| | 73 | 73 | 46 | 72 | 78 | 71 | 56 | 73 | 80 | | | | |

NOTE: On the first line of data for every program, raw values for each measure are reported; on the second line values are reported in standardized form, with mean = 50 and standard deviation = 10. "NA" indicates that the value for a measure is not available. Since the scale used to compute measure (16) is entirely arbitrary, only values in standardized form are reported for this measure.

TABLE 7.1 Program Measures (Raw and Standardized Values) in Physics

| Prog No. | University - Department/Academic Unit | Program Size | | | Characteristics of Program Graduates | | | |
|---|---|---|---|---|---|---|---|---|
| | | (01) | (02) | (03) | (04) | (05) | (06) | (07) |
| 041. | Houston, University of | 18 | 15 | 36 | .15 | 7.8 | .39 | .31 |
| | *Physics* | *43* | *44* | *46* | *41* | *43* | *27* | *54* |
| 042. | Howard University | 9 | 11 | NA | NA | NA | NA | NA |
| | *Physics* | *37* | *43* | | | | | |
| 043. | Illinois Institute of Technology | 16 | 14 | 12 | .23 | 7.1 | .62 | .15 |
| | *Physics* | *42* | *44* | *42* | *48* | *50* | *47* | *41* |
| 044. | Illinois, University of-Chicago Circle | 28 | 17 | 34 | .11 | 8.8 | .47 | .16 |
| | *Physics* | *50* | *45* | *46* | *36* | *34* | *35* | *41* |
| 045. | Illinois, University-Urbana/Champaign | 68 | 162 | 287 | .13 | 6.0 | .92 | .35 |
| | *Physics* | *78* | *89* | *93* | *39* | *60* | *72* | *58* |
| 046. | Indiana University-Bloomington | 42 | 35 | 67 | .27 | 7.5 | .77 | .43 |
| | *Physics* | *60* | *50* | *52* | *51* | *46* | *60* | *65* |
| 047. | Iowa State University-Ames | 43 | 46 | 76 | .30 | 5.9 | .82 | .36 |
| | *Physics* | *61* | *53* | *54* | *53* | *61* | *64* | *58* |
| 048. | Iowa, University of-Iowa City | 22 | 35 | 24 | .23 | 6.4 | .53 | .29 |
| | *Physics and Astronomy* | *46* | *50* | *44* | *48* | *57* | *39* | *53* |
| 049. | Johns Hopkins University | 18 | 34 | 56 | .34 | 7.3 | .78 | .25 |
| | *Physics* | *43* | *50* | *50* | *58* | *48* | *61* | *49* |
| 050. | Kansas State University-Manhattan | 24 | 11 | 10 | .13 | 5.3 | .60 | .13 |
| | *Physics* | *48* | *43* | *41* | *39* | *67* | *45* | *39* |
| 051. | Kansas, University of | 24 | 15 | 20 | .33 | 6.9 | .44 | .28 |
| | *Physics and Astronomy* | *48* | *44* | *43* | *57* | *52* | *32* | *52* |
| 052. | Kent State University | 16 | 15 | 50 | .31 | 6.8 | .56 | .19 |
| | *Physics* | *42* | *44* | *49* | *55* | *53* | *42* | *44* |
| 053. | Kentucky, University of | 22 | 9 | 11 | .30 | 6.4 | NA | NA |
| | *Physics and Astronomy* | *46* | *42* | *42* | *54* | *57* | | |
| 054. | Louisiana State University-Baton Rouge | 30 | 13 | 32 | .25 | 6.9 | .67 | .24 |
| | *Physics and Astronomy* | *52* | *43* | *46* | *49* | *51* | *51* | *48* |
| 055. | Maryland, University of-College Park | 75 | 144 | 181 | .17 | 7.0 | .65 | .30 |
| | *Physics and Astronomy* | *82* | *84* | *73* | *42* | *50* | *49* | *54* |
| 056. | Massachusetts Institute of Technology | 87 | 195 | 300 | .25 | 5.6 | .72 | .36 |
| | *Physics* | *91* | *99* | *96* | *50* | *64* | *56* | *59* |
| 057. | Massachusetts, University of-Amherst | 35 | 29 | 42 | .35 | 7.2 | .68 | .32 |
| | *Physics and Astronomy* | *55* | *48* | *47* | *58* | *49* | *52* | *56* |
| 058. | Michigan State University-East Lansing | 48 | 26 | 68 | .16 | 5.8 | .67 | .30 |
| | *Physics* | *64* | *47* | *52* | *41* | *62* | *51* | *54* |
| 059. | Michigan, University of-Ann Arbor | 50 | 51 | 81 | .29 | 6.6 | .62 | .28 |
| | *Physics* | *65* | *55* | *55* | *53* | *54* | *47* | *52* |
| 060. | Minnesota, University of | 46 | 43 | 83 | .28 | 7.8 | .66 | .42 |
| | *Physics and Astronomy* | *63* | *53* | *55* | *52* | *43* | *50* | *63* |

* indicates program was initiated since 1970.

NOTE: On the first line of data for every program, raw values for each measure are reported; on the second line values are reported in standardized form, with mean = 50 and standard deviation = 10. "NA" indicates that the value for a measure is not available.

TABLE 7.1  Program Measures (Raw and Standardized Values) in Physics

| Prog No. | Survey Results (08) | (09) | (10) | (11) | University Library (12) | Research Support (13) | (14) | Published Articles (15) | (16) | Survey Ratings Standard Error (08) | (09) | (10) | (11) |
|---|---|---|---|---|---|---|---|---|---|---|---|---|---|
| 041. | 2.1 | 1.3 | 1.6 | 0.3 | -0.9 | .44 | 542 | 77 | | .17 | .16 | .12 | .06 |
|  | 45 | 42 | 72 | 39 | 40 | 55 | 44 | 47 | 47 | | | | |
| 042. | 1.1 | 1.0 | NA | 0.2 | -0.4 | NA | NA | 27 | | .23 | .18 | NA | .05 |
|  | 34 | 36 | | 37 | 45 | | | 42 | 42 | | | | |
| 043. | 1.7 | 1.1 | 1.0 | 0.4 | NA | .56 | NA | 42 | | .11 | .14 | .12 | .06 |
|  | 40 | 39 | 45 | 43 | | 63 | | 44 | 44 | | | | |
| 044. | 2.1 | 1.4 | 1.2 | 0.6 | NA | .32 | 1014 | 47 | | .13 | .11 | .12 | .08 |
|  | 44 | 45 | 52 | 47 | | 47 | 46 | 44 | 44 | | | | |
| 045. | 4.3 | 2.5 | 1.1 | 1.6 | 2.0 | .46 | 4815 | 363 | | .07 | .06 | .06 | .06 |
|  | 67 | 66 | 50 | 70 | 68 | 56 | 54 | 75 | 75 | | | | |
| 046. | 3.1 | 1.9 | 1.2 | 1.0 | 0.9 | .38 | 4224 | 112 | | .07 | .07 | .07 | .07 |
|  | 55 | 54 | 55 | 56 | 58 | 51 | 53 | 51 | 51 | | | | |
| 047. | 2.9 | 1.8 | 1.0 | 0.7 | -0.5 | .14 | NA | 106 | | .11 | .08 | .07 | .07 |
|  | 53 | 52 | 46 | 50 | 44 | 36 | | 50 | 50 | | | | |
| 048. | 2.5 | 1.8 | 1.0 | 0.5 | 0.3 | .50 | 3475 | 58 | | .12 | .09 | .04 | .07 |
|  | 49 | 52 | 43 | 45 | 52 | 59 | 51 | 45 | 45 | | | | |
| 049. | 3.1 | 1.9 | 0.8 | 0.8 | -0.4 | .39 | 2000 | 89 | | .09 | .06 | .07 | .07 |
|  | 55 | 56 | 36 | 51 | 45 | 52 | 48 | 48 | 49 | | | | |
| 050. | 2.0 | 1.2 | 0.9 | 0.3 | NA | .21 | NA | 66 | | .18 | .12 | .10 | .06 |
|  | 43 | 41 | 42 | 41 | | 40 | | 46 | 46 | | | | |
| 051. | 2.2 | 1.5 | 1.0 | 0.6 | 0.1 | .33 | NA | 21 | | .11 | .10 | .08 | .08 |
|  | 45 | 46 | 45 | 47 | 50 | 48 | | 42 | 42 | | | | |
| 052. | 1.7 | 1.2 | 1.2 | 0.4 | -1.8 | .44 | NA | 20 | | .13 | .14 | .13 | .07 |
|  | 41 | 41 | 53 | 42 | 32 | 55 | | 42 | 42 | | | | |
| 053. | 1.8 | 1.2 | 1.4 | 0.6 | -0.1 | .23 | NA | 27 | | .10 | .13 | .12 | .08 |
|  | 41 | 40 | 62 | 47 | 48 | 41 | | 42 | 42 | | | | |
| 054. | 2.5 | 1.4 | 1.3 | 0.7 | -0.3 | .37 | 1340 | 65 | | .10 | .11 | .08 | .06 |
|  | 49 | 44 | 59 | 50 | 46 | 50 | 46 | 46 | 46 | | | | |
| 055. | 3.7 | 2.1 | 0.9 | 1.4 | 0.2 | .31 | 8341 | 290 | | .07 | .06 | .06 | .07 |
|  | 61 | 59 | 38 | 67 | 51 | 47 | 62 | 68 | 68 | | | | |
| 056. | 4.8 | 2.7 | 1.0 | 1.7 | -0.3 | .31 | 31429 | 559 | | .04 | .05 | .06 | .05 |
|  | 72 | 71 | 45 | 73 | 46 | 47 | 99 | 94 | 97 | | | | |
| 057. | 2.7 | 1.7 | 1.4 | 0.8 | -0.7 | .37 | 1267 | 68 | | .09 | .07 | .10 | .07 |
|  | 50 | 51 | 61 | 53 | 42 | 51 | 46 | 46 | 48 | | | | |
| 058. | 3.2 | 1.9 | 1.3 | 1.1 | 0.3 | .50 | 2612 | 144 | | .10 | .06 | .09 | .08 |
|  | 56 | 55 | 59 | 60 | 52 | 59 | 49 | 54 | 55 | | | | |
| 059. | 3.7 | 2.1 | 1.0 | 1.1 | 1.8 | .42 | 2912 | 169 | | .07 | .05 | .08 | .08 |
|  | 60 | 60 | 43 | 60 | 66 | 54 | 50 | 56 | 56 | | | | |
| 060. | 3.5 | 2.0 | 1.0 | 1.0 | 1.2 | .28 | 2664 | 197 | | .07 | .06 | .08 | .07 |
|  | 59 | 57 | 44 | 56 | 60 | 45 | 49 | 59 | 59 | | | | |

NOTE: On the first line of data for every program, raw values for each measure are reported; on the second line values are reported in standardized form, with mean = 50 and standard deviation = 10. "NA" indicates that the value for a measure is not available. Since the scale used to compute measure (16) is entirely arbitrary, only values in standardized form are reported for this measure.

TABLE 7.1  Program Measures (Raw and Standardized Values) in Physics

| Prog No. | University - Department/Academic Unit | Program Size (01) | (02) | (03) | Characteristics of Program Graduates (04) | (05) | (06) | (07) |
|---|---|---|---|---|---|---|---|---|
| 061. | Missouri, University of-Columbia | 19 | 17 | 20 | .06 | 6.6 | .59 | .18 |
| | *Physics* | *44* | *45* | *43* | *32* | *54* | *44* | *43* |
| 062. | Missouri, University of-Rolla | 22 | 21 | 23 | .27 | 6.0 | .73 | .13 |
| | *Physics* | *46* | *46* | *44* | *51* | *60* | *56* | *39* |
| 063. | Montana State University-Bozeman | 13 | 12 | 25 | .50 | 8.8 | .50 | .10 |
| | *Physics* | *40* | *43* | *44* | *71* | *33* | *37* | *36* |
| 064. | Nebraska, University of-Lincoln | 30 | 15 | 37 | .27 | 9.0 | .68 | .32 |
| | *Physics and Astronomy* | *52* | *44* | *47* | *51* | *32* | *52* | *55* |
| 065. | Nevada, University of-Reno | 12 | 15 | 9 | NA | NA | NA | NA |
| | *Physics* | *39* | *44* | *41* | | | | |
| 066. | New Hampshire, University of | 17 | 17 | 20 | .24 | 6.2 | .56 | .25 |
| | *Physics* | *43* | *45* | *43* | *48* | *59* | *42* | *49* |
| 067. | New Mexico State University-Las Cruces | 13 | 15 | 24 | .50 | 8.8 | .50 | .17 |
| | *Physics* | *40* | *44* | *44* | *71* | *34* | *37* | *42* |
| 068. | New Mexico, University of-Albuquerque | 22 | 15 | 45 | .23 | 7.0 | .67 | .14 |
| | *Physics and Astronomy* | *46* | *44* | *48* | *47* | *51* | *51* | *40* |
| 069. | New York University | 26 | 50 | 59 | .48 | 8.6 | .68 | .20 |
| | *Physics* | *49* | *55* | *51* | *69* | *36* | *52* | *45* |
| 070. | North Carolina State University-Raleigh | 31 | 18 | 38 | .40 | 7.5 | .60 | .15 |
| | *Physics* | *52* | *45* | *47* | *62* | *46* | *45* | *41* |
| 071. | North Carolina, University of-Chapel Hill | 30 | 37 | 43 | .12 | 6.3 | .71 | .32 |
| | *Physics and Astronomy* | *52* | *51* | *48* | *38* | *58* | *54* | *56* |
| 072. | North Texas State University-Denton | 12 | 19 | 23 | .30 | 9.0 | .74 | .21 |
| | *Physics* | *39* | *45* | *44* | *54* | *32* | *57* | *46* |
| 073. | Northeastern University | 32 | 18 | 64 | .16 | 7.5 | .37 | .21 |
| | *Physics* | *53* | *45* | *52* | *41* | *46* | *26* | *46* |
| 074. | Northwestern University | 24 | 26 | 45 | .26 | 6.5 | .71 | .50 |
| | *Physics and Astronomy* | *48* | *47* | *48* | *50* | *56* | *54* | *71* |
| 075. | Notre Dame, University of | 21 | 30 | 37 | .16 | 6.3 | .80 | .34 |
| | *Physics* | *45* | *49* | *47* | *41* | *57* | *62* | *57* |
| 076. | Ohio State University-Columbus | 36 | 57 | 105 | .12 | 6.7 | .60 | .26 |
| | *Physics* | *56* | *57* | *59* | *38* | *54* | *45* | *50* |
| 077. | Ohio University-Athens | 19 | 26 | 58 | .08 | 7.8 | .71 | .24 |
| | *Physics* | *44* | *47* | *50* | *34* | *43* | *55* | *48* |
| 078. | Oklahoma State University-Stillwater | 22 | 25 | 26 | .33 | 6.5 | .75 | .17 |
| | *Physics* | *46* | *47* | *44* | *57* | *56* | *58* | *42* |
| 079. | Oklahoma, University of-Norman | 22 | 16 | 16 | .24 | 9.0 | .71 | .24 |
| | *Physics and Astronomy* | *46* | *44* | *43* | *48* | *32* | *54* | *48* |
| 080. | Oregon State University-Corvallis | 17 | 6 | 17 | NA | 7.5 | NA | NA |
| | *Physics* | *43* | *41* | *43* | | *46* | | |

* indicates program was initiated since 1970.


NOTE: On the first line of data for every program, raw values for each measure are reported; on the second line values are reported in standardized form, with mean = 50 and standard deviation = 10. "NA" indicates that the value for a measure is not available.

TABLE 7.1  Program Measures (Raw and Standardized Values) in Physics

| Prog No. | Survey Results (08) | (09) | (10) | (11) | University Library (12) | Research Support (13) | (14) | Published Articles (15) | (16) | Survey Ratings Standard Error (08) | (09) | (10) | (11) |
|---|---|---|---|---|---|---|---|---|---|---|---|---|---|
| 061. | 2.0 | 1.3 | 1.5 | 0.3 | -0.2 | .26 | NA | 44 | | .15 | .15 | .11 | .06 |
| | 43 | 43 | 66 | 41 | 47 | 44 | | 44 | 43 | | | | |
| 062. | 2.0 | 1.6 | 1.3 | 0.4 | NA | .46 | 1287 | 70 | | .15 | .13 | .13 | .06 |
| | 43 | 49 | 60 | 43 | | 56 | 46 | 46 | 45 | | | | |
| 063. | 1.8 | 1.3 | 1.4 | 0.4 | NA | .31 | 565 | 28 | | .16 | .15 | .13 | .07 |
| | 41 | 42 | 62 | 42 | | 47 | 45 | 42 | 42 | | | | |
| 064. | 2.4 | 1.6 | 1.0 | 0.6 | -0.5 | .33 | 1203 | 68 | | .11 | .11 | .08 | .07 |
| | 48 | 48 | 45 | 46 | 44 | 48 | 46 | 46 | 47 | | | | |
| 065. | 0.9 | NA | NA | 0.1 | NA | .00 | NA | 15 | | .19 | NA | NA | .03 |
| | 32 | | | 35 | | 27 | | 41 | 41 | | | | |
| 066. | 1.7 | NA | 1.3 | 0.2 | NA | .47 | NA | 16 | | .11 | NA | .11 | .04 |
| | 40 | | 56 | 38 | | 57 | | 41 | 42 | | | | |
| 067. | NA | NA | NA | NA | NA | .08 | 392 | 25 | | NA | NA | NA | NA |
| | | | | | | 32 | 44 | 42 | 42 | | | | |
| 068. | 2.1 | 1.3 | 1.7 | 0.4 | -1.0 | .18 | 406 | 21 | | .15 | .17 | .13 | .07 |
| | 44 | 43 | 75 | 41 | 40 | 38 | 44 | 42 | 42 | | | | |
| 069. | 3.2 | 1.9 | 0.9 | 0.9 | 0.5 | .54 | 1659 | 132 | | .08 | .05 | .07 | .07 |
| | 55 | 55 | 41 | 55 | 54 | 62 | 47 | 53 | 55 | | | | |
| 070. | 2.2 | 1.4 | 1.6 | 0.5 | NA | .23 | 400 | 57 | | .11 | .12 | .09 | .07 |
| | 45 | 45 | 72 | 44 | | 41 | 44 | 45 | 44 | | | | |
| 071. | 3.0 | 1.9 | 1.1 | 0.9 | 1.0 | .30 | 546 | 75 | | .08 | .07 | .08 | .06 |
| | 53 | 55 | 50 | 53 | 59 | 46 | 44 | 47 | 47 | | | | |
| 072. | 1.3 | 0.7 | 1.2 | 0.4 | NA | .08 | 382 | 23 | | .16 | .20 | .11 | .07 |
| | 36 | 30 | 52 | 41 | | 32 | 44 | 42 | 43 | | | | |
| 073. | 2.7 | 1.7 | 1.1 | 0.8 | NA | .44 | 1223 | 58 | | .10 | .09 | .08 | .07 |
| | 50 | 50 | 47 | 51 | | 55 | 46 | 45 | 46 | | | | |
| 074. | 3.0 | 1.9 | 1.0 | 0.9 | 0.3 | .38 | 1254 | 154 | | .09 | .06 | .08 | .08 |
| | 54 | 55 | 44 | 54 | 52 | 51 | 46 | 55 | 55 | | | | |
| 075. | 2.4 | 1.6 | 1.0 | 0.7 | -1.3 | .67 | 850 | 82 | | .10 | .08 | .09 | .08 |
| | 47 | 48 | 45 | 50 | 36 | 70 | 45 | 48 | 49 | | | | |
| 076. | 3.0 | 1.8 | 1.2 | 0.9 | 0.9 | .50 | 1675 | 118 | | .09 | .06 | .08 | .07 |
| | 53 | 53 | 52 | 55 | 57 | 59 | 47 | 51 | 51 | | | | |
| 077. | 2.0 | 1.6 | 1.1 | 0.5 | NA | .16 | NA | 16 | | .12 | .11 | .09 | .07 |
| | 44 | 48 | 48 | 45 | | 37 | | 41 | 41 | | | | |
| 078. | 1.8 | 1.2 | NA | 0.2 | -1.9 | .27 | NA | 45 | | .16 | .14 | NA | .05 |
| | 41 | 40 | | 37 | 30 | 44 | | 44 | 44 | | | | |
| 079. | 2.1 | 1.4 | 1.5 | 0.3 | -0.6 | .27 | NA | 34 | | .12 | .12 | .12 | .05 |
| | 44 | 44 | 68 | 40 | 44 | 44 | | 43 | 43 | | | | |
| 080. | 1.8 | 1.2 | 1.0 | 0.3 | NA | .29 | NA | 47 | | .15 | .13 | .09 | .06 |
| | 41 | 40 | 45 | 40 | | 46 | | 44 | 45 | | | | |

NOTE: On the first line of data for every program, raw values for each measure are reported; on the second line values are reported in standardized form, with mean = 50 and standard deviation = 10. "NA" indicates that the value for a measure is not available. Since the scale used to compute measure (16) is entirely arbitrary, only values in standardized form are reported for this measure.

TABLE 7.1  Program Measures (Raw and Standardized Values) in Physics

| Prog No. | University - Department/Academic Unit | Program Size | | | Characteristics of Program Graduates | | | |
|---|---|---|---|---|---|---|---|---|
| | | (01) | (02) | (03) | (04) | (05) | (06) | (07) |
| 081. | Oregon, University of-Eugene | 19 | 38 | 64 | .27 | 7.2 | .42 | .13 |
| | *Physics* | *44* | *51* | *52* | *51* | *49* | *30* | *39* |
| 082. | Pennsylvania State University | 32 | 51 | 67 | .15 | 7.9 | .63 | .17 |
| | *Physics* | *53* | *55* | *52* | *41* | *43* | *48* | *43* |
| 083. | Pennsylvania, University of | 46 | 52 | 74 | .29 | 6.0 | .74 | .46 |
| | *Physics* | *63* | *55* | *53* | *53* | *61* | *57* | *67* |
| 084. | Pittsburgh, University of | 42 | 60 | 80 | .08 | 7.7 | .73 | .36 |
| | *Physics and Astronomy* | *60* | *58* | *55* | *34* | *44* | *56* | *58* |
| 085. | Polytech Institute of New York | 11 | 12 | 19 | .33 | 7.8 | .56 | .17 |
| | *Physics* | *39* | *43* | *43* | *57* | *43* | *42* | *42* |
| 086. | Princeton University | 51 | 90 | 83 | .45 | 5.2 | .81 | .46 |
| | *Physics* | *66* | *67* | *55* | *67* | *68* | *63* | *67* |
| 087. | Rensselaer Polytechnic Institute | 25 | 23 | 30 | .23 | 7.0 | .59 | .27 |
| | *Physics* | *48* | *46* | *45* | *47* | *51* | *45* | *51* |
| 088. | Rice University | 21 | 38 | 42 | .46 | 5.9 | .71 | .29 |
| | *Physics* | *45* | *51* | *47* | *67* | *61* | *55* | *53* |
| 089. | Rochester, University of | 30 | 68 | 68 | .15 | 7.1 | .77 | .33 |
| | *Physics and Astronomy* | *52* | *60* | *52* | *40* | *50* | *59* | *56* |
| 090. | Rockefeller University | 15 | 6 | 9 | NA | NA | NA | NA |
| | *Physics* | *41* | *41* | *41* | | | | |
| 091. | Rutgers, The State University-New Brunswick | 41 | 34 | 84 | .18 | 6.5 | .74 | .45 |
| | *Physics* | *59* | *50* | *55* | *43* | *56* | *57* | *66* |
| 092. | SUNY at Albany | 14 | 20 | 44 | .31 | 9.2 | .69 | .31 |
| | *Physics* | *41* | *45* | *48* | *55* | *30* | *53* | *55* |
| 093. | SUNY at Buffalo | 12 | 26 | 40 | .16 | 8.4 | .79 | .38 |
| | *Physics and Astronomy* | *39* | *47* | *47* | *41* | *38* | *61* | *60* |
| 094. | SUNY at Stony Brook | 52 | 82 | 141 | .14 | 5.8 | .74 | .33 |
| | *Physics* | *67* | *65* | *66* | *39* | *62* | *57* | *56* |
| 095. | South Carolina, University of-Columbia | 17 | 20 | 30 | .20 | 7.7 | .50 | .13 |
| | *Physics and Astronomy* | *43* | *45* | *45* | *45* | *44* | *37* | *39* |
| 096. | Southern California, University of | 20 | 31 | 34 | .14 | 6.8 | .58 | .11 |
| | *Physics* | *45* | *49* | *46* | *40* | *53* | *44* | *37* |
| 097. | Southern Illinois University-Carbondale | 61 | 8 | 12 | .00 | 10.5 | .42 | .17 |
| | *Molecular Science* | *73* | *42* | *42* | *27* | *17* | *30* | *42* |
| 098. | Stanford University | 14 | 52 | 100 | .41 | 5.8 | .69 | .16 |
| | *Applied Physics* | *41* | *55* | *58* | *63* | *62* | *53* | *41* |
| 099. | Stanford University | 23 | 68 | 121 | .58 | 6.2 | .74 | .49 |
| | *Physics* | *47* | *60* | *62* | *79* | *58* | *57* | *70* |
| 100. | Stevens Institute of Technology | 19 | 21 | 37 | .30 | 8.6 | .63 | .21 |
| | *Physics and Engineering Physics* | *44* | *46* | *47* | *54* | *35* | *48* | *46* |

\* indicates program was initiated since 1970.

NOTE: On the first line of data for every program, raw values for each measure are reported; on the second line values are reported in standardized form, with mean = 50 and standard deviation = 10. "NA" indicates that the value for a measure is not available.

TABLE 7.1 Program Measures (Raw and Standardized Values) in Physics

| Prog No. | Survey Results (08) | (09) | (10) | (11) | University Library (12) | Research Support (13) | (14) | Published Articles (15) | (16) | Survey Ratings Standard Error (08) | (09) | (10) | (11) |
|---|---|---|---|---|---|---|---|---|---|---|---|---|---|
| 081. | 2.8 | 1.6 | 1.1 | 0.8 | -0.9 | .53 | 918 | 74 | | .10 | .08 | .09 | .06 |
| | 51 | 50 | 47 | 52 | 40 | 61 | 45 | 47 | 47 | | | | |
| 082. | 2.5 | 1.8 | 1.1 | 0.8 | 0.7 | .44 | NA | 151 | | .09 | .07 | .08 | .07 |
| | 48 | 52 | 47 | 52 | 56 | 55 | | 54 | 51 | | | | |
| 083. | 4.0 | 2.3 | 0.8 | 1.4 | 0.7 | .30 | 5386 | 295 | | .07 | .05 | .08 | .07 |
| | 64 | 62 | 37 | 66 | 56 | 46 | 56 | 68 | 69 | | | | |
| 084. | 3.2 | 1.9 | 1.1 | 1.1 | 0.1 | .71 | 2549 | 144 | | .08 | .07 | .07 | .07 |
| | 55 | 55 | 49 | 58 | 50 | 73 | 49 | 54 | 54 | | | | |
| 085. | 1.9 | 1.2 | 1.0 | 0.3 | NA | .27 | NA | 30 | | .19 | .18 | .18 | .06 |
| | 42 | 40 | 43 | 40 | | 44 | | 43 | 43 | | | | |
| 086. | 4.9 | 2.8 | 0.9 | 1.6 | 0.9 | .28 | 3761 | 239 | | .03 | .05 | .06 | .06 |
| | 72 | 72 | 42 | 70 | 58 | 44 | 52 | 63 | 62 | | | | |
| 087. | 2.5 | 1.9 | 1.0 | 0.6 | NA | .40 | NA | 73 | | .08 | .05 | .09 | .07 |
| | 48 | 55 | 45 | 47 | | 53 | | 47 | 46 | | | | |
| 088. | 2.9 | 1.9 | 1.0 | 0.8 | -1.4 | .29 | 1031 | 87 | | .11 | .08 | .09 | .08 |
| | 52 | 55 | 43 | 51 | 35 | 45 | 46 | 48 | 48 | | | | |
| 089. | 3.5 | 2.0 | 0.8 | 1.2 | -0.6 | .43 | 4092 | 243 | | .07 | .06 | .07 | .06 |
| | 58 | 57 | 37 | 62 | 43 | 55 | 53 | 63 | 64 | | | | |
| 090. | 3.9 | 1.9 | 0.8 | 1.0 | NA | .27 | 1609 | 66 | | .09 | .10 | .07 | .08 |
| | 63 | 55 | 35 | 57 | | 44 | 47 | 46 | 49 | | | | |
| 091. | 3.1 | 1.9 | 1.1 | 1.1 | 0.8 | .49 | 2608 | 133 | | .09 | .06 | .09 | .08 |
| | 55 | 55 | 50 | 60 | 57 | 58 | 49 | 53 | 52 | | | | |
| 092. | 1.8 | 1.2 | 1.1 | 0.6 | -1.0 | .43 | 483 | 60 | | .13 | .11 | .08 | .07 |
| | 41 | 41 | 48 | 47 | 40 | 55 | 44 | 45 | 46 | | | | |
| 093. | 1.7 | 1.0 | 0.8 | 0.4 | 0.3 | .50 | NA | 74 | | .14 | .11 | .11 | .06 |
| | 40 | 37 | 35 | 43 | 52 | 59 | | 47 | 46 | | | | |
| 094. | 4.1 | 2.2 | 1.3 | 1.4 | -0.6 | .31 | 3469 | 230 | | .08 | .06 | .07 | .06 |
| | 65 | 62 | 58 | 66 | 43 | 47 | 51 | 62 | 64 | | | | |
| 095. | 1.7 | 1.1 | 1.3 | 0.3 | -0.4 | .29 | NA | 35 | | .14 | .15 | .11 | .06 |
| | 40 | 38 | 56 | 41 | 45 | 46 | | 43 | 43 | | | | |
| 096. | 2.7 | 1.7 | 1.1 | 0.6 | 0.4 | .35 | 900 | 196 | | .11 | .09 | .10 | .07 |
| | 50 | 51 | 49 | 48 | 52 | 49 | 45 | 59 | 56 | | | | |
| 097. | 1.4 | 0.7 | NA | 0.1 | -0.2 | .13 | NA | 6 | | .16 | .11 | NA | .03 |
| | 38 | 31 | | 36 | 47 | 35 | | 40 | 40 | | | | |
| 098. | 4.2 | 2.6 | 1.2 | 0.9 | 2.0 | .64 | 14262 | 365 | | .11 | .08 | .09 | .09 |
| | 65 | 69 | 52 | 53 | 69 | 68 | 76 | 75 | 71 | | | | |
| 099. | 4.6 | 2.7 | 1.0 | 1.5 | 2.0 | .65 | 14262 | 365 | | .07 | .06 | .05 | .06 |
| | 70 | 71 | 43 | 70 | 69 | 69 | 76 | 75 | 71 | | | | |
| 100. | 2.4 | 1.5 | 0.8 | 0.6 | NA | .21 | 508 | 23 | | .12 | .13 | .10 | .07 |
| | 47 | 46 | 36 | 46 | | 40 | 44 | 42 | 41 | | | | |

NOTE: On the first line of data for every program, raw values for each measure are reported; on the second line values are reported in standardized form, with mean = 50 and standard deviation = 10. "NA" indicates that the value for a measure is not available. Since the scale used to compute measure (16) is entirely arbitrary, only values in standardized form are reported for this measure.

TABLE 7.1  Program Measures (Raw and Standardized Values) in Physics

| Prog No. | University - Department/Academic Unit | Program Size | | | Characteristics of Program Graduates | | | |
|---|---|---|---|---|---|---|---|---|
| | | (01) | (02) | (03) | (04) | (05) | (06) | (07) |
| 101. | Syracuse University | 20 | 29 | 38 | .16 | 7.4 | .52 | .24 |
| | *Physics* | *45* | *48* | *47* | *41* | *47* | *38* | *49* |
| 102. | Temple University | 24 | 14 | 34 | .21 | 8.8 | .37 | .05 |
| | *Physics* | *48* | *44* | *46* | *46* | *33* | *26* | *32* |
| 103. | Tennessee, University of-Knoxville | 43 | 33 | 36 | .36 | 7.1 | .48 | .16 |
| | *Physics and Astronomy* | *61* | *49* | *46* | *59* | *50* | *36* | *42* |
| 104. | Texas A & M University | 35 | 29 | 34 | .35 | 7.5 | .56 | .24 |
| | *Physics* | *55* | *48* | *46* | *58* | *46* | *42* | *48* |
| 105. | Texas, U of-Health Science Center, Houston | 19 | 9 | 7 | NA | NA | NA | NA |
| | *Physics (M D Anderson Hospital)* | *44* | *42* | *41* | | | | |
| 106. | Texas, University of-Austin | 58 | 89 | 202 | .32 | 6.0 | .67 | .31 |
| | *Physics* | *71* | *67* | *77* | *56* | *60* | *51* | *54* |
| 107. | Texas, University of-Dallas | 21 | 18 | 46 | .00 | 6.6 | .71 | .47 |
| | *Physics** | *45* | *45* | *48* | *27* | *55* | *54* | *68* |
| 108. | Tufts University | 20 | 7 | 21 | NA | NA | NA | NA |
| | *Physics* | *45* | *41* | *44* | | | | |
| 109. | Tulane University | 10 | 4 | 10 | NA | NA | NA | NA |
| | *Physics* | *38* | *41* | *41* | | | | |
| 110. | Utah State University-Logan | 15 | 11 | 15 | .27 | 6.5 | .60 | .00 |
| | *Physics* | *41* | *43* | *42* | *51* | *56* | *45* | *28* |
| 111. | Utah, University of-Salt Lake City | 17 | 20 | 24 | .22 | 7.0 | .67 | .22 |
| | *Physics* | *43* | *45* | *44* | *47* | *51* | *51* | *47* |
| 112. | Vanderbilt University | 18 | 17 | 17 | .50 | 6.8 | .75 | .31 |
| | *Physics and Astronomy* | *43* | *45* | *43* | *71* | *52* | *58* | *55* |
| 113. | Virginia Polytechnic Institute & State Univ | 36 | 11 | 21 | .55 | 6.0 | .91 | .18 |
| | *Physics* | *56* | *43* | *44* | *75* | *60* | *71* | *43* |
| 114. | Virginia, University of | 31 | 53 | 58 | .23 | 5.8 | .66 | .26 |
| | *Physics* | *52* | *56* | *50* | *48* | *62* | *50* | *50* |
| 115. | Washington State University-Pullman | 16 | 16 | 20 | .29 | 7.8 | .64 | .14 |
| | *Physics* | *42* | *44* | *43* | *52* | *44* | *49* | *40* |
| 116. | Washington University-Saint Louis | 26 | 31 | 61 | .32 | 6.1 | .68 | .32 |
| | *Physics* | *49* | *49* | *51* | *55* | *59* | *52* | *55* |
| 117. | Washington, University of-Seattle | 45 | 45 | 37 | .18 | 7.0 | .71 | .41 |
| | *Physics* | *62* | *53* | *47* | *43* | *51* | *55* | *63* |
| 118. | Wayne State University | 24 | 16 | 14 | .25 | 6.8 | .73 | .33 |
| | *Physics and Astronomy* | *48* | *44* | *42* | *49* | *53* | *56* | *56* |
| 119. | William & Mary, College of | 24 | 21 | 32 | .42 | 6.4 | .58 | .12 |
| | *Physics* | *48* | *46* | *46* | *65* | *57* | *43* | *38* |
| 120. | Wisconsin, University of-Madison | 54 | 29 | 163 | .17 | 7.3 | .60 | .26 |
| | *Physics* | *68* | *48* | *70* | *42* | *48* | *45* | *50* |

* indicates program was initiated since 1970.


NOTE: On the first line of data for every program, raw values for each measure are reported; on the second line values are reported in standardized form, with mean = 50 and standard deviation = 10. "NA" indicates that the value for a measure is not available.

TABLE 7.1  Program Measures (Raw and Standardized Values) in Physics

| Prog No. | Survey Results (08) | (09) | (10) | (11) | University Library (12) | Research Support (13) | (14) | Published Articles (15) | (16) | Survey Ratings Standard Error (08) | (09) | (10) | (11) |
|---|---|---|---|---|---|---|---|---|---|---|---|---|---|
| 101. | 2.9 | 1.9 | 1.1 | 0.7 | -0.3 | .40 | 1217 | 73 | | .11 | .08 | .06 | .08 |
| | 52 | 55 | 47 | 49 | 46 | 53 | 46 | 47 | 47 | | | | |
| 102. | 2.3 | 1.5 | 0.9 | 0.5 | -0.4 | .29 | 405 | 39 | | .12 | .11 | .09 | .06 |
| | 46 | 46 | 42 | 46 | 45 | 46 | 44 | 43 | 44 | | | | |
| 103. | 2.5 | 1.6 | 1.3 | 0.5 | -0.4 | .02 | 710 | 65 | | .11 | .10 | .09 | .06 |
| | 49 | 48 | 56 | 44 | 45 | 28 | 45 | 46 | 46 | | | | |
| 104. | 2.6 | 1.6 | 1.4 | 0.7 | -0.5 | .31 | 663 | 121 | | .10 | .09 | .09 | .08 |
| | 50 | 48 | 63 | 50 | 44 | 47 | 45 | 51 | 53 | | | | |
| 105. | NA | NA | NA | 0.1 | NA | .37 | NA | NA | | NA | NA | NA | .03 |
| | | | | 34 | | 51 | | | NA | | | | |
| 106. | 3.9 | 1.9 | 1.9 | 1.1 | 1.6 | .31 | 8726 | 293 | | .08 | .07 | .04 | .05 |
| | 62 | 55 | 83 | 60 | 65 | 47 | 63 | 68 | 66 | | | | |
| 107. | 1.9 | 1.1 | NA | 0.2 | NA | .24 | 3407 | NA | | .16 | .16 | NA | .06 |
| | 43 | 39 | | 39 | | 42 | 51 | | NA | | | | |
| 108. | 2.2 | 1.5 | 1.0 | 0.4 | NA | .25 | NA | 29 | | .14 | .13 | .05 | .06 |
| | 45 | 48 | 43 | 42 | | 43 | | 42 | 42 | | | | |
| 109. | 1.2 | 0.7 | 1.1 | 0.3 | -1.0 | .30 | NA | 9 | | .18 | .15 | .17 | .05 |
| | 35 | 30 | 51 | 40 | 39 | 46 | | 40 | 41 | | | | |
| 110. | 1.4 | 1.0 | NA | 0.1 | NA | .33 | NA | 11 | | .17 | .16 | NA | .04 |
| | 37 | 37 | | 36 | | 48 | | 41 | 41 | | | | |
| 111. | 2.4 | 1.6 | 1.4 | 0.7 | -0.6 | .82 | 1171 | 112 | | .11 | .12 | .10 | .08 |
| | 47 | 48 | 60 | 48 | 43 | 80 | 46 | 51 | 50 | | | | |
| 112. | 2.3 | 1.6 | 1.1 | 0.5 | -0.7 | .33 | 657 | 29 | | .13 | .10 | .15 | .06 |
| | 46 | 49 | 49 | 44 | 42 | 48 | 45 | 42 | 43 | | | | |
| 113. | 2.9 | 1.6 | 1.6 | 0.8 | -0.0 | .33 | 801 | 74 | | .11 | .09 | .09 | .08 |
| | 52 | 48 | 73 | 53 | 49 | 48 | 45 | 47 | 45 | | | | |
| 114. | 2.9 | 1.9 | 1.3 | 1.0 | 0.7 | .45 | 1424 | 119 | | .08 | .07 | .09 | .07 |
| | 53 | 54 | 59 | 57 | 56 | 56 | 47 | 51 | 52 | | | | |
| 115. | 1.2 | 1.1 | NA | 0.2 | -0.3 | .25 | 424 | 31 | | .15 | .14 | NA | .04 |
| | 36 | 38 | | 37 | 46 | 43 | 44 | 43 | 42 | | | | |
| 116. | 2.8 | 1.8 | 1.0 | 0.8 | -0.4 | .31 | 1579 | 64 | | .09 | .10 | .09 | .06 |
| | 51 | 53 | 44 | 51 | 45 | 47 | 47 | 46 | 46 | | | | |
| 117. | 3.9 | 2.1 | 1.5 | 1.3 | 1.5 | .51 | 3339 | 176 | | .06 | .06 | .07 | .06 |
| | 62 | 60 | 65 | 64 | 63 | 60 | 51 | 57 | 57 | | | | |
| 118. | 1.8 | 1.1 | 0.7 | 0.4 | -0.4 | .38 | NA | 77 | | .10 | .12 | .10 | .06 |
| | 42 | 39 | 31 | 43 | 45 | 51 | | 47 | 47 | | | | |
| 119. | 2.4 | 1.6 | 1.2 | 0.6 | NA | .50 | 468 | 34 | | .11 | .09 | .08 | .06 |
| | 47 | 48 | 53 | 46 | | 59 | 44 | 43 | 43 | | | | |
| 120. | 3.8 | 2.1 | 0.9 | 1.4 | 1.6 | .26 | 5082 | 279 | | .08 | .06 | .06 | .07 |
| | 61 | 60 | 42 | 65 | 64 | 43 | 55 | 67 | 65 | | | | |

NOTE: On the first line of data for every program, raw values for each measure are reported; on the second line values are reported in standardized form, with mean = 50 and standard deviation = 10. "NA" indicates that the value for a measure is not available. Since the scale used to compute measure (16) is entirely arbitrary, only values in standardized form are reported for this measure.

TABLE 7.1  Program Measures (Raw and Standardized Values) in Physics

| Prog No. | University - Department/Academic Unit | Program Size | | | Characteristics of Program Graduates | | | |
|---|---|---|---|---|---|---|---|---|
| | | (01) | (02) | (03) | (04) | (05) | (06) | (07) |
| 121. | Wisconsin, University of-Milwaukee | 17 | 13 | 15 | .24 | 7.3 | .75 | .13 |
| | *Physics* | *43* | *43* | *42* | *48* | *48* | *58* | *39* |
| 122. | Wyoming, University of | 18 | 16 | 21 | .23 | 6.7 | .55 | .14 |
| | *Physics and Astronomy* | *43* | *44* | *44* | *47* | *54* | *41* | *40* |
| 123. | Yale University | 39 | 54 | 103 | .24 | 5.6 | .77 | .39 |
| | *Physics* | *58* | *56* | *59* | *49* | *64* | *60* | *61* |

* indicates program was initiated since 1970.

NOTE: On the first line of data for every program, raw values for each measure are reported; on the second line values are reported in standardized form, with mean = 50 and standard deviation = 10. "NA" indicates that the value for a measure is not available.

TABLE 7.1  Program Measures (Raw and Standardized Values) in Physics

| Prog No. | Survey Results | | | | University Library | Research Support | | Published Articles | | Survey Ratings Standard Error | | | |
|---|---|---|---|---|---|---|---|---|---|---|---|---|---|
| | (08) | (09) | (10) | (11) | (12) | (13) | (14) | (15) | (16) | (08) | (09) | (10) | (11) |
| 121. | 1.8 | 1.3 | 1.0 | 0.4 | NA | .35 | 415 | NA | | .17 | .11 | .14 | .06 |
| | 42 | 42 | 45 | 42 | | 50 | 44 | | NA | | | | |
| 122. | 1.7 | 1.2 | 1.1 | 0.3 | NA | .28 | 936 | 11 | | .14 | .17 | .12 | .06 |
| | 40 | 40 | 51 | 40 | | 45 | 45 | 41 | 41 | | | | |
| 123. | 4.2 | 2.4 | 0.9 | 1.4 | 2.1 | .15 | 4215 | 178 | | .08 | .06 | .07 | .07 |
| | 65 | 65 | 42 | 66 | 69 | 37 | 53 | 57 | 57 | | | | |

NOTE: On the first line of data for every program, raw values for each measure are reported; on the second line values are reported in standardized form, with mean = 50 and standard deviation = 10. "NA" indicates that the value for a measure is not available. Since the scale used to compute measure (16) is entirely arbitrary, only values in standardized form are reported for this measure.

TABLE 7.2  Summary Statistics Describing Each Program Measure--Physics

| Measure | Number of Programs Evaluated | Mean | Standard Deviation | DECILES | | | | | | | | |
|---|---|---|---|---|---|---|---|---|---|---|---|---|
| | | | | 1 | 2 | 3 | 4 | 5 | 6 | 7 | 8 | 9 |
| **Program Size** | | | | | | | | | | | | |
| 01 Raw Value | 123 | 28 | 15 | 12 | 17 | 19 | 21 | 23 | 25 | 31 | 37 | 46 |
| Std Value | 123 | 50 | 10 | 39 | 43 | 44 | 45 | 47 | 48 | 52 | 56 | 63 |
| 02 Raw Value | 123 | 35 | 33 | 11 | 15 | 16 | 19 | 24 | 29 | 35 | 51 | 69 |
| Std Value | 123 | 50 | 10 | 43 | 44 | 44 | 45 | 47 | 48 | 50 | 55 | 61 |
| 03 Raw Value | 122 | 56 | 53 | 14 | 20 | 25 | 34 | 38 | 44 | 58 | 81 | 111 |
| Std Value | 122 | 50 | 10 | 42 | 43 | 44 | 46 | 47 | 48 | 50 | 55 | 60 |
| **Program Graduates** | | | | | | | | | | | | |
| 04 Raw Value | 115 | .26 | .11 | .13 | .16 | .20 | .23 | .26 | .28 | .31 | .35 | .41 |
| Std Value | 115 | 50 | 10 | 38 | 41 | 45 | 47 | 50 | 52 | 55 | 58 | 64 |
| 05 Raw Value | 115 | 7.1 | 1.1 | 8.8 | 7.9 | 7.5 | 7.2 | 7.0 | 6.6 | 6.5 | 6.2 | 5.9 |
| Std Value | 115 | 50 | 10 | 34 | 42 | 46 | 49 | 51 | 55 | 56 | 58 | 61 |
| 06 Raw Value | 114 | .66 | .12 | .48 | .56 | .60 | .64 | .68 | .70 | .72 | .75 | .78 |
| Std Value | 114 | 50 | 10 | 35 | 42 | 45 | 48 | 52 | 53 | 55 | 58 | 60 |
| 07 Raw Value | 114 | .26 | .12 | .11 | .15 | .18 | .22 | .26 | .30 | .33 | .36 | .41 |
| Std Value | 114 | 50 | 10 | 38 | 41 | 43 | 47 | 50 | 53 | 56 | 58 | 63 |
| **Survey Results** | | | | | | | | | | | | |
| 08 Raw Value | 121 | 2.7 | 1.0 | 1.5 | 1.8 | 2.0 | 2.3 | 2.5 | 2.8 | 3.0 | 3.5 | 4.1 |
| Std Value | 121 | 50 | 10 | 38 | 41 | 43 | 46 | 48 | 52 | 54 | 59 | 65 |
| 09 Raw Value | 118 | 1.7 | .5 | 1.1 | 1.2 | 1.4 | 1.6 | 1.6 | 1.8 | 1.9 | 2.0 | 2.3 |
| Std Value | 118 | 50 | 10 | 39 | 41 | 45 | 49 | 49 | 53 | 55 | 57 | 63 |
| 10 Raw Value | 109 | 1.1 | .2 | .9 | .9 | 1.0 | 1.0 | 1.1 | 1.1 | 1.2 | 1.3 | 1.4 |
| Std Value | 109 | 50 | 10 | 41 | 41 | 45 | 45 | 50 | 50 | 54 | 58 | 63 |
| 11 Raw Value | 122 | .7 | .4 | .2 | .3 | .4 | .5 | .7 | .8 | .9 | 1.1 | 1.4 |
| Std Value | 122 | 50 | 10 | 38 | 40 | 42 | 45 | 50 | 52 | 54 | 59 | 66 |
| **University Library** | | | | | | | | | | | | |
| 12 Raw Value | 83 | .1 | 1.0 | -1.1 | -.7 | -.5 | -.4 | -.2 | .2 | .6 | .9 | 1.7 |
| Std Value | 83 | 50 | 10 | 38 | 42 | 44 | 45 | 47 | 51 | 55 | 58 | 65 |
| **Research Support** | | | | | | | | | | | | |
| 13 Raw Value | 121 | .36 | .15 | .16 | .24 | .28 | .31 | .33 | .39 | .44 | .49 | .54 |
| Std Value | 121 | 50 | 10 | 37 | 42 | 45 | 47 | 48 | 52 | 55 | 59 | 62 |
| 14 Raw Value | 88 | 2943 | 4353 | 415 | 591 | 876 | 1206 | 1372 | 1696 | 2610 | 3893 | 6502 |
| Std Value | 88 | 50 | 10 | 44 | 45 | 45 | 46 | 46 | 47 | 49 | 52 | 58 |
| **Publication Records** | | | | | | | | | | | | |
| 15 Raw Value | 120 | 106 | 102 | 19 | 28 | 42 | 57 | 68 | 82 | 119 | 169 | 269 |
| Std Value | 120 | 50 | 10 | 41 | 42 | 44 | 45 | 46 | 48 | 51 | 56 | 66 |
| 16 Std Value | 120 | 50 | 10 | 42 | 42 | 44 | 45 | 46 | 48 | 51 | 56 | 66 |

NOTE: Standardized values reported in the preceding table have been computed from exact values of the mean and standard deviation and not the rounded values reported here.  Since the scale used to compute measure 16 is entirely arbitrary, only data in standardized form are reported for this measure.

TABLE 7.3  Intercorrelations Among Program Measures on 123 Programs in Physics

|  | 01 | 02 | 03 | 04 | 05 | 06 | 07 | 08 | 09 | 10 | 11 | 12 | 13 | 14 | 15 | 16 |
|---|---|---|---|---|---|---|---|---|---|---|---|---|---|---|---|---|
| **Program Size** | | | | | | | | | | | | | | | | |
| 01 | | .77 | .80 | -.13 | .21 | .19 | .38 | .68 | .60 | .02 | .71 | .44 | -.07 | .54 | .71 | .72 |
| 02 | | | .92 | -.02 | .32 | .40 | .41 | .76 | .73 | -.17 | .78 | .47 | .13 | .66 | .85 | .86 |
| 03 | | | | -.05 | .28 | .34 | .39 | .75 | .70 | -.09 | .77 | .54 | .09 | .68 | .86 | .85 |
| **Program Graduates** | | | | | | | | | | | | | | | | |
| 04 | | | | | .11 | .17 | .01 | .15 | .19 | .06 | .11 | .05 | -.03 | .04 | .07 | .05 |
| 05 | | | | | | .34 | .29 | .42 | .44 | .00 | .38 | .23 | .21 | .31 | .37 | .38 |
| 06 | | | | | | | .51 | .42 | .44 | -.14 | .43 | .25 | .13 | .25 | .41 | .43 |
| 07 | | | | | | | | .58 | .56 | -.18 | .56 | .34 | .14 | .31 | .47 | .48 |
| **Survey Results** | | | | | | | | | | | | | | | | |
| 08 | | | | | | | | | .96 | -.15 | .96 | .67 | .24 | .61 | .85 | .86 |
| 09 | | | | | | | | | | -.17 | .92 | .65 | .23 | .61 | .82 | .82 |
| 10 | | | | | | | | | | | -.18 | -.11 | -.05 | -.08 | -.13 | -.14 |
| 11 | | | | | | | | | | | | .66 | .27 | .58 | .85 | .86 |
| **University Library** | | | | | | | | | | | | | | | | |
| 12 | | | | | | | | | | | | | .17 | .33 | .63 | .61 |
| **Research Support** | | | | | | | | | | | | | | | | |
| 13 | | | | | | | | | | | | | | .07 | .20 | .21 |
| 14 | | | | | | | | | | | | | | | .80 | .80 |
| **Publication Records** | | | | | | | | | | | | | | | | |
| 15 | | | | | | | | | | | | | | | | .99 |
| 16 | | | | | | | | | | | | | | | | |

NOTE: Since in computing correlation coefficients program data must be available for both of the measures being correlated, the actual number of programs on which each coefficient is based varies.

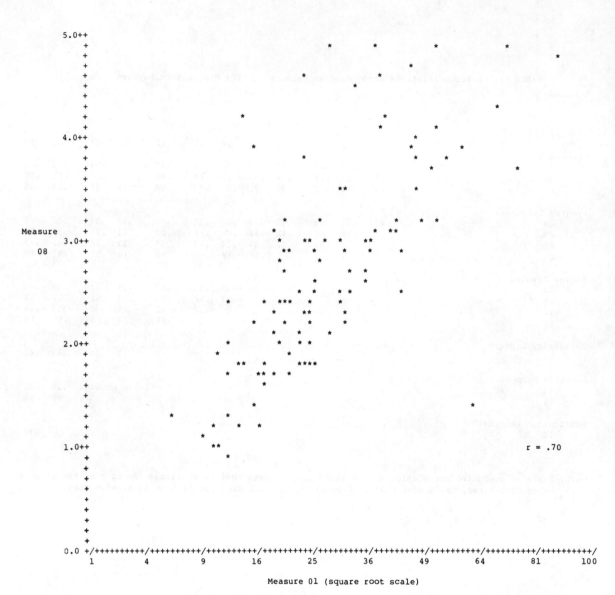

FIGURE 7.1  Mean rating of scholarly quality of faculty (measure 08) versus number of faculty members (measure 01)—121 programs in physics.

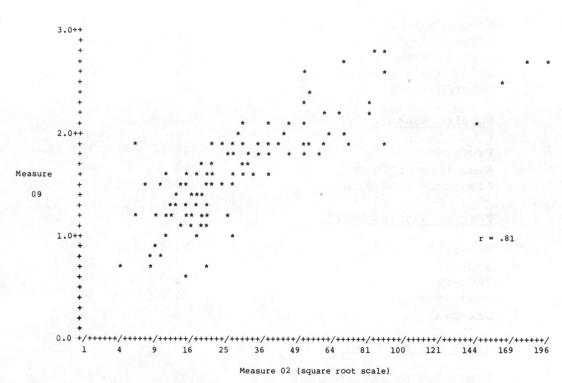

FIGURE 7.2 Mean rating of program effectiveness in educating research scholars/scientists (measure 09) versus number of graduates in last five years (measure 02)—118 programs in physics.

TABLE 7.4  Characteristics of Survey Participants in Physics

|  | Respondents | |
|---|---|---|
|  | N | % |
| **Field of Specialization** | | |
| Atomic/Molecular Physics | 29 | 14 |
| Elementary Particles | 56 | 27 |
| Nuclear Structure | 30 | 14 |
| Solid State Physics | 55 | 26 |
| Other/Unknown | 41 | 19 |
| **Faculty Rank** | | |
| Professor | 148 | 70 |
| Associate Professor | 50 | 24 |
| Assistant Professor | 13 | 6 |
| **Year of Highest Degree** | | |
| Pre-1950 | 13 | 6 |
| 1950-59 | 62 | 29 |
| 1960-69 | 103 | 49 |
| Post-1969 | 32 | 15 |
| Unknown | 1 | 1 |
| **Evaluator Selection** | | |
| Nominated by Institution | 169 | 80 |
| Other | 42 | 20 |
| **Survey Form** | | |
| With Faculty Names | 191 | 91 |
| Without Names | 20 | 10 |
| **Total Evaluators** | 211 | 100 |

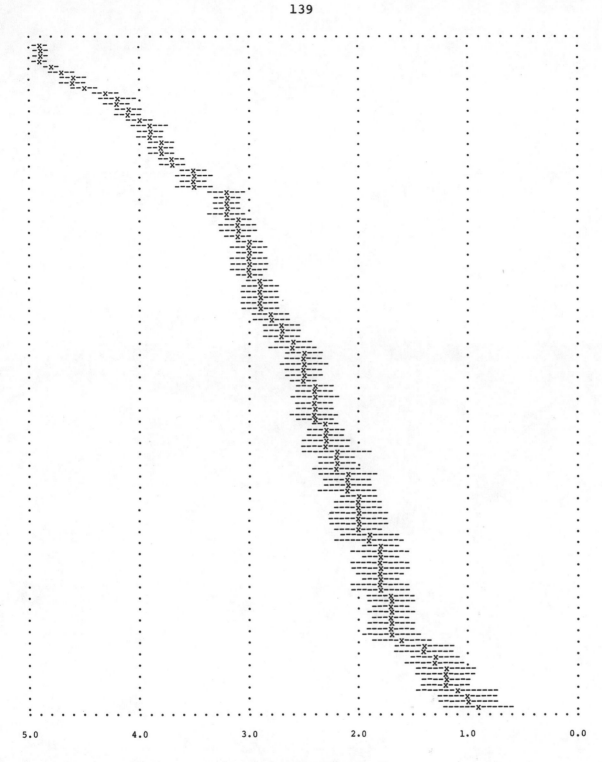

5.0       4.0       3.0       2.0       1.0       0.0

Mean Survey Rating (Measure 08)

FIGURE 7.3  Mean rating of scholarly quality of faculty in 121 programs in physics.

NOTE:  Programs are listed in sequence of mean rating, with the highest-rated program appearing at the top of the page.  The broken lines (---) indicate a confidence interval of ±1.5 standard errors around the reported mean (x) of each program.

# VIII
# Statistics/Biostatistics Programs

In this chapter 64 research-doctorate programs in statistics/ biostatistics are assessed. These programs, according to the information supplied by their universities, have accounted for 906 doctoral degrees awarded during the FY1976-80 period--approximately 87 percent of the aggregate number of statistics/biostatistics doctorates earned from U.S. universities in this five-year span.[1] On the average, 22 full-time and part-time students intending to earn doctorates were enrolled in a program in December 1980, with an average faculty size of 12 members.[2] The 64 programs, listed in Table 8.1, represent 58 different universities. The University of Michigan (Ann Arbor), University of Minnesota, North Carolina State University (Raleigh), University of North Carolina (Chapel Hill), University of Pittsburgh, and Yale University each have one statistics program and one biomedical program (i.e., biostatistics, biometry, biomathematics, epidemiology, or public health) included in the assessment. All but nine of the programs were initiated prior to 1970. In addition to the 58 universities represented in this discipline, another 5 were initially identified as meeting the criteria[3] for inclusion in the assessment:

> Dartmouth College
> University of Illinois--Chicago Circle
> New York University
> University of Northern Colorado
> University of South Carolina

Statistics/biostatistics programs at these five institutions have not been included in the evaluations in this discipline, since in each

---

[1] Data from the NRC's Survey of Earned Doctorates indicate that 1,038 research doctorates in statistics/biostatistics were awarded by U.S. universities between FY1976 and FY1980.
[2] See the reported means for measures 03 and 01 in Table 8.2.
[3] As mentioned in Chapter I, the primary criterion for inclusion was that a university had awarded at least 5 doctorates in statistics/ biostatistics during the FY1976-78 period.

142

case the study coordinator either indicated that the institution did not at that time have a research-doctorate program in statistics/ biostatistics or failed to provide the information requested by the committee.

Before examining the individual program results presented in Table 8.1, the reader is urged to refer to Chapter II, in which each of the 16 measures used in the assessment is discussed. Summary statistics describing every measure are given in Table 8.2. For nine of the measures, data are reported for at least 61 of the 64 statistics/ biostatistics programs. For measures 04-07, which pertain to characteristics of the program graduates, data are presented for only 36 of the programs; the other 28 had too few graduates on which to base statistics.[4] For measure 12, a composite index of the size of a university library, data are available for 50 programs; for measure 13, the fraction of faculty with research support from the National Science Foundation, the National Institutes of Health, or the Alcohol, Drug Abuse, and Mental Health Administration, data are reported for 37 programs that had at least 10 faculty members. As mentioned in Chapter II, data are not available on total university expenditures for research in the area of statistics and biostatistics--measure 14.

Intercorrelations among the 15 measures (Pearson product-moment coefficients) are given in Table 8.3. Of particular note are the high positive correlations of the measures of the numbers of faculty and doctoral graduates (01, 02) with measures of publication records (15, 16) and reputational survey ratings (08, 09, and 11). Figure 8.1 illustrates the relation between the mean rating of the scholarly quality of faculty (measure 08) and the number of faculty members (measure 01) for each of the 63 statistics/biostatistics programs. Figure 8.2 plots the mean rating of program effectiveness (measure 09) against the total number of FY1976-80 program graduates (measure 02). Although in both figures there is a significant positive correlation between program size and reputational rating, it is quite apparent that some of the smaller programs received high mean ratings and some of the larger programs received low mean ratings.

Table 8.4 describes the 135 faculty members who participated in the evaluation of statistics/biostatistics programs. These individuals constituted 71 percent of those asked to respond to the survey in this discipline and 17 percent of the faculty population in the 64 research-doctorate programs being evaluated.[5] More than one-third of the survey participants were mathematical statisticians and 16 percent were biostatisticians. More than half of them held the rank of full professor; almost two-thirds had earned their highest degree prior to 1970.

---

[4] As mentioned in Chapter II, data for measures 04-07 are not reported if they are based on the survey responses of fewer than 10 FY1975-79 program graduates.
[5] See Table 2.3 in Chapter II.

Two exceptions should be noted with regard to the survey evaluations in this discipline. The biostatistics program in the School of Public Health at the University of Michigan (Ann Arbor) was omitted on the survey form because at the time of the survey mailing no information on this program had been provided by the institution. At the request of the study coordinator at this university, the program has been included in all other aspects of the assessment. Shortly after the survey mailing it was called to the committee's attention that the name of a faculty member in the statistics program at the University of Rochester was not included on the survey form. The department chairman at this university contacted other department chairmen in the discipline informing them of this omission.

To assist the reader in interpreting results of the survey evaluations, estimated standard errors have been computed for mean ratings of the scholarly quality of faculty in 63 statistics/ biostatistics programs (and are given in Table 8.1). For each program the mean rating and an associated "confidence interval" of 1.5 standard errors are illustrated in Figure 8.3 (listed in order of highest to lowest mean rating). In comparing two programs, if their confidence intervals do not overlap, one may conclude that there is a significant difference in their mean ratings at a .05 level of significance.[6] From this figure it is also apparent that one should have somewhat more confidence in the accuracy of the mean ratings of higher-rated programs than lower-rated programs. This generalization results primarily from the fact that evaluators are not as likely to be familiar with the less prestigious programs, and consequently the mean ratings of these programs are usually based on fewer survey responses.

---

[6] See pp. 29-31 for a discussion of the interpretation of mean ratings and associated confidence intervals.

TABLE 8.1  Program Measures (Raw and Standardized Values) in Statistics/Biostatistics

| Prog No. | University - Department/Academic Unit | Program Size (01) | (02) | (03) | Characteristics of Program Graduates (04) | (05) | (06) | (07) |
|---|---|---|---|---|---|---|---|---|
| 001. | American University | 6 | 12 | 21 | NA | NA | NA | NA |
|  | *Mathematics, Statistics, and Computer Sci* | *40* | *47* | *50* | | | | |
| 002. | Boston University | 6 | 9 | 15 | NA | NA | NA | NA |
|  | *Mathematics* | *40* | *44* | *46* | | | | |
| 003. | California, University of-Berkeley | 24 | 50 | 69 | .12 | 6.3 | .80 | .43 |
|  | *Statistics* | *68* | *85* | *75* | *40* | *53* | *52* | *50* |
| 004. | California, University of-Los Angeles | 19 | 15 | 27 | .63 | 6.5 | .67 | .50 |
|  | *Public Health/Mathematics* | *60* | *50* | *53* | *65* | *51* | *40* | *55* |
| 005. | California, University of-Riverside | 9 | 13 | 21 | NA | NA | NA | NA |
|  | *Statistics** | *45* | *48* | *50* | | | | |
| 006. | Carnegie-Mellon University | 14 | 7 | 23 | NA | NA | NA | NA |
|  | *Statistics* | *52* | *42* | *51* | | | | |
| 007. | Case Western Reserve University | 9 | NA | 3 | NA | NA | NA | NA |
|  | *Biometry** | *45* | | *40* | | | | |
| 008. | Chicago, University of | 12 | 11 | 7 | .50 | 6.0 | NA | NA |
|  | *Statistics* | *49* | *46* | *42* | *59* | *55* | | |
| 009. | Colorado State University-Fort Collins | 15 | 12 | 13 | NA | NA | NA | NA |
|  | *Statistics* | *54* | *47* | *45* | | | | |
| 010. | Columbia University | 17 | 17 | 34 | .19 | 6.5 | .80 | .47 |
|  | *Mathematical Statistics* | *57* | *52* | *57* | *44* | *51* | *52* | *53* |
| 011. | Connecticut, University of-Storrs | 8 | 2 | 9 | NA | NA | NA | NA |
|  | *Statistics* | *43* | *37* | *43* | | | | |
| 012. | Cornell University-Ithaca | 24 | 3 | 8 | .32 | 5.3 | .96 | .55 |
|  | *Statistics* | *68* | *38* | *43* | *50* | *61* | *66* | *59* |
| 013. | Delaware, University of-Newark | 6 | NA | 8 | NA | NA | NA | NA |
|  | *Applied Sciences** | *40* | | *43* | | | | |
| 014. | Florida State University-Tallahassee | 14 | 28 | 38 | .55 | 6.7 | .77 | .57 |
|  | *Statistics* | *52* | *63* | *59* | *61* | *50* | *49* | *60* |
| 015. | Florida, University of-Gainesville | 18 | 8 | 39 | .09 | 5.0 | .73 | .46 |
|  | *Statistics* | *58* | *43* | *59* | *39* | *63* | *45* | *52* |
| 016. | George Washington University | 3 | 12 | 42 | .33 | 12.0 | .67 | .08 |
|  | *Statistics* | *36* | *47* | *61* | *51* | *8* | *40* | *24* |
| 017. | Georgia, University of-Athens | 7 | 22 | 15 | .10 | 8.5 | .81 | .29 |
|  | *Statistics and Computer Science* | *42* | *57* | *46* | *39* | *36* | *53* | *39* |
| 018. | Harvard University | 7 | 16 | 14 | .50 | 6.1 | .94 | .59 |
|  | *Statistics* | *42* | *51* | *46* | *59* | *54* | *64* | *62* |
| 019. | Illinois, University-Urbana/Champaign | 13 | 1 | 3 | NA | NA | NA | NA |
|  | *Mathematics* | *51* | *36* | *40* | | | | |
| 020. | Indiana University-Bloomington | 2 | 6 | 9 | .20 | 6.0 | .70 | .60 |
|  | *Mathematics* | *34* | *41* | *43* | *44* | *55* | *43* | *63* |

* indicates program was initiated since 1970.

NOTE: On the first line of data for every program, raw values for each measure are reported;
on the second line values are reported in standardized form, with mean = 50 and
standard deviation = 10. "NA" indicates that the value for a measure is not available.

TABLE 8.1 Program Measures (Raw and Standardized Values) in Statistics/Biostatistics

| Prog No. | Survey Results (08) | (09) | (10) | (11) | University Library (12) | Research Support (13) | (14) | Published Articles (15) | (16) | Survey Ratings Standard Error (08) | (09) | (10) | (11) |
|---|---|---|---|---|---|---|---|---|---|---|---|---|---|
| 001. | 0.9 | 0.4 | 0.9 | 0.2 | NA | NA | NA | 0 | | .12 | .10 | .10 | .04 |
| | 30 | 28 | 45 | 31 | | | | 39 | 40 | | | | |
| 002. | 1.4 | 0.8 | 1.0 | 0.4 | -0.4 | NA | NA | 2 | | .14 | .13 | .07 | .06 |
| | 35 | 37 | 48 | 38 | 42 | | | 40 | 43 | | | | |
| 003. | 4.9 | 2.6 | 1.3 | 1.8 | 2.2 | .54 | NA | 52 | | .04 | .06 | .06 | .04 |
| | 72 | 70 | 59 | 70 | 65 | 66 | | 86 | 87 | | | | |
| 004. | 3.7 | 2.0 | 1.1 | 1.3 | 2.0 | .32 | NA | 3 | | .08 | .06 | .06 | .06 |
| | 59 | 58 | 52 | 58 | 63 | 53 | | 41 | 41 | | | | |
| 005. | 2.4 | 1.3 | 1.6 | 0.8 | -1.0 | NA | NA | 4 | | .08 | .08 | .08 | .06 |
| | 46 | 46 | 69 | 47 | 36 | | | 42 | 43 | | | | |
| 006. | 3.5 | 2.0 | 1.6 | 1.3 | NA | .29 | NA | 9 | | .07 | .06 | .06 | .06 |
| | 57 | 58 | 71 | 58 | | 52 | | 47 | 46 | | | | |
| 007. | 1.6 | 0.9 | 0.6 | 0.4 | -1.3 | NA | NA | 0 | | .14 | .11 | .12 | .06 |
| | 37 | 37 | 34 | 36 | 33 | | | 39 | 40 | | | | |
| 008. | 4.7 | 2.3 | 1.2 | 1.5 | 0.9 | .42 | NA | 25 | | .05 | .06 | .05 | .06 |
| | 70 | 65 | 56 | 65 | 53 | 59 | | 61 | 58 | | | | |
| 009. | 3.2 | 1.9 | 1.1 | 1.2 | -1.1 | .33 | NA | 15 | | .07 | .06 | .06 | .06 |
| | 54 | 56 | 52 | 55 | 35 | 54 | | 52 | 49 | | | | |
| 010. | 3.8 | 2.0 | 1.0 | 1.3 | 1.7 | .47 | NA | 16 | | .07 | .06 | .07 | .07 |
| | 61 | 58 | 48 | 59 | 61 | 62 | | 53 | 57 | | | | |
| 011. | 2.1 | 1.1 | 1.0 | 0.9 | -0.5 | NA | NA | 3 | | .10 | .10 | .06 | .06 |
| | 43 | 42 | 49 | 49 | 41 | | | 41 | 41 | | | | |
| 012. | 4.0 | 2.2 | 1.0 | 1.5 | 1.6 | .25 | NA | 30 | | .08 | .07 | .07 | .06 |
| | 62 | 62 | 48 | 63 | 60 | 50 | | 66 | 66 | | | | |
| 013. | 1.8 | 0.7 | 1.3 | 0.6 | NA | NA | NA | 8 | | .13 | .10 | .10 | .06 |
| | 39 | 34 | 57 | 41 | | | | 46 | 46 | | | | |
| 014. | 3.8 | 2.2 | 1.0 | 1.4 | -0.4 | .14 | NA | 18 | | .07 | .05 | .07 | .06 |
| | 61 | 62 | 48 | 61 | 41 | 44 | | 55 | 55 | | | | |
| 015. | 2.4 | 1.4 | 1.1 | 0.9 | 0.8 | .00 | NA | 23 | | .09 | .08 | .08 | .06 |
| | 45 | 47 | 53 | 49 | 52 | 36 | | 60 | 56 | | | | |
| 016. | 2.6 | 1.3 | 0.4 | 1.0 | NA | NA | NA | 3 | | .11 | .09 | .07 | .06 |
| | 48 | 46 | 26 | 51 | | | | 41 | 41 | | | | |
| 017. | 1.7 | 1.0 | 1.1 | 0.7 | 0.4 | NA | NA | 11 | | .11 | .09 | .10 | .06 |
| | 39 | 39 | 52 | 44 | 49 | | | 49 | 46 | | | | |
| 018. | 4.0 | 1.9 | 1.0 | 1.4 | 3.0 | NA | NA | 13 | | .08 | .06 | .06 | .05 |
| | 62 | 56 | 49 | 62 | 73 | | | 50 | 49 | | | | |
| 019. | 3.4 | 1.6 | 1.2 | 1.0 | 2.0 | .46 | NA | 20 | | .11 | .11 | .06 | .07 |
| | 57 | 50 | 53 | 51 | 63 | 62 | | 57 | 64 | | | | |
| 020. | 1.7 | 0.8 | 0.6 | 0.9 | 0.9 | NA | NA | 10 | | .15 | .10 | .08 | .07 |
| | 39 | 36 | 34 | 49 | 54 | | | 48 | 49 | | | | |

NOTE: On the first line of data for every program, raw values for each measure are reported; on the second line values are reported in standardized form, with mean = 50 and standard deviation = 10. "NA" indicates that the value for a measure is not available. Since the scale used to compute measure (16) is entirely arbitrary, only values in standardized form are reported for this measure.

TABLE 8.1  Program Measures (Raw and Standardized Values) in Statistics/Biostatistics

| Prog No. | University - Department/Academic Unit | Program Size | | | Characteristics of Program Graduates | | | |
|---|---|---|---|---|---|---|---|---|
| | | (01) | (02) | (03) | (04) | (05) | (06) | (07) |
| 021. | Iowa State University-Ames | 19 | 38 | 40 | .16 | 6.5 | .84 | .36 |
| | *Statistics* | *60* | *73* | *60* | *42* | *51* | *55* | *45* |
| 022. | Iowa, University of-Iowa City | 13 | 13 | 15 | .40 | 6.5 | .60 | .30 |
| | *Statistics* | *51* | *48* | *46* | *54* | *51* | *34* | *41* |
| 023. | Johns Hopkins University | 9 | 15 | 10 | .82 | 7.0 | .82 | .47 |
| | *Biostatistics* | *45* | *50* | *44* | *75* | *48* | *54* | *53* |
| 024. | Kansas State University-Manhattan | 13 | 12 | 12 | .21 | 7.5 | .71 | .36 |
| | *Statistics* | *51* | *47* | *45* | *45* | *44* | *44* | *45* |
| 025. | Kentucky, University of | 9 | 16 | 19 | .27 | 6.3 | .80 | .40 |
| | *Statistics* | *45* | *51* | *49* | *48* | *53* | *52* | *48* |
| 026. | Maryland, University of-College Park | 8 | 5 | 14 | NA | NA | NA | NA |
| | *Mathematics** | *43* | *40* | *46* | | | | |
| 027. | Michigan State University-East Lansing | 15 | 12 | 14 | NA | NA | NA | NA |
| | *Statistics and Probability* | *54* | *47* | *46* | | | | |
| 028. | Michigan, University of-Ann Arbor | 13 | 11 | 28 | NA | NA | NA | NA |
| | *Biostatistics (School of Public Health)* | *51* | *46* | *53* | | | | |
| 029. | Michigan, University of-Ann Arbor | 13 | 14 | 20 | .36 | 6.0 | 1.00 | .55 |
| | *Statistics* | *51* | *49* | *49* | *52* | *55* | *70* | *59* |
| 030. | Minnesota, University of | 7 | 11 | 7 | NA | 8.0 | .91 | .46 |
| | *Biometry* | *42* | *46* | *42* | | *40* | *61* | *52* |
| 031. | Minnesota, University of | 15 | 12 | 12 | .22 | 6.5 | .72 | .50 |
| | *Statistics* | *54* | *47* | *45* | *45* | *51* | *45* | *55* |
| 032. | Missouri, University of-Columbia | 11 | 9 | 10 | NA | NA | NA | NA |
| | *Statistics* | *48* | *44* | *44* | | | | |
| 033. | Missouri, University of-Rolla | 4 | 4 | 5 | NA | NA | NA | NA |
| | *Statistics* | *37* | *39* | *41* | | | | |
| 034. | New Mexico, University of-Albuquerque | 8 | 10 | 6 | NA | NA | NA | NA |
| | *Mathematics and Statistics* | *43* | *45* | *42* | | | | |
| 035. | North Carolina State University-Raleigh | 12 | 5 | 11 | NA | NA | NA | NA |
| | *Biomathematics* | *49* | *40* | *44* | | | | |
| 036. | North Carolina State University-Raleigh | 31 | 19 | 32 | .43 | 7.5 | .86 | .43 |
| | *Statistics* | *78* | *54* | *56* | *55* | *44* | *57* | *50* |
| 037. | North Carolina, University of-Chapel Hill | 23 | 35 | 44 | .62 | 7.0 | .79 | .43 |
| | *Biostatistics* | *66* | *70* | *62* | *65* | *48* | *50* | *50* |
| 038. | North Carolina, University of-Chapel Hill | 15 | 22 | 23 | .28 | 5.8 | .72 | .56 |
| | *Statistics* | *54* | *57* | *51* | *48* | *57* | *44* | *60* |
| 039. | Ohio State University-Columbus | 10 | 17 | 28 | .00 | 5.4 | .83 | .44 |
| | *Statistics/Biostatistics** | *46* | *52* | *53* | *34* | *60* | *55* | *51* |
| 040. | Oklahoma State University-Stillwater | 9 | 19 | 20 | .07 | 6.4 | .67 | .20 |
| | *Statistics* | *45* | *54* | *49* | *38* | *52* | *40* | *33* |

* indicates program was initiated since 1970.

NOTE: On the first line of data for every program, raw values for each measure are reported; on the second line values are reported in standardized form, with mean = 50 and standard deviation = 10. "NA" indicates that the value for a measure is not available.

TABLE 8.1  Program Measures (Raw and Standardized Values) in Statistics/Biostatistics

| Prog No. | Survey Results (08) | (09) | (10) | (11) | University Library (12) | Research Support (13) | (14) | Published Articles (15) | (16) | Survey Ratings Standard Error (08) | (09) | (10) | (11) |
|---|---|---|---|---|---|---|---|---|---|---|---|---|---|
| 021. | 4.0 | 2.3 | 1.2 | 1.5 | -0.5 | .21 | NA | 29 | | .07 | .06 | .04 | .06 |
| | 62 | 64 | 53 | 64 | 41 | 48 | | 65 | 60 | | | | |
| 022. | 3.0 | 1.8 | 1.1 | 1.2 | 0.3 | .08 | NA | 9 | | .08 | .06 | .06 | .06 |
| | 52 | 55 | 51 | 57 | 48 | 40 | | 47 | 44 | | | | |
| 023. | 2.8 | 1.7 | 0.9 | 0.7 | -0.4 | NA | NA | 1 | | .10 | .08 | .06 | .06 |
| | 50 | 53 | 44 | 44 | 42 | | | 40 | 40 | | | | |
| 024. | 1.8 | 1.2 | 1.0 | 0.4 | NA | .00 | NA | 5 | | .10 | .09 | .09 | .05 |
| | 40 | 43 | 49 | 37 | | 36 | | 43 | 43 | | | | |
| 025. | 2.6 | 1.4 | 1.0 | 1.1 | -0.1 | NA | NA | 15 | | .08 | .08 | .09 | .06 |
| | 48 | 47 | 49 | 53 | 44 | | | 52 | 52 | | | | |
| 026. | 1.6 | 0.7 | 1.1 | 0.4 | 0.2 | NA | NA | 7 | | .12 | .10 | .06 | .06 |
| | 37 | 34 | 52 | 38 | 47 | | | 45 | 49 | | | | |
| 027. | 3.1 | 1.8 | 0.9 | 1.0 | 0.3 | .60 | NA | 14 | | .08 | .06 | .05 | .06 |
| | 53 | 54 | 44 | 51 | 48 | 69 | | 51 | 55 | | | | |
| 028. | NA | NA | NA | NA | 1.8 | .08 | NA | 2 | | NA | NA | NA | NA |
| | | | | | 62 | 40 | | 40 | 41 | | | | |
| 029. | 3.1 | 1.8 | 1.0 | 0.8 | 1.8 | .39 | NA | 13 | | .09 | .05 | .07 | .07 |
| | 53 | 55 | 48 | 48 | 62 | 57 | | 50 | 50 | | | | |
| 030. | 2.2 | 1.2 | 1.0 | 0.4 | 1.2 | NA | NA | 4 | | .15 | .12 | .05 | .06 |
| | 43 | 44 | 46 | 36 | 56 | | | 42 | 43 | | | | |
| 031. | 3.7 | 2.1 | 1.1 | 1.4 | 1.2 | .53 | NA | 31 | | .07 | .05 | .06 | .06 |
| | 60 | 60 | 51 | 60 | 56 | 66 | | 67 | 65 | | | | |
| 032. | 2.3 | 1.4 | 0.9 | 0.7 | -0.2 | .27 | NA | 7 | | .09 | .07 | .04 | .07 |
| | 44 | 46 | 44 | 45 | 43 | 51 | | 45 | 46 | | | | |
| 033. | 1.5 | 0.7 | 1.0 | 0.5 | NA | NA | NA | 12 | | .15 | .11 | .12 | .07 |
| | 36 | 34 | 46 | 39 | | | | 50 | 49 | | | | |
| 034. | 2.1 | 1.3 | 0.9 | 0.6 | -1.0 | NA | NA | 6 | | .10 | .07 | .10 | .06 |
| | 43 | 45 | 45 | 42 | 36 | | | 44 | 43 | | | | |
| 035. | 2.4 | 1.4 | 0.7 | 0.3 | NA | .08 | NA | 7 | | .17 | .12 | .11 | .05 |
| | 45 | 47 | 35 | 34 | | 40 | | 45 | 45 | | | | |
| 036. | 3.1 | 1.9 | 0.9 | 1.0 | NA | .10 | NA | 7 | | .10 | .07 | .05 | .07 |
| | 53 | 57 | 45 | 52 | | 41 | | 45 | 45 | | | | |
| 037. | 3.7 | 2.2 | 1.1 | 1.3 | 1.0 | .13 | NA | 4 | | .09 | .07 | .05 | .07 |
| | 60 | 62 | 52 | 59 | 54 | 43 | | 42 | 43 | | | | |
| 038. | 4.0 | 2.2 | 0.9 | 1.4 | 1.0 | .27 | NA | 41 | | .07 | .06 | .06 | .06 |
| | 63 | 62 | 43 | 62 | 54 | 51 | | 76 | 77 | | | | |
| 039. | 2.5 | 1.6 | 1.1 | 1.0 | 0.9 | .30 | NA | 19 | | .08 | .07 | .06 | .06 |
| | 47 | 50 | 51 | 51 | 53 | 53 | | 56 | 56 | | | | |
| 040. | 1.7 | 1.3 | 0.9 | 0.5 | -1.9 | NA | NA | 3 | | .10 | .08 | .07 | .06 |
| | 39 | 46 | 45 | 40 | 27 | | | 41 | 42 | | | | |

NOTE: On the first line of data for every program, raw values for each measure are reported; on the second line values are reported in standardized form, with mean = 50 and standard deviation = 10. "NA" indicates that the value for a measure is not available. Since the scale used to compute measure (16) is entirely arbitrary, only values in standardized form are reported for this measure.

TABLE 8.1  Program Measures (Raw and Standardized Values) in Statistics/Biostatistics

| Prog No. | University - Department/Academic Unit | Program Size | | | Characteristics of Program Graduates | | | |
|---|---|---|---|---|---|---|---|---|
| | | (01) | (02) | (03) | (04) | (05) | (06) | (07) |
| 041. | Oregon State University-Corvallis | 16 | 17 | 20 | .44 | 8.5 | .63 | .25 |
| | *Statistics* | 55 | 52 | 49 | 56 | 36 | 36 | 37 |
| 042. | Pennsylvania State University | 13 | 15 | 16 | .27 | 8.0 | .80 | .40 |
| | *Statistics* | 51 | 50 | 47 | 47 | 40 | 52 | 48 |
| 043. | Pennsylvania, University of | 12 | 6 | 10 | NA | NA | NA | NA |
| | *Statistics* | 49 | 41 | 44 | | | | |
| 044. | Pittsburgh, University of | 8 | 6 | 16 | NA | NA | NA | NA |
| | *Biostatistics (Grad Schl of Public Health)* | 43 | 41 | 47 | | | | |
| 045. | Pittsburgh, University of | 9 | 3 | 20 | NA | NA | NA | NA |
| | *Mathematics and Statistics** | 45 | 38 | 49 | | | | |
| 046. | Princeton University | 7 | 13 | 17 | .07 | 4.5 | .86 | .36 |
| | *Statistics* | 42 | 48 | 48 | 38 | 67 | 57 | 45 |
| 047. | Purdue University-West Lafayette | 23 | 25 | 33 | .04 | 6.2 | .79 | .58 |
| | *Statistics* | 66 | 60 | 56 | 36 | 54 | 51 | 62 |
| 048. | Rochester, University of | 11 | 13 | 19 | .15 | 7.8 | 1.00 | .58 |
| | *Statistics* | 48 | 48 | 49 | 42 | 42 | 70 | 62 |
| 049. | Rutgers, The State University-New Brunswick | 14 | 8 | 126 | NA | NA | NA | NA |
| | *Statistics* | 52 | 43 | 99 | | | | |
| 050. | SUNY at Buffalo | 7 | 15 | 8 | .59 | 5.8 | .71 | .41 |
| | *Statistics* | 42 | 50 | 43 | 63 | 57 | 43 | 49 |
| 051. | SUNY at Stony Brook | 5 | 17 | 36 | NA | NA | NA | NA |
| | *Applied Mathematics and Statistics* | 39 | 52 | 58 | | | | |
| 052. | South Florida, University of-Tampa | 3 | NA | 2 | NA | NA | NA | NA |
| | *Mathematics** | 36 | | 39 | | | | |
| 053. | Southern Methodist University | 11 | 16 | 25 | .39 | 5.5 | .83 | .56 |
| | *Statistics* | 48 | 51 | 52 | 53 | 59 | 55 | 60 |
| 054. | Stanford University | 20 | 40 | 48 | .57 | 5.9 | .83 | .53 |
| | *Statistics* | 61 | 75 | 64 | 62 | 56 | 55 | 57 |
| 055. | Temple University | 16 | 12 | 36 | NA | NA | NA | NA |
| | *Statistics** | 55 | 47 | 58 | | | | |
| 056. | Texas A & M University | 18 | 25 | 25 | .10 | 7.2 | .91 | .43 |
| | *Statistics* | 58 | 60 | 52 | 39 | 46 | 61 | 50 |
| 057. | Texas, U of-Health Science Center, Houston | 14 | 2 | 2 | NA | NA | NA | NA |
| | *Biomathematics (M D Anderson Hospital)* | 52 | 37 | 39 | | | | |
| 058. | Virginia Commonwealth University/Medical Col | 5 | 8 | 5 | NA | NA | NA | NA |
| | *Biostatistics* | 39 | 43 | 41 | | | | |
| 059. | Virginia Polytechnic Institute & State Univ | 19 | 26 | 25 | .30 | 7.5 | .65 | .30 |
| | *Statistics and Statistical Lab* | 60 | 61 | 52 | 49 | 44 | 38 | 41 |
| 060. | Washington, University of-Seattle | 34 | 24 | 36 | .70 | 7.0 | .86 | .64 |
| | *Biomathematics Group and Biostatistics* | 83 | 59 | 58 | 68 | 48 | 57 | 66 |

* indicates program was initiated since 1970.


NOTE: On the first line of data for every program, raw values for each measure are reported;
on the second line values are reported in standardized form, with mean = 50 and
standard deviation = 10. "NA" indicates that the value for a measure is not available.

TABLE 8.1  Program Measures (Raw and Standardized Values) in Statistics/Biostatistics

| Prog No. | Survey Results (08) | (09) | (10) | (11) | University Library (12) | Research Support (13) | (14) | Published Articles (15) | (16) | Survey Ratings Standard Error (08) | (09) | (10) | (11) |
|---|---|---|---|---|---|---|---|---|---|---|---|---|---|
| 041. | 2.8 | 1.7 | 1.0 | 0.8 | NA | .13 | NA | 8 | | .08 | .08 | .05 | .06 |
| | 50 | 53 | 49 | 48 | | 43 | | 46 | 48 | | | | |
| 042. | 2.7 | 1.6 | 1.1 | 1.0 | 0.7 | .31 | NA | 13 | | .08 | .06 | .05 | .07 |
| | 49 | 52 | 52 | 52 | 52 | 53 | | 50 | 51 | | | | |
| 043. | 2.6 | 1.4 | 1.2 | 0.8 | 0.7 | .00 | NA | 4 | | .09 | .09 | .06 | .07 |
| | 47 | 48 | 54 | 46 | 51 | 36 | | 42 | 41 | | | | |
| 044. | 1.9 | 1.0 | 1.0 | 0.3 | 0.1 | NA | NA | 0 | | .17 | .14 | .10 | .05 |
| | 40 | 39 | 46 | 34 | 46 | | | 39 | 40 | | | | |
| 045. | 3.1 | 1.3 | 1.8 | 1.1 | 0.1 | NA | NA | 13 | | .10 | .10 | .05 | .07 |
| | 53 | 45 | 78 | 53 | 46 | | | 50 | 50 | | | | |
| 046. | 4.0 | 1.9 | 0.9 | 1.5 | 0.9 | NA | NA | 11 | | .10 | .08 | .04 | .05 |
| | 62 | 56 | 43 | 65 | 53 | | | 49 | 45 | | | | |
| 047. | 3.9 | 2.1 | 1.2 | 1.4 | -0.5 | .48 | NA | 26 | | .06 | .06 | .05 | .05 |
| | 61 | 60 | 56 | 61 | 40 | 63 | | 62 | 62 | | | | |
| 048. | 3.0 | 1.7 | 1.1 | 1.0 | -0.6 | .46 | NA | 18 | | .08 | .07 | .06 | .07 |
| | 52 | 53 | 50 | 52 | 39 | 61 | | 55 | 54 | | | | |
| 049. | 3.2 | 1.8 | 0.9 | 1.0 | 0.8 | .50 | NA | 31 | | .09 | .06 | .08 | .07 |
| | 54 | 54 | 44 | 52 | 53 | 64 | | 67 | 71 | | | | |
| 050. | 1.9 | 1.1 | 0.1 | 0.7 | 0.3 | NA | NA | 7 | | .10 | .09 | .05 | .07 |
| | 40 | 42 | 15 | 44 | 48 | | | 45 | 44 | | | | |
| 051. | 2.2 | 1.1 | 1.5 | 0.8 | -0.6 | NA | NA | 3 | | .14 | .13 | .10 | .07 |
| | 44 | 42 | 66 | 46 | 39 | | | 41 | 41 | | | | |
| 052. | 2.0 | 0.6 | 1.3 | 0.7 | NA | NA | NA | 2 | | .15 | .10 | .14 | .07 |
| | 41 | 32 | 59 | 44 | | | | 40 | 43 | | | | |
| 053. | 2.6 | 1.6 | 1.0 | 0.9 | NA | .00 | NA | 13 | | .09 | .06 | .05 | .06 |
| | 48 | 51 | 48 | 50 | | 36 | | 50 | 49 | | | | |
| 054. | 4.9 | 2.8 | 1.2 | 1.7 | 2.0 | .35 | NA | 36 | | .04 | .04 | .05 | .05 |
| | 72 | 73 | 55 | 69 | 64 | 55 | | 71 | 71 | | | | |
| 055. | 1.8 | 1.1 | 1.0 | 0.5 | -0.4 | .00 | NA | 10 | | .10 | .10 | .07 | .06 |
| | 39 | 41 | 48 | 39 | 41 | 36 | | 48 | 48 | | | | |
| 056. | 3.1 | 1.8 | 1.4 | 1.3 | -0.5 | .06 | NA | 14 | | .07 | .07 | .08 | .05 |
| | 53 | 55 | 63 | 59 | 41 | 39 | | 51 | 49 | | | | |
| 057. | 2.4 | 1.3 | 1.2 | 0.4 | NA | .29 | NA | NA | | .16 | .12 | .12 | .06 |
| | 46 | 45 | 57 | 37 | | 52 | | | NA | | | | |
| 058. | 1.1 | 0.8 | 0.8 | 0.3 | NA | NA | NA | 2 | | .12 | .14 | .13 | .05 |
| | 33 | 36 | 41 | 34 | | | | 40 | 41 | | | | |
| 059. | 2.9 | 1.8 | 1.2 | 1.1 | -0.0 | .05 | NA | 17 | | .09 | .06 | .05 | .06 |
| | 51 | 54 | 56 | 55 | 45 | 39 | | 54 | 52 | | | | |
| 060. | 3.8 | 2.2 | 1.8 | 1.1 | 1.5 | .35 | NA | 9 | | .07 | .07 | .06 | .07 |
| | 61 | 62 | 77 | 55 | 59 | 56 | | 47 | 48 | | | | |

NOTE: On the first line of data for every program, raw values for each measure are reported; on the second line values are reported in standardized form, with mean = 50 and standard deviation = 10. "NA" indicates that the value for a measure is not available. Since the scale used to compute measure (16) is entirely arbitrary, only values in standardized form are reported for this measure.

TABLE 8.1  Program Measures (Raw and Standardized Values) in Statistics/Biostatistics

| Prog No. | University – Department/Academic Unit | Program Size | | | Characteristics of Program Graduates | | | |
|---|---|---|---|---|---|---|---|---|
| | | (01) | (02) | (03) | (04) | (05) | (06) | (07) |
| 061. | Wisconsin, University of–Madison | 23 | 41 | 40 | .21 | 5.7 | .67 | .29 |
| | *Statistics* | 66 | 76 | 60 | 45 | 58 | 40 | 40 |
| 062. | Wyoming, University of | 7 | 9 | 9 | NA | NA | NA | NA |
| | *Statistics** | 42 | 44 | 43 | | | | |
| 063. | Yale University | 6 | 9 | 6 | NA | NA | NA | NA |
| | *Epidemiology and Public Health* | 40 | 44 | 42 | | | | |
| 064. | Yale University | 6 | 13 | 18 | .30 | 6.5 | .50 | .10 |
| | *Statistics* | 40 | 48 | 48 | 49 | 51 | 25 | 26 |

* indicates program was initiated since 1970.

NOTE: On the first line of data for every program, raw values for each measure are reported; on the second line values are reported in standardized form, with mean = 50 and standard deviation = 10. "NA" indicates that the value for a measure is not available.

TABLE 8.1  Program Measures (Raw and Standardized Values) in Statistics/Biostatistics

| Prog No. | Survey Results (08) | (09) | (10) | (11) | University Library (12) | Research Support (13) | (14) | Published Articles (15) | (16) | Survey Ratings Standard Error (08) | (09) | (10) | (11) |
|---|---|---|---|---|---|---|---|---|---|---|---|---|---|
| 061. | 4.3 | 2.4 | 1.2 | 1.6 | 1.6 | .22 | NA | 35 | | .06 | .05 | .05 | .05 |
| | 66 | 67 | 54 | 67 | 59 | 48 | | 70 | 70 | | | | |
| 062. | 1.9 | 1.1 | 1.0 | 0.5 | NA | NA | NA | 7 | | .10 | .10 | .04 | .06 |
| | 40 | 41 | 48 | 40 | | | | 45 | 41 | | | | |
| 063. | 2.0 | 1.4 | 1.1 | 0.4 | 2.1 | NA | NA | 2 | | .16 | .13 | .08 | .06 |
| | 42 | 47 | 51 | 37 | 64 | | | 40 | 41 | | | | |
| 064. | 3.7 | 2.0 | 1.0 | 1.3 | 2.1 | NA | NA | 5 | | .08 | .06 | .04 | .07 |
| | 60 | 57 | 46 | 60 | 64 | | | 43 | 46 | | | | |

NOTE: On the first line of data for every program, raw values for each measure are reported;
on the second line values are reported in standardized form, with mean = 50 and
standard deviation = 10. "NA" indicates that the value for a measure is not available.
Since the scale used to compute measure (16) is entirely arbitrary, only values in
standardized form are reported for this measure.

TABLE 8.2 Summary Statistics Describing Each Program Measure—Statistics/Biostatistics

| Measure | Number of Programs Evaluated | Mean | Standard Deviation | DECILES | | | | | | | | |
|---|---|---|---|---|---|---|---|---|---|---|---|---|
| | | | | 1 | 2 | 3 | 4 | 5 | 6 | 7 | 8 | 9 |
| **Program Size** | | | | | | | | | | | | |
| 01 Raw Value | 64 | 12 | 7 | 5 | 7 | 8 | 9 | 12 | 13 | 15 | 17 | 22 |
| Std Value | 64 | 50 | 10 | 39 | 42 | 43 | 45 | 49 | 51 | 54 | 57 | 65 |
| 02 Raw Value | 61 | 15 | 10 | 4 | 7 | 9 | 12 | 13 | 15 | 16 | 19 | 26 |
| Std Value | 61 | 50 | 10 | 39 | 42 | 44 | 47 | 48 | 50 | 51 | 54 | 61 |
| 03 Raw Value | 64 | 22 | 19 | 5 | 8 | 10 | 14 | 17 | 20 | 25 | 33 | 40 |
| Std Value | 64 | 50 | 10 | 41 | 43 | 44 | 46 | 48 | 49 | 52 | 56 | 60 |
| **Program Graduates** | | | | | | | | | | | | |
| 04 Raw Value | 36 | .32 | .20 | .07 | .10 | .18 | .22 | .28 | .33 | .41 | .50 | .60 |
| Std Value | 36 | 50 | 10 | 38 | 39 | 43 | 45 | 48 | 51 | 55 | 59 | 64 |
| 05 Raw Value | 37 | 6.7 | 1.3 | 8.2 | 7.5 | 7.0 | 6.5 | 6.5 | 6.3 | 6.0 | 5.8 | 5.5 |
| Std Value | 37 | 50 | 10 | 38 | 44 | 48 | 51 | 51 | 53 | 55 | 57 | 59 |
| 06 Raw Value | 36 | .78 | .11 | .64 | .67 | .71 | .74 | .80 | .81 | .83 | .86 | .92 |
| Std Value | 36 | 50 | 10 | 37 | 40 | 44 | 46 | 52 | 53 | 55 | 57 | 63 |
| 07 Raw Value | 36 | .43 | .13 | .23 | .30 | .36 | .42 | .43 | .46 | .51 | .55 | .58 |
| Std Value | 36 | 50 | 10 | 35 | 40 | 45 | 49 | 50 | 52 | 56 | 59 | 62 |
| **Survey Results** | | | | | | | | | | | | |
| 08 Raw Value | 63 | 2.8 | 1.0 | 1.6 | 1.8 | 2.1 | 2.4 | 2.7 | 3.1 | 3.2 | 3.8 | 4.0 |
| Std Value | 63 | 50 | 10 | 37 | 40 | 43 | 46 | 49 | 53 | 54 | 61 | 63 |
| 09 Raw Value | 63 | 1.6 | .5 | .8 | 1.1 | 1.3 | 1.4 | 1.6 | 1.8 | 1.9 | 2.0 | 2.2 |
| Std Value | 63 | 50 | 10 | 36 | 41 | 45 | 47 | 51 | 55 | 56 | 58 | 62 |
| 10 Raw Value | 63 | 1.1 | .3 | .8 | .9 | 1.0 | 1.0 | 1.0 | 1.1 | 1.1 | 1.2 | 1.3 |
| Std Value | 63 | 50 | 10 | 40 | 44 | 48 | 48 | 48 | 52 | 52 | 55 | 59 |
| 11 Raw Value | 63 | .9 | .4 | .4 | .5 | .7 | .8 | 1.0 | 1.0 | 1.2 | 1.3 | 1.5 |
| Std Value | 63 | 50 | 10 | 37 | 40 | 44 | 47 | 52 | 52 | 57 | 59 | 64 |
| **University Library** | | | | | | | | | | | | |
| 12 Raw Value | 50 | .5 | 1.1 | -1.0 | -.5 | -.4 | .1 | .4 | .9 | 1.0 | 1.6 | 2.0 |
| Std Value | 50 | 50 | 10 | 36 | 40 | 41 | 46 | 49 | 53 | 54 | 60 | 64 |
| **Research Support** | | | | | | | | | | | | |
| 13 Raw Value | 37 | .25 | .18 | .00 | .06 | .10 | .20 | .27 | .30 | .35 | .44 | .49 |
| Std Value | 37 | 50 | 10 | 36 | 39 | 42 | 47 | 51 | 53 | 56 | 61 | 63 |
| **Publication Records** | | | | | | | | | | | | |
| 15 Raw Value | 63 | 12 | 11 | 2 | 3 | 5 | 7 | 9 | 13 | 14 | 18 | 30 |
| Std Value | 63 | 50 | 10 | 41 | 41 | 43 | 45 | 47 | 50 | 51 | 55 | 66 |
| 16 Std Value | 63 | 50 | 10 | 41 | 41 | 43 | 45 | 47 | 49 | 51 | 56 | 65 |

NOTE: Standardized values reported in the preceding table have been computed from exact values of the mean and standard deviation and not the rounded values reported here. Since the scale used to compute measure 16 is entirely arbitrary, only data in standardized form are reported for this measure.

TABLE 8.3  Intercorrelations Among Program Measures on 64 Programs in Statistics/Biostatistics

Measure

| | 01 | 02 | 03 | 04 | 05 | 06 | 07 | 08 | 09 | 10 | 11 | 12 | 13 | 14 | 15 | 16 |
|---|---|---|---|---|---|---|---|---|---|---|---|---|---|---|---|---|
| **Program Size** | | | | | | | | | | | | | | | | |
| 01 | | .53 | .42 | .16 | .15 | .16 | .29 | .63 | .72 | .35 | .54 | .24 | .05 | N/A | .50 | .49 |
| 02 | | | .48 | .00 | .04 | .00 | -.03 | .55 | .63 | .17 | .59 | .11 | .06 | N/A | .52 | .48 |
| 03 | | | | -.07 | -.05 | -.03 | -.03 | .40 | .42 | .12 | .40 | .14 | .17 | N/A | .48 | .50 |
| **Program Graduates** | | | | | | | | | | | | | | | | |
| 04 | | | | | -.11 | -.02 | .22 | .19 | .24 | -.06 | .00 | .24 | -.11 | N/A | -.32 | -.29 |
| 05 | | | | | | .14 | .45 | .32 | .30 | .15 | .35 | .25 | .11 | N/A | .39 | .37 |
| 06 | | | | | | | .62 | .15 | .13 | .24 | .04 | .06 | .31 | N/A | .14 | .11 |
| 07 | | | | | | | | .25 | .24 | .26 | .17 | .20 | .44 | N/A | .28 | .30 |
| **Survey Results** | | | | | | | | | | | | | | | | |
| 08 | | | | | | | | | .95 | .30 | .93 | .53 | .53 | N/A | .70 | .67 |
| 09 | | | | | | | | | | .24 | .87 | .45 | .41 | N/A | .67 | .63 |
| 10 | | | | | | | | | | | .27 | .04 | .09 | N/A | .16 | .15 |
| 11 | | | | | | | | | | | | .43 | .38 | N/A | .71 | .66 |
| **University Library** | | | | | | | | | | | | | | | | |
| 12 | | | | | | | | | | | | | .26 | N/A | .32 | .36 |
| **Research Support** | | | | | | | | | | | | | | | | |
| 13 | | | | | | | | | | | | | | N/A | .48 | .56 |
| 14 | | | | | | | | | | | | | | | N/A | N/A |
| **Publication Records** | | | | | | | | | | | | | | | | |
| 15 | | | | | | | | | | | | | | | | .98 |
| 16 | | | | | | | | | | | | | | | | |

NOTE: Since in computing correlation coefficients program data must be available for both of the measures being correlated, the actual number of programs on which each coefficient is based varies.

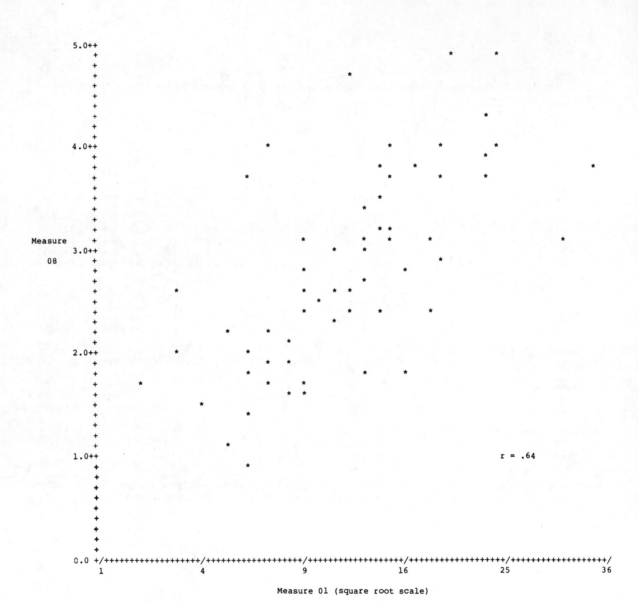

FIGURE 8.1  Mean rating of scholarly quality of faculty (measure 08) versus number of faculty members (measure 01)--63 programs in statistics/biostatistics.

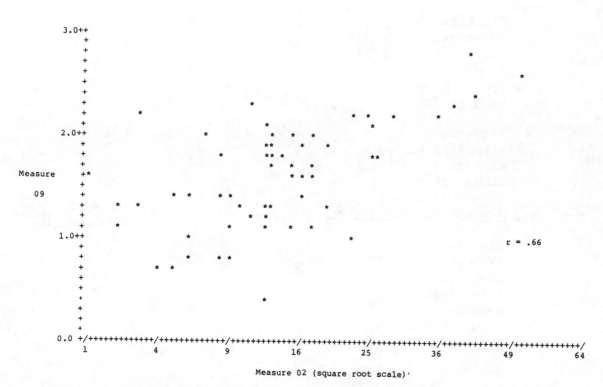

FIGURE 8.2  Mean rating of program effectiveness in educating research scholars/scientists (measure 09) versus number of graduates in last five years (measure 02)--60 programs in statistics/biostatistics.

TABLE 8.4  Characteristics of Survey Participants in Statistics/Biostatistics

|  | Respondents | |
|---|---|---|
|  | N | % |
| **Field of Specialization** | | |
| Biostatistics/Biometrics | 22 | 16 |
| Mathematical Statistics | 51 | 38 |
| Statistics, General | 48 | 36 |
| Other/Unknown | 14 | 10 |
| **Faculty Rank** | | |
| Professor | 78 | 58 |
| Associate Professor | 32 | 24 |
| Assistant Professor | 24 | 18 |
| Other/Unknown | 1 | 1 |
| **Year of Highest Degree** | | |
| Pre-1950 | 3 | 2 |
| 1950-59 | 28 | 21 |
| 1960-69 | 55 | 41 |
| Post-1969 | 47 | 35 |
| Unknown | 2 | 2 |
| **Evaluator Selection** | | |
| Nominated by Institution | 112 | 83 |
| Other | 23 | 17 |
| **Survey Form** | | |
| With Faculty Names | 119 | 88 |
| Without Names | 16 | 12 |
| **Total Evaluators** | 135 | 100 |

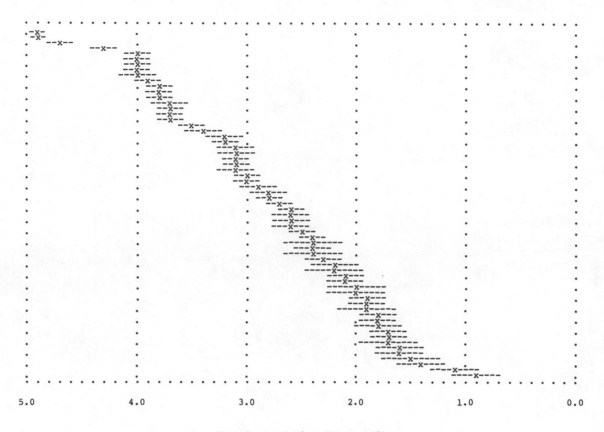

5.0     4.0     3.0     2.0     1.0     0.0

Mean Survey Rating (Measure 08)

FIGURE 8.3  Mean rating of scholarly quality of faculty in 63 programs in statistics/biostatistics.

NOTE:  Programs are listed in sequence of mean rating, with the highest-rated program appearing at the top of the page.  The broken lines (---) indicate a confidence interval of $\pm 1.5$ standard errors around the reported mean (x) of each program.

# IX
# Summary and Discussion

In the six preceding chapters results are presented of the assessment of 596 research-doctorate programs in chemistry, computer sciences, geosciences, mathematics, physics, and statistics/biostatistics. Included in each chapter are summary data describing the means and intercorrelations of the program measures in a particular discipline. In this chapter a comparison is made of the summary data reported for the six disciplines. Also presented are an analysis of the reliability (consistency) of the reputational survey ratings and an examination of some factors that might possibly have influenced the survey results. This chapter concludes with suggestions for improving studies of this kind--with particular attention given to the types of measures one would like to have available for an assessment of research-doctorate programs.

This chapter necessarily involves a detailed discussion of various statistics (means, standard deviations, correlation coefficients) describing the measures. Throughout, the reader should bear in mind that all these statistics and measures are necessarily imperfect attempts to describe the real quality of research-doctorate programs. Quality and some differences in quality are real, but these differences cannot be subsumed completely under any one quantitative measure. For example, no single numerical ranking--by measure 08 or by any weighted average of measures--can rank the quality of different programs with precision.

However, the evidence for reliability indicates considerable stability in the assessment of quality. For instance, a program that comes out in the first decile of a ranking is quite unlikely to "really" belong in the third decile, or vice versa. If numerical ranks of programs were replaced by groupings (distinguished, strong, etc.), these groupings again would not fully capture actual differences in quality since there would likely be substantial ambiguity about the borderline between adjacent groups. Furthermore, any attempt at linear ordering (best, next best, . . .) also may be inaccurate. Programs of roughly comparable quality may be better in different ways, so that there simply is no one best program--as will also be indicated in some of the numerical analyses. However, these difficulties of formulating ranks should not hide the underlying

159

reality of differences in quality or the importance of high quality for effective doctoral education.

## SUMMARY OF THE RESULTS

Displayed in Table 9.1 are the numbers of programs evaluated (bottom line) and the mean values for each measure in the six mathematical and physical science disciplines.[1]  As can be seen, the mean values reported for individual measures vary considerably among disciplines.  The pattern of means on each measure is summarized below, but the reader interested in a detailed comparison of the distribution of a measure should refer to the second table in each of the preceding six chapters.[2]

Program Size (Measures 01-03).  Based on the information provided to the committee by the study coordinator at each university, mathematics programs had, on the average, the largest number of faculty members (33 in December 1980), followed by physics (28) and chemistry (23). Chemistry programs graduated the most students (51 Ph.D. recipients in the FY1975-79 period) and had the largest enrollment (75 doctoral students in December 1980).  In contrast, statistics and biostatistics programs were reported to have an average of only 12 faculty members, 15 graduates, and 22 doctoral students.

Program Graduates (Measures 04-07).  The mean fraction of FY1975-79 doctoral recipients who as graduate students had received some national fellowship or training grant support (measure 04) ranges from .17 for graduates of computer science programs to .32 for graduates in statistics/biostatistics.  (The relatively high figure for the latter group may be explained by the availability of National Institutes of Health (NIH) training grant support for students in biostatistics.) With respect to the median number of years from first enrollment in a graduate program to receipt of the doctorate (measure 05), chemistry graduates typically earned their degrees more than half a year sooner than graduates in any of the other disciplines.  Graduates in physics and geosciences report the longest median times to the Ph.D.  In terms of employment status at graduation (measure 06), an average of 80 percent of the Ph.D. recipients from computer science programs reported that they had made firm job commitments by the time they had completed the requirements for their degrees, contrasted with 61 percent of the program graduates in mathematics.  A mean of 43 percent of the statistics/biostatistics graduates reported that they had made

---

[1] Means for measure 16, "influence" of publication, are omitted since arbitrary scaling of this measure prevents meaningful comparisons across disciplines.
[2] The second table in each of the six preceding chapters presents the standard deviation and decile values for each measure.

TABLE 9.1  Mean Values for Each Program Measure, by Discipline

| | Chemistry | Computer Sciences | Geo- sciences | Math | Physics | Statistics/ Biostat. |
|---|---|---|---|---|---|---|
| **Program Size** | | | | | | |
| 01 | 23 | 16 | 16 | 33 | 28 | 12 |
| 02 | 51 | 20 | 19 | 24 | 35 | 15 |
| 03 | 75 | 41 | 25 | 35 | 56 | 22 |
| **Program Graduates** | | | | | | |
| 04 | .23 | .17 | .26 | .25 | .26 | .32 |
| 05 | 5.9 | 6.5 | 7.0 | 6.6 | 7.1 | 6.7 |
| 06 | .76 | .80 | .77 | .61 | .66 | .78 |
| 07 | .33 | .38 | .22 | .25 | .26 | .43 |
| **Survey Results** | | | | | | |
| 08 | 2.5 | 2.5 | 2.9 | 2.7 | 2.7 | 2.8 |
| 09 | 1.6 | 1.5 | 1.8 | 1.6 | 1.7 | 1.6 |
| 10 | 1.1 | 1.1 | 1.1 | 1.2 | 1.1 | 1.1 |
| 11 | .9 | .9 | .9 | .8 | .7 | .9 |
| **University Library** | | | | | | |
| 12 | .1 | .4 | .4 | .1 | .1 | .5 |
| **Research Support** | | | | | | |
| 13 | .48 | .36 | .47 | .32 | .36 | .25 |
| 14 | 1788 | 1171 | 3996 | 616 | 2943 | NA |
| **Publication Records** | | | | | | |
| 15 | 78 | 34 | 44 | 39 | 106 | 12 |
| **Total Programs** | 145 | 58 | 91 | 115 | 123 | 64 |

firm commitments to take positions in Ph.D.-granting institutions (measure 07), while only 22 percent of those in the geosciences had made such plans. This difference may be due, to a great extent, to the availability of employment opportunities for geoscientists outside the academic sector.

Survey Results (Measures 08-11). Differences in the mean ratings derived from the reputational survey are small. In all six disciplines the mean rating of scholarly quality of program faculty (measure 08) is slightly below 3.0 ("good"), and programs were judged to be, on the average, a bit below "moderately" effective (2.0) in educating research scholars/scientists (measure 09). In the opinions of the survey respondents, there has been "little or no change" (approximately 1.0 on measure 10) in the last five years in the overall average quality of programs. The mean rating of an evaluator's familiarity with the work of program faculty (measure 11) is close to 1.0 ("some familiarity") in every discipline--about which more will be said later in this chapter.

University Library (Measure 12). Measure 12, based on a composite index of the size[3] of the library at the university in which a program resides, is calculated on a scale from -2.0 to 3.0, with means ranging from .1 in chemistry, mathematics, and physics to .4 in computer sciences and geosciences, and .5 in statistics/biostatistics. These differences may be explained, in large part, by the number of programs evaluated in each discipline. In the disciplines with the fewest doctoral programs (statistics/biostatistics, computer sciences, and geosciences), programs included are typically found in the larger institutions, which are likely to have high scores on the library size index. Ph.D. programs in chemistry, physics, and mathematics are found in a much broader spectrum of universities that includes the smaller institutions as well as the larger ones.

Research Support (Measures 13-14). Measure (13), the proportion of program faculty who had received NSF, NIH, or ADAMHA[4] research grant awards during the FY1978-80 period, has mean values ranging from as high as .48 and .47 in chemistry and geosciences, respectively, to .25 in statistics/biostatistics. It should be emphasized that this measure does not take into account research support that faculty members have received from sources other than these three federal

---

[3] The index, derived by the Association of Research Libraries, reflects a number of different measures, including number of volumes, fiscal expenditures, and other factors relevant to the size of a university library. See the description of this measure presented in Appendix D.

[4] Very few faculty members in mathematical and physical science programs received any research support from the Alcohol, Drug Abuse, and Mental Health Administration.

agencies.  In terms of total university expenditures for R&D in a particular discipline (measure 14), the mean values are reported to range from $616,000 in mathematics to $3,996,000 in the geosciences. (R&D expenditure data are not available for statistics/biostatistics.) The large differences in reported expenditures are likely to be related to three factors:  the differential availability of research support in the six disciplines, the differential average cost of doing research, and the differing numbers of individuals involved in a research effort.

Publication Records (Measures 15 and 16).  Considerable diversity is found in the mean number of articles associated with a research-doctorate program (measure 15).  An average of 106 articles published in the 1978-79 period is reported for programs in physics and 75 articles for programs in chemistry; in each of the other four disciplines the mean number of articles is fewer than 40.  These large differences reflect both the program size in a particular discipline (i.e., the total number of faculty and other staff members involved in research) and the frequency with which scientists in that discipline publish; it may also depend on the length of a typical paper in a discipline.  Mean scores are not reported on measure 16, the estimated "overall influence" of the articles attributed to a program.  Since this measure is calculated from an average of journal influence weights,[5] normalized for the journals covered in a particular discipline, mean differences among disciplines are uninterpretable.

Correlations with Measure 02.  Relations among the program measures are of intrinsic interest and are relevant to the issue of validity of the measures as indices of the quality of a research-doctorate program.  Measures that are logically related to program quality are expected to be related to each other.  To the extent that they are, a stronger case might be made for the validity of each as a quality measure.

A reasonable index of the relationship between any two measures is the Pearson product-moment correlation coefficient.  A table of correlation coefficients between all possible pairs of measures has been presented in each of the six preceding chapters.  In this chapter selected correlations to determine the extent to which coefficients are comparable in the six disciplines are presented.  Special attention is given to the correlations involving the number of FY1975-79 program graduates (measure 02), the survey rating of the scholarly quality of program faculty (measure 08), university R&D expenditures in a particular discipline (measure 14), and the influence-weighted number of publications (measure 16).

Table 9.2 presents the correlations of measure 02 with each of the other measures used in the assessment.  As might be expected, correlations of this measure with the other two measures of program size--number of faculty (01) and doctoral student enrollment (03)--are

---

[5] See Appendix F for a description of the derivation of this measure.

TABLE 9.2  Correlations of the Number of Program Graduates (Measure 02) with Other Measures, by Discipline

|  | Chemistry | Computer Sciences | Geo-sciences | Math | Physics | Statistics/ Biostat. |
|---|---|---|---|---|---|---|
| **Program Size** |  |  |  |  |  |  |
| 01 | .68 | .62 | .42 | .50 | .77 | .53 |
| 03 | .92 | .52 | .72 | .85 | .92 | .48 |
| **Program Graduates** |  |  |  |  |  |  |
| 04 | .02 | .05 | -.01 | .08 | -.02 | .00 |
| 05 | .38 | -.07 | .29 | .31 | .32 | .04 |
| 06 | .23 | .12 | .05 | .18 | .40 | .00 |
| 07 | .13 | -.05 | .36 | .46 | .41 | -.03 |
| **Survey Results** |  |  |  |  |  |  |
| 08 | .83 | .66 | .64 | .70 | .76 | .55 |
| 09 | .81 | .68 | .67 | .68 | .73 | .63 |
| 10 | .23 | -.02 | .06 | .01 | -.17 | .17 |
| 11 | .83 | .61 | .67 | .72 | .78 | .59 |
| **University Library** |  |  |  |  |  |  |
| 12 | .61 | .44 | .43 | .45 | .47 | .11 |
| **Research Support** |  |  |  |  |  |  |
| 13 | .57 | .34 | .40 | .35 | .13 | .06 |
| 14 | .72 | .58 | .25 | .41 | .66 | N/A |
| **Publication Records** |  |  |  |  |  |  |
| 15 | .83 | .85 | .73 | .75 | .85 | .52 |
| 16 | .86 | .84 | .74 | .81 | .86 | .48 |

quite high in all six disciplines.  Of greater interest are the strong positive correlations between measure 02 and measures derived from either reputational survey ratings or publication records.  The coefficients describing the relationship of measure 02 with measures 15 and 16 are greater than .70 in all disciplines except statistics/biostatistics.  This result is not surprising, of course, since both of the publication measures reflect total productivity and have not been adjusted for program size.  The correlations of measure 02 with measures 08, 09, and 11 are almost as strong.  It is quite apparent that the programs that received high survey ratings and with which evaluators were more likely to be familiar were also ones that had larger numbers of graduates.  Although the committee gave serious consideration to presenting an alternative set of survey measures that were adjusted for program size, a satisfactory algorithm for making such an adjustment was not found.  In attempting such an adjustment on the basis of the regression of survey ratings on measures of program size, it was found that some exceptionally large programs appeared to be unfairly penalized and that some very small programs received unjustifiably high adjusted scores.

Measure 02 also has positive correlations in most disciplines with measure 12, an index of university library size, and with measures 13 and 14, which pertain to the level of support for research in a program.  Of particular note are the moderately large coefficients--in disciplines other than statistics/biostatistics and physics--for measure 13, the fraction of faculty members receiving federal research grants.  Unlike measure 14, this measure has been adjusted for the number of program faculty.  The correlations of measure 02 with measures 05, 06, and 07 are smaller but still positive in most of the disciplines.  From this analysis it is apparent that the number of program graduates tends to be positively correlated with all other variables except measure 04--the fraction of students with national fellowship support.  It is also apparent that the relationship of measure 02 with the other variables tends to be weakest for programs in statistics/biostatistics.

Correlations with Measure 08.  Table 9.3 shows the correlation coefficients for measure 08, the mean rating of the scholarly quality of program faculty, with each of the other variables.  The correlations of measure 08 with measures of program size (01, 02, and 03) are .40 or greater for all six disciplines.  Not surprisingly, the larger the program, the more likely its faculty is to be rated high in quality.  However, it is interesting to note that in all disciplines except statistics/biostatistics the correlation with the number of program graduates (measure 02) is larger than that with the number of faculty or the number of enrolled students.

Correlations of measure 08 with measure 04, the fraction of students with national fellowship awards, are positive but close to zero in all disciplines except computer sciences and mathematics.  For programs in the biological and social sciences, the corresponding coefficients (not reported in this volume) are found to be greater, typically in the range of .40 to .70.  Perhaps in the mathematical and

TABLE 9.3   Correlations of the Survey Ratings of Scholarly Quality of Program Faculty
(Measure 08) with Other Measures, by Discipline

| | Chemistry | Computer Sciences | Geo-sciences | Math | Physics | Statistics/ Biostat. |
|---|---|---|---|---|---|---|
| **Program Size** | | | | | | |
| 01 | .64 | .54 | .45 | .48 | .68 | .63 |
| 02 | .83 | .66 | .64 | .70 | .76 | .55 |
| 03 | .81 | .50 | .61 | .64 | .75 | .40 |
| **Program Graduates** | | | | | | |
| 04 | .11 | .35 | .08 | .30 | .15 | .19 |
| 05 | .47 | .14 | .50 | .57 | .42 | .32 |
| 06 | .28 | .21 | .24 | .19 | .42 | .15 |
| 07 | .30 | .17 | .58 | .63 | .58 | .25 |
| **Survey Results** | | | | | | |
| 09 | .98 | .98 | .97 | .98 | .96 | .95 |
| 10 | .35 | .29 | .29 | -.01 | -.15 | .30 |
| 11 | .96 | .97 | .87 | .96 | .96 | .93 |
| **University Library** | | | | | | |
| 12 | .66 | .58 | .58 | .65 | .67 | .53 |
| **Research Support** | | | | | | |
| 13 | .77 | .59 | .72 | .70 | .24 | .53 |
| 14 | .79 | .63 | .27 | .42 | .61 | N/A |
| **Publication Records** | | | | | | |
| 15 | .80 | .70 | .75 | .75 | .85 | .70 |
| 16 | .86 | .77 | .77 | .83 | .86 | .67 |

physical sciences, the departments with highly regarded faculty are more likely to provide support to doctoral students as teaching assistants or research assistants on faculty research grants—thereby reducing dependency on national fellowships. (The low correlation of rated faculty quality with the fraction of students with national fellowships is not, of course, inconsistent with the thesis that programs with large numbers of students are programs with large numbers of fellowship holders.)

Correlations of rated faculty quality with measure 05, shortness of time from matriculation in graduate school to award of the doctorate, are notably high for programs in mathematics, geosciences, and chemistry and still sizeable for physics and statistics/bio-statistics programs. Thus, those programs producing graduates in shorter periods of time tended to receive higher survey ratings. This finding is surprising in view of the smaller correlations in these disciplines between measures of program size and shortness of time-to-Ph.D. It seems there is a tendency for programs that produce doctoral graduates in a shorter time to have more highly rated faculty, and this tendency is relatively independent of the number of faculty members.

Correlations of ratings of faculty quality with measure 06, the fraction of program graduates with definite employment plans, are moderately high in physics and somewhat lower, but still positive, in the other disciplines. In every discipline except computer sciences the correlation of measure 08 is higher with measure 07, the fraction of graduates having agreed to employment at a Ph.D.-granting institution. These coefficients are greater than .50 in mathematics, geosciences, and physics.

The correlations of measure 08 with measure 09, the rated effectiveness of doctoral education, are uniformly very high, at or above .95 in every discipline. This finding is consistent with results from the Cartter and Roose-Andersen studies.[6] The coefficients describing the relationship between measure 08 and measure 11, familiarity with the work of program faculty, are also very high, ranging from .87 to .97. In general, evaluators were more likely to have high regard for the quality of faculty in those programs with which they were most familiar. That the correlation coefficients are as large as observed may simply reflect the fact that "known" programs tend to be those that have earned strong reputations.

Correlations of ratings of faculty quality with measure 10, the ratings of perceived improvement in program quality, are near zero for mathematics and physics programs and range from .29 to .35 in other disciplines. One might have expected that a program judged to have improved in quality would have been somewhat more likely to receive high ratings on measure 08 than would a program judged to have declined—thereby imposing a small positive correlation between these two variables.

---

[6] Roose and Andersen, p. 19.

Moderately high correlations are observed in most disciplines between measure 08 and university library size (measure 12), support for research (measures 13 and 14), and publication records (measures 15 and 16). With few exceptions these coefficients are .50 or greater in all disciplines. Of particular note are the strong correlations with the two publication measures--ranging from .70 to .86. In all disciplines except statistics/biostatistics the correlations with measure 16 are higher than those with measure 15; the "weighted influence" of journals in which articles are published yields an index that tends to relate more closely to faculty reputation than does an unadjusted count of the number of articles published. Although the observed differences between the coefficients for measures 15 and 16 are not large, this result is consistent with earlier findings of Anderson et al.[7]

Correlations with Measure 14. Correlations of measure 14, reported dollars of support for R&D, with other measures are shown in Table 9.4. (Data on research expenditures in statistics/biostatistics are not available.) The pattern of relations is quite similar for programs in chemistry, computer sciences, and physics: moderately high correlations with measures of program size and somewhat higher correlations with both reputational survey results (except measure 10) and publication measures. For programs in mathematics many of these relations are positive but not as strong. For geoscience programs, measure 14 is related more closely to faculty size (measure 01) than to any other measure, and the correlations with rated quality of faculty and program effectiveness are lower than in any other discipline. In interpreting these relationships one must keep in mind the fact that the research expenditure data have not been adjusted for the number of faculty and other staff members involved in research in a program.

Correlations with Measure 16. Measure 16 is the number of published articles attributed to a program and adjusted for the "average influence" of the journals in which the articles appear. The correlations of this measure with all others appear in Table 9.5. Of particular interest are the high correlations with all three measures of program size and with the reputational survey results (excluding measure 10). Most of those coefficients exceed .70, although for programs in statistics/biostatistics they are below this level. Moderately high correlations are also observed between measure 16 and measures 12, 13, and 14. With the exception of computer science programs, the correlations between the adjusted publication measure and measure 05, time-to-Ph.D., range from .31 to .41. It should be pointed out that the exceptionally large coefficients reported for measure 15 result from the fact that the two publication measures are empirically as well as logically interdependent.

---

[7]Anderson et al., p. 95.

TABLE 9.4  Correlations of the University Research Expenditures in a Discipline
(Measure 14) with Other Measures, by Discipline

| | Chemistry | Computer Sciences | Geosciences | Math | Physics | Statistics/ Biostat. |
|---|---|---|---|---|---|---|
| **Program Size** | | | | | | |
| 01 | .43 | .44 | .61 | .18 | .54 | N/A |
| 02 | .72 | .58 | .25 | .41 | .66 | N/A |
| 03 | .66 | .43 | .28 | .44 | .68 | N/A |
| **Program Graduates** | | | | | | |
| 04 | .18 | .22 | .22 | .29 | .04 | N/A |
| 05 | .35 | -.21 | -.05 | .17 | .31 | N/A |
| 06 | .31 | -.03 | -.04 | .23 | .25 | N/A |
| 07 | .20 | -.16 | .06 | .22 | .31 | N/A |
| **Survey Results** | | | | | | |
| 08 | .79 | .63 | .27 | .42 | .61 | N/A |
| 09 | .74 | .61 | .25 | .42 | .61 | N/A |
| 10 | .14 | -.02 | .13 | -.12 | -.08 | N/A |
| 11 | .77 | .64 | .18 | .43 | .58 | N/A |
| **University Library** | | | | | | |
| 12 | .45 | .16 | .33 | .33 | .33 | N/A |
| **Research Support** | | | | | | |
| 13 | .55 | .10 | .20 | .18 | .07 | N/A |
| **Publication Records** | | | | | | |
| 15 | .70 | .66 | .42 | .35 | .80 | N/A |
| 16 | .78 | .73 | .35 | .42 | .80 | N/A |

TABLE 9.5  Correlations of the Influence-Weighted Number of Publications
(Measure 16) with Other Measures, by Discipline

|  | Chemistry | Computer Sciences | Geo-sciences | Math | Physics | Statistics/ Biostat. |
|---|---|---|---|---|---|---|
| **Program Size** | | | | | | |
| 01 | .65 | .61 | .36 | .63 | .72 | .49 |
| 02 | .86 | .84 | .74 | .81 | .86 | .48 |
| 03 | .84 | .52 | .64 | .78 | .85 | .50 |
| **Program Graduates** | | | | | | |
| 04 | .03 | .20 | .07 | .15 | .05 | -.29 |
| 05 | .41 | -.04 | .31 | .40 | .38 | .37 |
| 06 | .22 | .14 | .00 | .16 | .43 | .11 |
| 07 | .23 | -.01 | .39 | .50 | .48 | .30 |
| **Survey Results** | | | | | | |
| 08 | .86 | .77 | .77 | .83 | .86 | .67 |
| 09 | .82 | .75 | .75 | .80 | .82 | .63 |
| 10 | .33 | .05 | .09 | .05 | -.14 | .15 |
| 11 | .88 | .74 | .70 | .83 | .86 | .66 |
| **University Library** | | | | | | |
| 12 | .56 | .52 | .66 | .59 | .61 | .36 |
| **Research Support** | | | | | | |
| 13 | .60 | .35 | .51 | .51 | .21 | .56 |
| 14 | .78 | .73 | .35 | .42 | .80 | N/A |
| **Publication Records** | | | | | | |
| 15 | .95 | .98 | .97 | .90 | .99 | .98 |

Despite the appreciable correlations between reputational ratings of quality and program size measures, the functional relations between the two probably are complex. If there is a minimum size for a high quality program, this size is likely to vary from discipline to discipline. Increases in size beyond the minimum may represent more high quality faculty, or a greater proportion of inactive faculty, or a faculty with heavy teaching responsibilities. In attempting to select among these alternative interpretations, a single correlation coefficient provides insufficient guidance. Nonetheless, certain similarities may be seen in the pattern of correlations among the measures. High correlations consistently appear among measures 08, 09, and 11 from the reputational survey, and these measures also are prominently related to program size (measures 01, 02, and 03), to publication productivity (measures 15 and 16), to R&D expenditures (measure 14), and to library size (measure 12). These results show that for all disciplines the reputational rating measures (08, 09, and 11) tend to be associated with program size and with other correlates of size--publication volume, R&D expenditures, and library size. Furthermore, for most disciplines the reputational measures 08, 09, and 11 tend to be positively related to shortness of time-to-Ph.D. (measure 05), to employment prospects of program graduates (measures 06 and 07), and to fraction of faculty holding research grants (measure 13). These latter measures are not consistently correlated highly with the size measures or with any other measures besides reputational ratings.

## ANALYSIS OF THE SURVEY RESPONSE

Measures 08-11, derived from the reputational survey, may be of particular interest to many readers since measures of this type have been the most widely used (and frequently criticized) indices of quality of graduate programs. In designing the survey instrument for this assessment the committee made several changes in the form that had been used in the Roose-Andersen study. The modifications served two purposes: to provide the evaluators with a clearer understanding of the programs that they were asked to judge and to provide the committee with supplemental information for the analysis of the survey response. One change was to restrict to 50 the number of programs that any individual was asked to evaluate. Probably the most important change was the inclusion of lists of names and ranks of individual faculty members involved in the research-doctorate programs to be evaluated on the survey form, together with the number of doctoral degrees awarded in the previous five years. Ninety percent of the evaluators were sent forms with faculty names and numbers of degrees awarded; the remaining ten percent were given forms without this information so that an analysis could be made of the effect of this modification on survey results. Another change was the addition of a question concerning an evaluator's familiarity with each of the programs. In addition to providing an index of program recognition (measure 11), the inclusion of this question permits a comparison of the ratings furnished by individuals who had considerable familiarity

TABLE 9.6  Distribution of Responses to Each Survey Item, by Discipline

| Survey Measure | Total | Chem-istry | Computer Sciences | Geo-sciences | Math | Physics | Statistics/ Biostat. |
|---|---|---|---|---|---|---|---|
| **08 SCHOLARLY QUALITY OF PROGRAM FACULTY** | | | | | | | |
| Distinguished | 7.2 | 6.3 | 7.5 | 6.5 | 7.7 | 7.9 | 8.3 |
| Strong | 15.9 | 15.1 | 12.5 | 19.1 | 15.5 | 13.6 | 20.3 |
| Good | 21.2 | 22.4 | 20.4 | 22.8 | 19.2 | 19.6 | 22.7 |
| Adequate | 16.3 | 19.5 | 19.4 | 13.4 | 14.5 | 14.6 | 16.2 |
| Marginal | 7.8 | 10.4 | 9.8 | 4.7 | 6.9 | 6.9 | 7.3 |
| Not Sufficient for Doctoral Education | 2.2 | 3.0 | 3.0 | .8 | 2.5 | 1.3 | 2.7 |
| Don't Know Well Enough to Evaluate | 29.4 | 23.3 | 27.4 | 32.7 | 33.8 | 36.1 | 22.4 |
| TOTAL | 100.0 | 100.0 | 100.0 | 100.0 | 100.0 | 100.0 | 100.0 |
| **09 EFFECTIVENESS OF PROGRAM IN EDUCATING SCIENTISTS** | | | | | | | |
| Extremely Effective | 8.0 | 8.7 | 7.9 | 8.3 | 7.4 | 7.8 | 7.2 |
| Reasonably Effective | 28.7 | 32.5 | 25.7 | 34.1 | 22.1 | 27.0 | 29.0 |
| Minimally Effective | 13.2 | 15.0 | 15.7 | 12.1 | 11.3 | 11.1 | 15.1 |
| Not Effective | 3.1 | 3.6 | 4.6 | 1.7 | 3.4 | 2.0 | 3.8 |
| Don't Know Well Enough to Evaluate | 47.0 | 40.2 | 46.1 | 43.8 | 55.8 | 52.1 | 45.0 |
| TOTAL | 100.0 | 100.0 | 100.0 | 100.0 | 100.0 | 100.0 | 100.0 |
| **10 CHANGE IN PROGRAM QUALITY IN LAST FIVE YEARS** | | | | | | | |
| Better | 11.5 | 12.7 | 15.7 | 14.2 | 9.3 | 9.2 | 9.2 |
| Little or No Change | 29.4 | 33.9 | 25.9 | 27.1 | 25.8 | 28.4 | 32.5 |
| Poorer | 6.2 | 8.4 | 8.2 | 6.6 | 3.5 | 5.1 | 5.1 |
| Don't Know Well Enough to Evaluate | 52.9 | 44.9 | 50.1 | 52.1 | 61.5 | 57.3 | 53.2 |
| TOTAL | 100.0 | 100.0 | 100.0 | 100.0 | 100.0 | 100.0 | 100.0 |
| **11 FAMILIARITY WITH WORK OF PROGRAM FACULTY** | | | | | | | |
| Considerable | 20.0 | 20.9 | 20.2 | 22.3 | 17.9 | 16.3 | 24.0 |
| Some | 41.1 | 43.1 | 42.8 | 40.7 | 38.8 | 38.2 | 43.6 |
| Little or None | 37.2 | 34.6 | 34.6 | 35.4 | 41.8 | 43.0 | 31.1 |
| No Response | 1.7 | 1.4 | 2.3 | 1.6 | 1.5 | 2.5 | 1.3 |
| TOTAL | 100.0 | 100.0 | 100.0 | 100.0 | 100.0 | 100.0 | 100.0 |

NOTE: For survey measures 08, 09, and 10 the "don't know" category includes a small number of cases for which the respondents provided no response to the survey item.

with a particular program and the ratings by those not as familiar
with the program. Each evaluator was also asked to identify his or
her own institution of highest degree and current field of specialization. This information enables the committee to compare, for each
program, the ratings furnished by alumni of a particular institution
with the ratings by other evaluators as well as to examine differences
in the ratings supplied by evaluators in certain specialty fields.

Before examining factors that may have influenced the survey
results, some mention should be made of the distributions of responses
to the four survey items and the reliability (consistency) of the
ratings. As Table 9.6 shows, the response distribution for each
survey item does not vary greatly from discipline to discipline. For
example, in judging the scholarly quality of faculty (measure 08),
survey respondents in each discipline rated between 6 and 8 percent of
the programs as being "distinguished" and between 1 and 3 percent as
"not sufficient for doctoral education." In evaluating the
effectiveness in educating research scholars/scientists, 7 to 9
percent of the programs were rated as being "extremely effective" and
approximatey 2 to 5 percent as "not effective." Of particular
interest in this table are the frequencies with which evaluators
failed to provide responses on survey measures 08, 09, and 10.
Approximately 30 percent of the total number of evaluations requested
for measure 08 were not furnished because survey respondents in the
mathematical and physical sciences felt that they were not familiar
enough with a particular program to evaluate it. The corresponding
percentages of "don't know" responses for measures 09 and 10 are
considerably larger--47 and 53 percent, respectively--suggesting that
survey respondents found it more difficult (or were less willing) to
judge program effectiveness and change than to judge the scholarly
quality of program faculty.

The large fractions of "don't know" responses are a matter of some
concern. However, given the broad coverage of research-doctorate
programs, it is not surprising that faculty members would be unfamiliar
with many of the less distinguished programs. As shown in Table 9.7,
survey respondents in each discipline were much more likely to furnish
evaluations for programs with high reputational standings than they
were for programs of lesser distinction. For example, for mathematical
and physical science programs that received mean ratings of 4.0 or
higher on measure 08, almost 95 percent of the evaluations requested
on measure 08 were provided; 85 and 77 percent were provided on
measures 09 and 10. In contrast, the corresponding response rates for
programs with mean ratings below 2.0 are much lower--52, 35, and 28
percent response on measures 08, 09, and 10, respectively.

Of great importance to the interpretation of the survey results is
the reliability of the response. How much confidence can one have in
the reliability of a mean rating reported for a particular program?
In the first table in each of the preceding six chapters, estimated
standard errors associated with the mean ratings of every program are
presented for all four survey items (measures 08-11). While there is
some variation in the magnitude of the standard errors reported in
every discipline, they rarely exceed .15 for any of the four measures
and typically range from .05 to .10. For programs with higher mean

TABLE 9.7  Survey Item Response Rates, by Discipline and Mean Rating on Measure 08

| Survey Measure | Total | Chem-istry | Computer Sciences | Geo-sciences | Math | Physics | Statistics/ Biostat. |
|---|---|---|---|---|---|---|---|
| 08  SCHOLARLY QUALITY OF PROGRAM FACULTY | | | | | | | |
| Mean Rating on Measure 08 | | | | | | | |
| 4.0 or Higher | 94.7 | 98.0 | 99.4 | 87.6 | 95.9 | 93.7 | 97.0 |
| 3.0 - 3.9 | 85.9 | 91.9 | 91.8 | 76.3 | 83.6 | 83.6 | 91.3 |
| 2.0 - 2.9 | 67.7 | 77.4 | 76.4 | 60.7 | 62.5 | 61.5 | 72.3 |
| Less than 2.0 | 51.7 | 61.2 | 51.6 | 40.1 | 45.8 | 39.9 | 58.5 |
| 09  EFFECTIVENESS OF PROGRAM IN EDUCATING SCIENTISTS | | | | | | | |
| Mean Rating on Measure 08 | | | | | | | |
| 4.0 or Higher | 85.2 | 92.6 | 90.9 | 79.4 | 80.9 | 85.4 | 85.4 |
| 3.0 - 3.9 | 68.1 | 77.1 | 72.4 | 66.2 | 56.3 | 65.6 | 68.8 |
| 2.0 - 2.9 | 47.5 | 57.4 | 53.1 | 47.6 | 37.8 | 42.8 | 47.1 |
| Less than 2.0 | 34.9 | 42.9 | 36.8 | 31.6 | 28.5 | 25.7 | 36.3 |
| 10  CHANGE IN PROGRAM QUALITY IN LAST FIVE YEARS | | | | | | | |
| Mean Rating on Measure 08 | | | | | | | |
| 4.0 or Higher | 76.8 | 88.3 | 85.6 | 69.0 | 70.0 | 76.7 | 74.7 |
| 3.0 - 3.9 | 62.3 | 74.0 | 67.5 | 56.8 | 51.9 | 61.9 | 60.3 |
| 2.0 - 2.9 | 43.1 | 54.4 | 52.2 | 40.9 | 33.7 | 38.5 | 39.9 |
| Less than 2.0 | 27.7 | 35.5 | 29.1 | 22.7 | 22.0 | 19.9 | 27.7 |

ratings the estimated errors associated with these means are generally smaller--a finding consistent with the fact that survey respondents were more likely to furnish evaluations for programs with high reputational standing.  The "split-half" correlations[8] presented in Table 9.8 give an indication of the overall reliability of the survey results in each discipline and for each measure.  In the derivation of these correlations, individual ratings of each program were randomly divided into two groups (A and B), and a separate mean rating was computed for each group.  The last column in Table 9.8 reports the correlations between the mean program ratings of the two groups and is not corrected for the fact that the mean ratings of each group are based on only half rather than a full set of the responses.[9]  As the reader will note, the coefficients reported for measure 08, the scholarly quality of program faculty, are in the range of .96 to

[8] For a discussion of the interpretation of "split-half" coefficients, see Robert L. Thorndike and Elizabeth Hagan, Measurement and Evaluation in Psychology and Education, John Wiley & Sons, New York, 1969, pp. 182-185.

[9] To compensate for the smaller sample size the "split-half" coefficient may be adjusted using the Spearman-Brown formula: $r' = 2r/(1 + r)$.  This adjustment would have the effect of increasing a correlation of .70, for example, to .82; a correlation of .80 to .89; a correlation of .90 to .95; and a correlation of .95 to .97.

TABLE 9.8  Correlations Between Two Sets of Average Ratings from Two Randomly Selected Groups of Evaluators in the Mathematical and Physical Sciences

### MEASURE 08: SCHOLARLY QUALITY OF PROGRAM FACULTY

| Discipline | Mean Rating | | Std. Deviation | | Correlation | |
|---|---|---|---|---|---|---|
| | Group A | Group B | Group A | Group B | N | r |
| Chemistry | 2.55 | 2.53 | 1.00 | 1.00 | 145 | .99 |
| Computer Sciences | 2.51 | 2.50 | .97 | 1.00 | 57 | .96 |
| Geosciences | 2.92 | 2.93 | .83 | .82 | 91 | .97 |
| Mathematics | 2.64 | 2.66 | 1.03 | 1.00 | 114 | .98 |
| Physics | 2.66 | 2.63 | .99 | 1.01 | 122 | .96 |
| Statistics/Biostat. | 2.80 | 2.79 | .94 | .97 | 63 | .98 |

### MEASURE 09: EFFECTIVENESS OF PROGRAM IN EDUCATING SCHOLARS

| Discipline | Mean Rating | | Std. Deviation | | Correlation | |
|---|---|---|---|---|---|---|
| | Group A | Group B | Group A | Group B | N | r |
| Chemistry | 1.63 | 1.64 | .54 | .54 | 145 | .95 |
| Computer Sciences | 1.52 | 1.50 | .56 | .56 | 57 | .95 |
| Geosciences | 1.74 | 1.76 | .44 | .45 | 91 | .94 |
| Mathematics | 1.54 | 1.55 | .57 | .59 | 114 | .91 |
| Physics | 1.63 | 1.65 | .52 | .51 | 122 | .89 |
| Statistics/Biostat. | 1.55 | 1.57 | .54 | .53 | 63 | .97 |

### MEASURE 10: IMPROVEMENT IN PROGRAM IN LAST FIVE YEARS

| Discipline | Mean Rating | | Std. Deviation | | Correlation | |
|---|---|---|---|---|---|---|
| | Group A | Group B | Group A | Group B | N | r |
| Chemistry | 1.05 | 1.06 | .22 | .23 | 145 | .76 |
| Computer Sciences | 1.14 | 1.11 | .28 | .29 | 57 | .82 |
| Geosciences | 1.15 | 1.13 | .28 | .30 | 91 | .77 |
| Mathematics | 1.12 | 1.14 | .22 | .22 | 114 | .62 |
| Physics | 1.10 | 1.11 | .26 | .25 | 122 | .64 |
| Statistics/Biostat. | 1.06 | 1.07 | .28 | .27 | 63 | .85 |

### MEASURE 11: FAMILIARITY WITH WORK OF PROGRAM FACULTY

| Discipline | Mean Rating | | Std. Deviation | | Correlation | |
|---|---|---|---|---|---|---|
| | Group A | Group B | Group A | Group B | N | r |
| Chemistry | .86 | .86 | .43 | .41 | 145 | .95 |
| Computer Sciences | .84 | .86 | .42 | .45 | 57 | .94 |
| Geosciences | .87 | .86 | .36 | .37 | 91 | .93 |
| Mathematics | .75 | .76 | .39 | .40 | 114 | .95 |
| Physics | .71 | .73 | .42 | .42 | 122 | .96 |
| Statistics/Biostat. | .92 | .94 | .42 | .40 | 63 | .95 |

TABLE 9.9  Comparison of Mean Ratings for 11 Mathematics Programs
Included in Two Separate Survey Administrations

| | Survey Measure | All Evaluators | | | | Evaluators Rating the Same Program in Both Surveys | | | |
| | | First | | Second | | First | | Second | |
| | | N | $\overline{X}$ | N | $\overline{X}$ | N | $\overline{X}$ | N | $\overline{X}$ |
|---|---|---|---|---|---|---|---|---|---|
| Program A | 08 | 100 | 4.9 | 114 | 4.9 | 50 | 4.9 | 50 | 4.9 |
| | 09 | 90 | 2.7 | 100 | 2.8 | 42 | 2.7 | 43 | 2.7 |
| | 10 | 74 | 1.2 | 83 | 1.2 | 38 | 1.1 | 34 | 1.2 |
| | 11 | 100 | 1.6 | 115 | 1.6 | 50 | 1.5 | 50 | 1.6 |
| Program B | 08 | 94 | 4.6 | 115 | 4.6 | 48 | 4.6 | 50 | 4.5 |
| | 09 | 81 | 2.6 | 91 | 2.5 | 40 | 2.6 | 39 | 2.5 |
| | 10 | 69 | 1.0 | 82 | 1.0 | 37 | 1.0 | 36 | 0.9 |
| | 11 | 98 | 1.4 | 116 | 1.4 | 50 | 1.5 | 50 | 1.5 |
| Program C | 08 | 86 | 3.4 | 103 | 3.6 | 42 | 3.4 | 44 | 3.5 |
| | 09 | 56 | 2.0 | 66 | 2.1 | 28 | 2.1 | 29 | 2.0 |
| | 10 | 55 | 1.1 | 62 | 1.3 | 30 | 1.2 | 27 | 1.4 |
| | 11 | 99 | 1.0 | 116 | 1.1 | 50 | 1.1 | 50 | 1.0 |
| Program D | 08 | 74 | 3.0 | 93 | 3.0 | 37 | 2.8 | 38 | 2.9 |
| | 09 | 50 | 1.8 | 48 | 1.6 | 27 | 1.7 | 16 | 1.6 |
| | 10 | 46 | 1.4 | 52 | 1.5 | 24 | 1.4 | 23 | 1.5 |
| | 11 | 90 | 1.0 | 113 | 0.9 | 46 | 1.0 | 46 | 0.9 |
| Program E | 08 | 69 | 3.0 | 95 | 3.1 | 39 | 3.0 | 46 | 3.1 |
| | 09 | 40 | 1.8 | 60 | 1.9 | 25 | 1.8 | 30 | 1.8 |
| | 10 | 36 | 0.8 | 58 | 0.9 | 24 | 0.8 | 29 | 0.9 |
| | 11 | 96 | 0.8 | 115 | 0.9 | 52 | 0.9 | 52 | 1.0 |
| Program F | 08 | 63 | 2.9 | 90 | 3.0 | 26 | 3.0 | 32 | 3.1 |
| | 09 | 35 | 1.8 | 46 | 1.7 | 10 | 1.6 | 13 | 1.8 |
| | 10 | 32 | 1.1 | 43 | 1.1 | 11 | 1.3 | 12 | 1.2 |
| | 11 | 95 | 0.7 | 115 | 0.8 | 43 | 0.7 | 44 | 0.7 |
| Program G | 08 | 69 | 2.7 | 92 | 2.8 | 39 | 2.7 | 39 | 3.0 |
| | 09 | 35 | 1.7 | 45 | 1.6 | 17 | 1.7 | 19 | 1.7 |
| | 10 | 36 | 1.1 | 43 | 1.2 | 17 | 1.1 | 19 | 1.2 |
| | 11 | 85 | 0.9 | 116 | 0.8 | 46 | 0.9 | 46 | 0.9 |
| Program H | 08 | 58 | 2.2 | 73 | 2.5 | 36 | 2.2 | 37 | 2.4 |
| | 09 | 32 | 1.3 | 43 | 1.3 | 22 | 1.2 | 19 | 1.3 |
| | 10 | 30 | 1.5 | 39 | 1.5 | 20 | 1.7 | 17 | 1.4 |
| | 11 | 90 | 0.7 | 116 | 0.6 | 51 | 0.7 | 52 | 0.6 |
| Program I | 08 | 55 | 2.0 | 74 | 1.9 | 30 | 1.9 | 30 | 2.0 |
| | 09 | 33 | 1.0 | 41 | 0.9 | 19 | 1.0 | 18 | 0.8 |
| | 10 | 27 | 1.2 | 31 | 1.1 | 15 | 1.1 | 13 | 1.2 |
| | 11 | 99 | 0.5 | 115 | 0.5 | 50 | 0.5 | 50 | 0.5 |
| Program J | 08 | 51 | 1.5 | 67 | 1.5 | 26 | 1.4 | 28 | 1.4 |
| | 09 | 31 | 0.8 | 36 | 0.7 | 14 | 0.6 | 14 | 0.7 |
| | 10 | 26 | 1.2 | 23 | 1.1 | 14 | 1.2 | 12 | 1.3 |
| | 11 | 96 | 0.5 | 113 | 0.3 | 49 | 0.4 | 48 | 0.4 |
| Program K | 08 | 33 | 1.2 | 48 | 1.2 | 17 | 1.1 | 21 | 1.4 |
| | 09 | 19 | 0.8 | 21 | 0.5 | 11 | 0.6 | 8 | 0.4 |
| | 10 | 12 | 0.8 | 15 | 0.9 | 5 | 1.0 | 5 | 0.8 |
| | 11 | 99 | 0.2 | 114 | 0.2 | 48 | 0.2 | 47 | 0.2 |

.98--indicating a high degree of consistency in evaluators' judgments. The correlations reported for measures 09 and 11, the rated effectiveness of a program and evaluators' familiarity with a program, are somewhat lower but still at a level of .92 or higher in each discipline. Not surprisingly, the reliability coefficients for ratings of change in program quality in the last five years (measure 10) are considerably lower, ranging from .67 to .88 in the six mathematical and physical science disciplines. While these coefficients represent tolerable reliability, it is quite evident that the responses to measure 10 are not as reliable as the responses to the other three items.

Further evidence of the reliability of the survey responses is presented in Table 9.9. As mentioned in Chapter VI, 11 mathematics programs, selected at random, were included on a second form sent to 178 survey respondents in this discipline, and 116 individuals (65 percent) furnished responses to the second survey. A comparison of the overall results of the two survey administrations (columns 2 and 4 in Table 9.9) demonstrates the consistency of the ratings provided for each of the 11 programs. The average, absolute observed difference in the two sets of mean ratings is less than 0.1 for each measure. Columns 6 and 8 in this table report the results based on the responses of only those evaluators who had been asked to consider a particular program in both administrations of the survey. (For a given program approximately 40-45 percent of the 116 respondents to the second survey were asked to evaluate that program in the prior survey.) It is not surprising to find comparable small differences in the mean ratings provided by this subgroup of evaluators.

Critics of past reputational studies have expressed concern about the credibility of reputational assessments when evaluators provide judgments of programs about which they may know very little. As already mentioned, survey participants in this study were offered the explicit alternative, "Don't know well enough to evaluate." This response option was quite liberally used for measures 08, 09, and 10, as shown in Table 9.6. In addition, evaluators were asked to indicate their degree of familiarity with each program. Respondents reported "considerable" familiarity with an average of only one program in every five. While this finding supports the conjecture that many program ratings are based on limited information, the availability of reported familiarity permits us to analyze how ratings vary as a function of familiarity.

This issue can be addressed in more than one way. It is evident from the data reported in Table 9.10 that mean ratings of the scholarly quality of program faculty tend to be higher if the evaluator has considerable familiarity with the program. There is nothing surprising or, for that matter, disconcerting about such an association. When a particular program fails to provoke more than vague images in the evaluator's mind, he or she is likely to take this as some indication that the program is not an extremely lustrous one on the national scene. While visibility and quality are scarcely the same, the world of research in higher education is structured to encourage high quality to achieve high visibility, so that any association of the two is far from spurious.

TABLE 9.10  Mean Ratings of Scholarly Quality of Program Faculty,
by Evaluator's Familiarity with Work of Faculty

|  | MEAN RATINGS | | CORRELATION | |
|  | Consid-erable | Some/Little | r | N |
|---|---|---|---|---|
| Chemistry | 2.81 | 2.46 | .93 | 145 |
| Computer Sciences | 2.83 | 2.47 | .89 | 55 |
| Geosciences | 3.24 | 2.80 | .89 | 91 |
| Mathematics | 3.05 | 2.55 | .92 | 114 |
| Physics | 3.00 | 2.64 | .87 | 116 |
| Statistics/Biostat. | 2.99 | 2.69 | .94 | 63 |

NOTE:  N reported in last column represents the number of programs
with a rating from at least one evaluator in each of the two groups.

From the data presented in Table 9.10 it is evident that if mean ratings were computed on the basis of the responses of only those most familiar with programs, the values reported for individual programs would be increased.  A largely independent question is whether a restriction of this kind would substantially change our sense of the relative standings of programs on this measure.  Quite naturally, the answer depends to some degree on the nature of the restriction imposed.  For example, if we exclude evaluations provided by those who confessed "little or no" familiarity with particular programs, then the revised mean ratings would be correlated at a level of at least .99 with the mean ratings computed using all of the data.[10]  (This similarity arises, in part, because only a small fraction of evaluations are given on the basis of no more than "little" familiarity with the program.)

The third column in Table 9.10 presents the correlation in each discipline between the array of mean ratings supplied by respondents claiming "considerable" familiarity and the mean ratings of those indicating "some" or "little or no" familiarity with particular programs.  This coefficient is a rather conservative estimate of agreement since there is not a sufficient number of ratings from those with "considerable" familiarity to provide highly stable means.  Were more such ratings available, one might expect the correlations to be higher.  However, even in the form presented, the correlations, which are at least .92 in all six disciplines, are high enough to suggest that the relative standing of programs on measure 08 is not greatly affected by the admixtures of ratings from evaluators who recognize that their knowledge of a given program is limited.

As mentioned previously, 90 percent of the survey sample members were supplied the names of faculty members associated with each program to be evaluated, along with the reported number of program

[10] These correlations, not reported here, were found to exceed .995 for program ratings in chemistry, geosciences, mathematics, and statistics/biostatistics.

TABLE 9.11   Item Response Rate on Measure 08, by Selected Characteristics
of Survey Evaluators in the Mathematical and Physical Sciences

| | Total | Chem-istry | Computer Sciences | Geo-sciences | Math | Physics | Statistics/ Biostat. |
|---|---|---|---|---|---|---|---|
| **EVALUATOR'S FAMILIARITY** | | | | | | | |
| **WITH PROGRAM** | | | | | | | |
| Considerable | 100.0 | 100.0 | 100.0 | 100.0 | 100.0 | 100.0 | 100.0 |
| Some | 98.2 | 98.8 | 97.2 | 98.1 | 98.0 | 98.4 | 98.2 |
| Little or None | 26.4 | 36.6 | 29.2 | 13.5 | 23.6 | 22.0 | 33.3 |
| | | | | | | | |
| **TYPE OF SURVEY FORM** | | | | | | | |
| Names | 70.6 | 77.0 | 72.4 | 67.9 | 65.1 | 63.3 | 78.7 |
| No Names | 70.8 | 73.6 | 74.2 | 62.6 | 74.7 | 69.3 | 69.8 |
| | | | | | | | |
| **INSTITUTION OF HIGHEST** | | | | | | | |
| **DEGREE** | | | | | | | |
| Alumni | 98.0 | 98.1 | 100.0 | 95.1 | 98.8 | 100.0 | 97.1 |
| Nonalumni | 70.4 | 76.5 | 72.3 | 67.0 | 65.9 | 63.6 | 77.3 |
| | | | | | | | |
| **EVALUATOR'S PROXIMITY** | | | | | | | |
| **TO PROGRAM** | | | | | | | |
| Same Region | 81.8 | 87.7 | 79.9 | 81.8 | 77.2 | 78.5 | 83.2 |
| Outside Region | 69.0 | 75.1 | 71.4 | 65.3 | 64.5 | 61.8 | 76.7 |

NOTE:  The item response rate is the percentage of the total ratings requested from survey
participants that included a response other than "don't know."

graduates (Ph.D. or equivalent degrees) in the previous five years.
Since earlier reputational surveys had not provided such information,
10 percent of the sample members, randomly selected, were given forms
without faculty names or doctoral data, as a "control group."
Although one might expect that those given faculty names would have
been more likely than other survey respondents to provide evaluations
of the scholarly quality of program faculty, no appreciable
differences were found (Table 9.11) between the two groups in their
frequency of response to this survey item.  (The reader may recall
that the provision of faculty names apparently had little effect on
survey sample members' willingness to complete and return their
questionnaires.[11])

The mean ratings provided by the group furnished faculty names are
lower than the mean ratings supplied by other respondents (Table
9.12).  Although the differences are small, they attract attention
because they are reasonably consistent from discipline to discipline
and because the direction of the differences was not anticipated.
After all, those programs more familiar to evaluators tended to
receive higher ratings, yet when steps were taken to enhance the
evaluator's familiarity, the resulting ratings are somewhat lower.
One post hoc interpretation of this finding is that a program may be
considered to have distinguished faculty if even only a few of its

---

[11]See Table 2.3.

members are considered by the evaluator to be outstanding in their field. However, when a full list of program faculty is provided, the evaluator may be influenced by the number of individuals whom he or she could not consider to be distinguished. Thus, the presentation of these additional, unfamiliar names may occasionally result in a lower rating of program faculty.

However interesting these effects may be, one should not lose sight of the fact that they are small at best and that their existence does not necessarily imply that a program's relative standing on measure 08 would differ much whichever type of survey form were used. Since only about 1 in 10 ratings was supplied without the benefit of faculty names, it is hard to establish any very stable picture of relative mean ratings of individual programs. However, the correlations between the mean ratings supplied by the two groups are reasonably high--ranging from .85 to .94 in the six disciplines (Table 9.12). Were these coefficients adjusted for the fact that the group furnished forms without names constituted only about 10 percent of the survey respondents, they would be substantially larger. From this result it seems reasonable to conclude that differences in the alternative survey forms used are not likely to be responsible for any large-scale reshuffling in the reputational ranking of programs on measure 08. It also suggests that the inclusion of faculty names in the committee's assessment need not prevent comparisons of the results with those obtained from the Roose-Andersen survey.

Another factor that might be thought to influence an evaluator's judgment about a particular program is the geographic proximity of that program to the evaluator. There is enough regional traffic in academic life that one might expect proximate programs to be better known than those in distant regions of the country. This hypothesis may apply especially to the smaller and less visible programs and is

TABLE 9.12  Mean Ratings of Scholarly Quality of Program Faculty, by Type of Survey Form Provided to Evaluator

|  | MEAN RATINGS | | CORRELATION | |
|  | Names | No Names | r | N |
|---|---|---|---|---|
| Chemistry | 2.53 | 2.66 | .93 | 145 |
| Computer Sciences | 2.49 | 2.61 | .93 | 57 |
| Geosciences | 2.93 | 3.01 | .88 | 90 |
| Mathematics | 2.62 | 2.72 | .94 | 113 |
| Physics | 2.62 | 2.88 | .85 | 122 |
| Statistics/Biostat. | 2.79 | 2.85 | .92 | 63 |

NOTE:  N reported in last column represents the number of programs with a rating from at least one evaluator in each of the two groups.

TABLE 9.13  Mean Ratings of Scholarly Quality of Program Faculty,
by Evaluator's Proximity to Region of Program

|  | MEAN RATINGS | | CORRELATION | |
|  | Nearby | Outside | r | N |
| --- | --- | --- | --- | --- |
| Chemistry | 2.59 | 2.54 | .95 | 144 |
| Computer Sciences | 2.51 | 2.52 | .95 | 55 |
| Geosciences | 3.00 | 2.94 | .93 | 87 |
| Mathematics | 2.74 | 2.64 | .94 | 114 |
| Physics | 2.75 | 2.65 | .88 | 120 |
| Statistics/Biostat. | 2.96 | 2.77 | .94 | 62 |

NOTE:  N reported in last column represents the number of programs
with a rating from at least one evaluator in each of the two groups.

confirmed by the survey results.  For purposes of analysis, programs
were assigned to one of nine geographic regions[12] in the United
States, and ratings of programs within an evaluator's own region are
categorized in Table 9.13 as "nearby."  Ratings of programs in any of
the other eight regions were put in the "outside" group.  Findings
reported elsewhere in this chapter confirm that evaluators were more
likely to provide ratings if a program was within their own region of
the country,[13] and it is reasonable to imagine that the smaller and
the less visible programs received a disproportionate share of their
ratings either from evaluators within their own region or from others
who for one reason or another were particularly familiar with programs
in that region.

Although the data in Table 9.13 suggest that "nearby" programs
were given higher ratings than those outside the evaluator's region,
the differences in reported means are quite small and probably
represent no more than a secondary effect that might be expected
because, as we have already seen, evaluators tended to rate higher
those programs with which they were more familiar.  Furthermore, the
high correlations found between the mean ratings of the two groups
indicate that the relative standings of programs are not dramatically
influenced by the geographic proximity of those evaluating it.

Another consideration that troubles some critics is that large
programs may be unfairly favored in a faculty survey because they are
likely to have more alumni contributing to their ratings who, it would
stand to reason, would be generous in the evaluations of their alma

---

[12] See Appendix I for a list of the states included in each region.
[13] See Table 9.11.

TABLE 9.14 Mean Ratings of Scholarly Quality of Program Faculty,
by Evaluator's Institution of Highest Degree

|  | MEAN RATINGS | | NUMBER OF PROGRAMS WITH ALUMNI RATINGS |
|  | Alumni | Nonalumni | N |
| --- | --- | --- | --- |
| Chemistry | 3.88 | 3.60 | 37 |
| Computer Sciences | 3.56 | 3.02 | 26 |
| Geosciences | 3.83 | 3.51 | 34 |
| Mathematics | 3.73 | 3.41 | 37 |
| Physics | 4.11 | 3.87 | 27 |
| Statistics/Biostat. | 3.90 | 3.32 | 35 |

NOTE: The pairs of means reported in each discipline are computed
for a subset of programs with a rating from at least one alumnus
and are substantially greater than the mean ratings for the full set
of programs in each discipline.

maters. Information collected in the survey on each evaluator's
institution of highest degree enables us to investigate this concern.
The findings presented in Table 9.14 support the hypothesis that
alumni provided generous ratings--with differences in the mean ratings
(for measure 08) of alumni and nonalumni ranging from .24 to .58 in
the six disciplines. It is interesting to note that the largest
differences are found in statistics/biostatistics and computer
sciences, the disciplines with the fewest programs. Given the
appreciable differences between the ratings furnished by program
alumni and other evaluators, one might ask how much effect this has
had on the overall results of the survey. The answer is "very
little." As shown in the table, in chemistry and physics only one
program in every four received ratings from any alumnus; in
statistics/biostatistics slightly more than half of the programs were
evaluated by one or more alumni.[14] Even in the latter discipline,
however, the fraction of alumni providing ratings of a program is
always quite small and should have had minimal impact on the overall
mean rating of any program. To be certain that this was the case,
mean ratings of the scholarly quality of faculty were recalculated for
every mathematical and physical science program--with the evaluations
provided by alumni excluded. The results were compared with the mean
scores based on a full set of evaluations. Out of the 592 mathematical and physical science programs evaluated in the survey, only 1

---

[14] Because of the small number of alumni ratings in every discipline,
the mean ratings for this group are unstable and therefore the correlations between alumni and nonalumni mean ratings are not reported.

program (in geosciences) had an observed difference as large as 0.2, and for 562 programs (95 percent) their mean ratings remain unchanged (to the nearest tenth of a unit). On the basis of these findings the committee saw no reason to exclude alumni ratings in the calculation of program means.

Another concern that some critics have is that a survey evaluation may be affected by the interaction of the research interests of the evaluator and the area(s) of focus of the research-doctorate program to be rated. It is said, for example, that some narrowly focused programs may be strong in a particular area of research but that this strength may not be recognized by a large fraction of evaluators who happen to be unknowledgeable in this area. This is a concern more difficult to address than those discussed in the preceding pages since little or no information is available about the areas of focus of the programs being evaluated (although in certain disciplines the title of a department or academic unit may provide a clue). To obtain a better understanding of the extent to which an evaluator's field of specialty may have influenced the ratings he or she has provided, evaluators in physics and in statistics/biostatistics were separated into groups according to their specialty fields (as reported on the survey questionnaire). In physics, Group A includes those specializing in elementary particles and nuclear structure, and Group B is made up of those in all other areas of physics. In statistics/biostatistics, Group A consists of evaluators who designated biostatistics or biomathematics as their specialty and Group B of those in all other specialty areas of statistics. The mean ratings of the two groups in each discipline are reported in Table 9.15. The program ratings

TABLE 9.15 Mean Ratings of Scholarly Quality of Program Faculty, by Evaluator's Field of Specialty Within Physics or Statistics/Biostatistics

PHYSICS: Group A includes evaluators in elementary particles and nuclear structure; Group B includes those in atomic/molecular, solid state, and other fields of physics.

STATISTICS/BIOSTATISTICS: Group A includes evaluators in biostatistics, biometrics, and epidemiology; Group B includes those in all other fields of statistics.

|  | MEAN RATINGS | | CORRELATION | |
|  | Group A | Group B | r | N |
| Physics | 2.58 | 2.68 | .95 | 122 |
| Statistics/Biostat. | 3.13 | 2.73 | .93 | 63 |

NOTE: N reported in last column represents the number of programs with a rating from at least one evaluator in each of the two groups.

supplied by evaluators in elementary particles and nuclear structure are, on the average, slightly below those provided by other physicists. The mean ratings of the biostatistics group are typically higher than those of other statisticians. Despite these differences there is a high degree of correlation in the mean ratings provided by the two groups in each discipline. Although the differences in the mean ratings of biostatisticians (Group A) and other statisticians (Group B) are comparatively large, a detailed inspection of the individual ratings reveals that biomedical evaluators rated programs appreciably higher <u>regardless</u> of whether a program was located in a department of biostatistics (and related fields) or in a department outside the biomedical area. Although one cannot conclude from these findings that an evaluator's specialty field has no bearing on how he or she rates a program, these findings do suggest that the relative standings of programs in physics and statistics/biostatistics would not be greatly altered if the ratings by either group were discarded.

## INTERPRETATION OF REPUTATIONAL SURVEY RATINGS

It is not hard to foresee that results from this survey will receive considerable attention through enthusiastic and uncritical reporting in some quarters and sharp castigation in others. The study committee understands the grounds for both sides of this polarized response but finds that both tend to be excessive. It is important to make clear how we view these ratings as fitting into the larger study of which they are a part.

The reputational results are likely to receive a disproportionate degree of attention for several reasons, including the fact that they reflect the opinions of a large group of faculty colleagues and that they form a bridge with earlier studies of graduate programs. But the results will also receive emphasis because they alone, among all of the measures, seem to address quality in an overall or global fashion. While most recognize that "objective" program characteristics (i.e., publication productivity, research funding, or library size) have some bearing on program quality, probably no one would contend that a single one of these measures encompasses all that need be known about the quality of research-doctorate programs. Each is obviously no more than an indicator of some aspect of program quality. In contrast, the reputational ratings are global from the start because the respondents are asked to take into account many objective characteristics and to arrive at a general assessment of the quality of the faculty and effectiveness of the program. This generality has self-evident appeal.

On the other hand, it is wise to keep in mind that these reputational ratings are measures of <u>perceived</u> program quality rather than of "quality" in some ideal or absolute sense. What this means is that, just as for all of the more objective measures, the reputational

ratings represent only a partial view of what most of us would consider quality to be; hence, they must be kept in careful perspective.

Some critics may argue that such ratings are positively misleading
because of a variety of methodological artifacts or because they are
supplied by "judges" who often know very little about the programs
they are rating.  The committee has conducted the survey in a way that
permits the empirical examination of a number of the alleged artifacts
and, although our analysis is by no means exhaustive, the general
conclusion is that their effects are slight.

Among the criticisms of reputational ratings from prior studies
are some that represent a perspective that may be misguided.  This
perspective assumes that one asks for ratings in order to find out
what quality really is and that to the degree that the ratings miss
the mark of "quintessential quality," they are unreal, although the
quality that they attempt to measure is real.  What this perspective
misses is the reality of quality and the fact that impressions of
quality, if widely shared, have an imposing reality of their own and
therefore are worth knowing about in their own right.  After all,
these perceptions govern a large-scale system of traffic around the
nation's graduate institutions--for example, when undergraduate
students seek the advice of professors concerning graduate programs
that they might attend.  It is possible that some professors put in
this position disqualify themselves on grounds that they are not well
informed about the relative merits of the programs being considered.
Most faculty members, however, surely attempt to be helpful on the
basis of impressions gleaned from their professional experience, and
these assessments are likely to have major impact on student
decision-making.  In short, the impressions are real and have very
real effects not only on students shopping for graduate schools but
also on other flows, such as job-seeking young faculty and the
distribution of research resources.  At the very least, the survey
results provide a snapshot of these impressions from discipline to
discipline.  Although these impressions may be far from ideally
informed, they certainly show a strong degree of consensus within each
discipline, and it seems safe to assume that they are more than
passingly related to what a majority of keen observers might agree
program quality is all about.

## COMPARISON WITH RESULTS OF THE ROOSE-ANDERSEN STUDY

An analysis of the response to the committee's survey would not be
complete without comparing the results with those obtained in the
survey by Roose and Andersen 12 years earlier.  Although there are
obvious similarities in the two surveys, there are also some important
differences that should be kept in mind in examining individual program
ratings of the scholarly quality of faculty.  Already mentioned in
this chapter is the inclusion, on the form sent to 90 percent of the
sample members in the committee's survey, of the names and academic
ranks of faculty and the numbers of doctoral graduates in the previous

five years. Other significant changes in the committee's form are the identification of the university department or academic unit in which each program may be found, the restriction of requesting evaluators to make judgments about no more than 50 research-doctorate programs in their discipline, and the presentation of these programs in random sequence on each survey form. The sampling frames used in the two surveys also differ. The sample selected in the earlier study included only individuals who had been nominated by the participating universities, while more than one-fourth of the sample in the committee's survey were chosen at random from full faculty lists. (Except for this difference the samples were quite similar--i.e., in terms of number of evaluators in each discipline and the fraction of senior scholars.[15])

Several dissimilarities in the coverage of the Roose-Andersen and this committee's reputational assessments should be mentioned. The former included a total of 130 institutions that had awarded at least 100 doctoral degrees in two or more disciplines during the FY1958-67 period. The institutional coverage in the committee's assessment was based on the number of doctorates awarded in each discipline (as described in Chapter I) and covered a total population of 228 universities. Most of the universities represented in the present study but not the earlier one are institutions that offered research-doctorate programs in a limited set of disciplines. In the Roose-Andersen study, programs in five mathematical and physical science disciplines were rated: astronomy, chemistry, geology, mathematics, and physics. In the committee's assessment, two disciplines were added to this list[16]--computer sciences and statistics/biostatistics--and programs in astronomy were not evaluated (for reasons explained in Chapter I). Finally, in the Roose-Andersen study only one set of ratings was compiled from each institution represented in a discipline, whereas in the committee's survey, separate ratings were requested if a university offered more than one research-doctorate program in a given discipline. The consequences of these differences in survey coverage are quite apparent: in the committee's survey, evaluations were requested for a total of 593 research-doctorate programs in the mathematical and physical sciences, compared with 444 programs in the Roose-Andersen study.

Figures 9.1-9.4 plot the mean ratings of scholarly quality of faculty in programs included in both surveys; sets of ratings are graphed for 103 programs in chemistry, 57 in geosciences, 86 in mathematics, and 90 in physics. Since in the Roose-Andersen study programs were identified by institution and discipline (but not by department), the matching of results from this survey with those from

---

[15] For a description of the sample group used in the earlier study, see Roose and Andersen, pp. 28-31.
[16] It should be emphasized that the committee's assessment of geoscience programs encompasses--in addition to geology--geochemistry, geophysics, and other earth sciences.

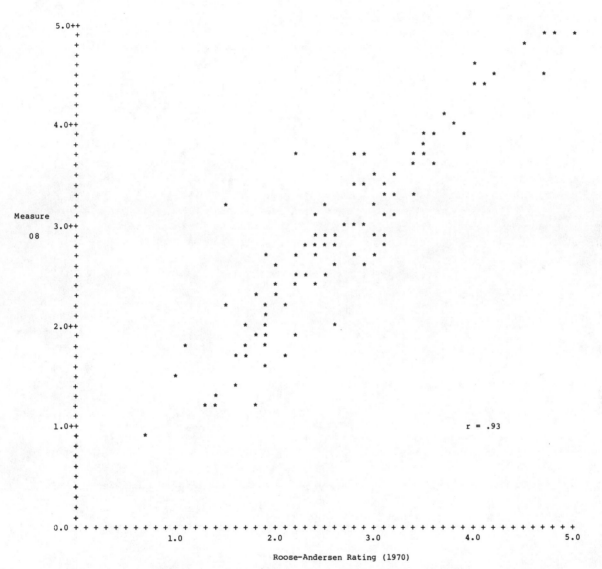

FIGURE 9.1  Mean rating of scholarly quality of faculty (measure 08) versus mean rating of faculty in the Roose-Andersen study--103 programs in chemistry.

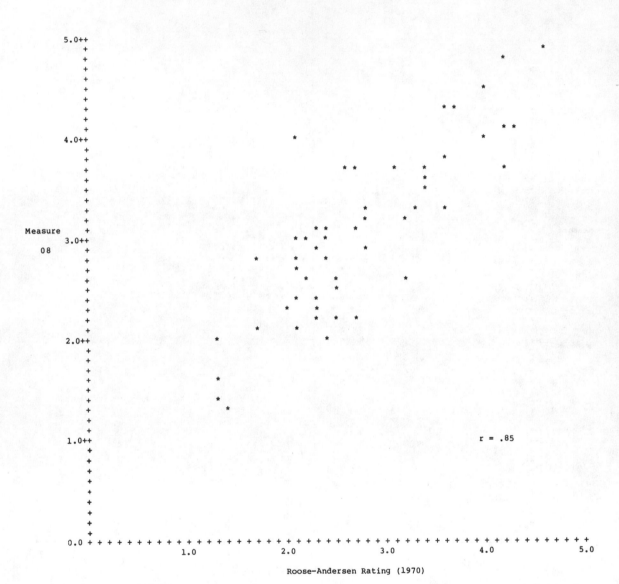

FIGURE 9.2 Mean rating of scholarly quality of faculty (measure 08) versus mean rating of faculty in the Roose-Andersen study--57 programs in geosciences.

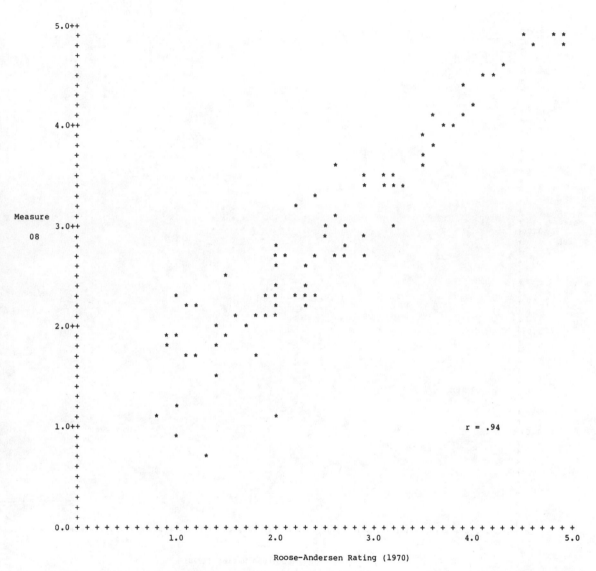

FIGURE 9.3  Mean rating of scholarly quality of faculty (measure 08) versus mean rating of faculty in the Roose-Andersen study--86 programs in mathematics.

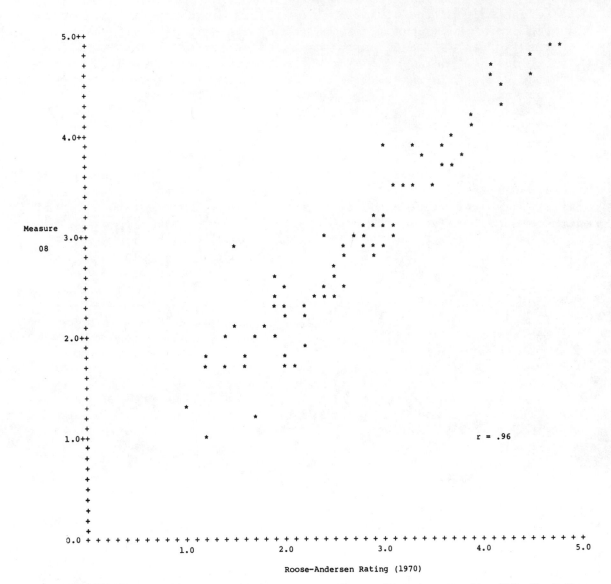

FIGURE 9.4  Mean rating of scholarly quality of faculty (measure 08) versus mean rating of faculty in the Roose-Andersen study--90 programs in physics.

the committee's survey is not precise. For universities represented in the latter survey by more than one program in a particular discipline, the mean rating for the program with the largest number of graduates (measure 02) is the only one plotted here. Although the results of both surveys are reported on identical scales, some caution must be taken in interpreting differences in the mean ratings a program received in the two evaluations. It is impossible to estimate what effect all of the differences described above may have had on the results of the two surveys. Furthermore, one must remember that the reported scores are based on the opinions of different groups of faculty members and were provided at different time periods. In 1969, when the Roose-Andersen survey was conducted, graduate departments in most universities were still expanding and not facing the enrollment and budget reductions that many departments have had to deal with in recent years. Consequently, a comparison of the overall findings from the two surveys reveals nothing about how much the quality of graduate education has improved (or declined) in the past decade. Nor should the reader place much stock in any small differences in the mean ratings that a particular program may have received in the two surveys. On the other hand, it is of particular interest to note the high correlations between the results of the evaluations. For programs in chemistry, mathematics, and physics the correlation coefficients range between .93 and .96; in the geosciences the coefficient is .85. The lower coefficient in geosciences may be explained, in part, by the difference, described in footnote 16, in the field coverage of the two surveys. The extraordinarily high correlations found in chemistry, mathematics, and physics may suggest to some readers that reputational standings of programs in these disciplines have changed very little in the last decade. However, differences are apparent for some institutions. Also, one must keep in mind that the correlations are based on the reputational ratings of only three-fourths of the programs evaluated in this assessment in these disciplines and do not take into account the emergence of many new programs that did not exist or were too small to be rated in the Roose-Andersen study.

## FUTURE STUDIES

One of the most important objectives in undertaking this assessment was to test new measures not used extensively in past evaluations of graduate programs. Although the committee believes that it has been successful in this effort, much more needs to be done. First and foremost, studies of this kind should be extended to cover other types of programs and other disciplines not included in this effort. As a consequence of budgeting limitations, the committee had to restrict its study to 32 disciplines, selected on the basis of the number of doctorates awarded in each. Among those omitted were programs in astronomy, which was included in the Roose-Andersen study; a multidimensional assessment of research-doctorate programs in this and many other important disciplines would be of value. Consideration should also be given to embarking on evaluations of programs offering other types of graduate and professional degrees. As a matter of fact, plans for including master's-degree programs in this assessment were originally contemplated, but because of a lack of available information about the resources and graduates of programs at the master's level, it was decided to focus on programs leading to the research doctorate.

Perhaps the most debated issue the committee has had to address concerned which measures should be reported in this assessment. In fact, there is still disagreement among some of its members about the relative merits of certain measures, and the committee fully recognizes a need for more reliable and valid indices of the quality of graduate programs. First on a list of needs is more precise and meaningful information about the product of research-doctorate programs--the graduates. For example, what fraction of the program graduates have gone on to be productive investigators--either in the academic setting or in government and industrial laboratories? What fraction have gone on to become outstanding investigators--as measured by receipt of major prizes, membership in academies, and other such distinctions? How do program graduates compare with regard to their publication records? Also desired might be measures of the quality of the students applying for admittance to a graduate program (e.g., Graduate Record Examination scores, undergraduate grade point averages). If reliable data of this sort were made available, they might provide a useful index of the reputational standings of programs, from the perspective of graduate students.

A number of alternative measures relevant to the quality of program faculty were considered by the committee but not included in the assessment because of the associated difficulties and costs of compiling the necessary data. For example, what fraction of the program faculty were invited to present papers at national meetings? What fraction had been elected to prestigious organizations/groups in their field? What fraction had received senior fellowships and other awards of distinction? In addition, it would be highly desirable to supplement the data presented on NSF, NIH, and ADAMHA research grant awards (measure 13) with data on awards from other federal agencies (e.g., Department of Defense, Department of Energy, National Aeronautics and Space Administration) as well as from major private foundations.

As described in the preceding pages, the committee was able to make several changes in the survey design and procedures, but further improvements could be made. Of highest priority in this regard is the expansion of the survey sample to include evaluators from outside the academic setting (in particular, those in government and industrial laboratories who regularly employ graduates of the programs to be evaluated). To add evaluators from these sectors would require a major effort in identifying the survey population from which a sample could be selected. Although such an effort is likely to involve considerable costs in both time and financial resources, the committee believes that the addition of evaluators from the government and industrial settings would be of value in providing a different perspective to the reputational assessment and that comparisons between the ratings supplied by academic and nonacademic evaluators would be of particular interest.

# Minority Statement

The inclusion of several different and independent possible measures reflecting the quality of graduate education in this report seems to us a substantial addition and a significant improvement to previous such studies.  However, we are concerned with the possibility that there are perhaps too many measures, some of which have little or no bearing on the objectives of the present study.  In particular, measures 06 and 07 (on the employment plans of graduates) are not informative, have little or nothing to do with the quality of the program, and yield numbers that are not very dependable.  Both measures come from data in the NRC's Survey of Earned Doctorates. Measure 06, the fraction of FY1975-79 program graduates with definite employment or study plans at time of doctorate, is vague because the "time of doctorate" may vary considerably from the time of year when, say, academic appointments are offered--and this in turn can vary substantially among institutions.  This measure may be associated with the prosperity of the program, but its connection with quality is tenuous.  Measure 07, the fraction of FY1975-79 program graduates planning to take positions in Ph.D.-granting universities, is even more nebulous.  <u>What</u> is meant by "planning"?  How firm are those plans?  (We can't know; all there is is a check somewhere on a questionnaire.)  What about the variation in quality among <u>different</u> Ph.D.-granting universities?  It can be considerable, and such considerable differences are precisely those that the whole study is attempting to measure.  Such data obscure the differences.  Further, measure 07 betrays the inherent bias of the present study and previous ones in that the "program graduates planning to take positions in Ph.D.-granting universities" is tacitly offered as a measure of the "goodness" of the program.  In the late 1970's and 1980's nothing can be farther from the truth.  The kindest evaluation of measures 06 and 07 is that they are irrelevant.

These two measures do not result from careful plans made by the committee for this study in order to find other useful new measures. Such plans were considered, but for various good reasons could not be carried out.  These two particular measures just happen to be available in the vast data collected and recorded (but not critically evaluated) over the years by the Commission on Human Resources of the

193

National Research Council.  Their inclusion in this report might be explained by bureaucratic inertia, but this inclusion adds nothing to the report.

SAUNDERS MAC LANE

C. K. N. PATEL

ERNEST S. KUH

# Appendixes

APPENDIX A

LETTER TO INSTITUTIONAL COORDINATORS

## COMMITTEE ON AN ASSESSMENT OF QUALITY-RELATED CHARACTERISTICS OF RESEARCH-DOCTORATE PROGRAMS IN THE UNITED STATES

*Established by the Conference Board of Associated Research Councils*

*Office of the Staff Director / National Research Council / 2101 Constitution Avenue, N.W. / Washington, D.C. 20418*

(202) 389-6552

December 5, 1980

Dear

We are pleased to learn that you have been designated to coordinate the efforts of your institution in assisting our committee with an assessment of the characteristics and effectiveness of research-doctorate programs in U.S. universities. A prospectus describing the goals and procedures for this study has already been distributed to university presidents and graduate deans. The cooperation of universities and their faculties is essential for the assessment to be carried out in an objective and accurate fashion.

The study is being conducted under the aegis of the Conference Board of Associated Research Councils and is housed administratively within the National Research Council. Financial support has been provided by the Andrew W. Mellon Foundation, the Ford Foundation, the National Science Foundation, and the National Institutes of Health. The study will examine more than 2,600 programs in 31 fields in the physical sciences, engineering, life sciences, social sciences, and humanities. Approximately 10,000 faculty members will be asked to evaluate programs in their own fields. In addition to the reputational evaluations by faculty, information will be compiled from national data banks on the achievements of both the faculty involved in each program and the program graduates.

The product of this study will be a series of reports with descriptive data on institutional programs in each of 31 fields to be covered. These reports will present several different measures of the quality-related characteristics of each program being evaluated. Some of the measures will be adjusted for program size. With the cooperation of your institution and that of other universities, we plan to produce these reports by late spring of 1982. At that time the detailed data that have

COMMITTEE MEMBERS

Lyle V. Jones, Co-Chairman
Gardner Lindzey, Co-Chairman
Paul A. Albrecht

Marcus Alexis
Robert M. Bock
Philip E. Converse
James H. M. Henderson
Ernest S. Kuh

Winfred P. Lehmann
Saunders Mac Lane
Nancy S. Milburn
Lincoln E. Moses
James C. Olson

Kumar Patel
Michael J. Pelczar, Jr.
Jerome B. Schneewind
Duane C. Spriestersbach
Harriet A. Zuckerman

197

been compiled on research-doctorate programs within your
institution will be made available to you for a nominal cost.
These data should prove to be quite valuable for an assessment of
the particular strengths and weaknesses of individual programs
at your institution.

For the past three months the committee has deliberated over
what fields are to be covered in the study and which programs
within each field are to be evaluated.  The financial resources
available limit us to an assessment of approximately 2,600
programs in 31 fields.  The fields to be included have been
determined on the basis of the total number of doctorates awarded
by U.S. universities during the FY1976-78 period and the feasibility of identifying and evaluating comparable programs in a
particular field.  Within each of the 31 fields, programs which
awarded more than a specified number of doctorates during the
period have been designated for inclusion in the study.

For each of the programs at your institution that are to be
evaluated, we ask that you furnish the names and ranks of all
faculty members who participate significantly in education toward
the research doctorate, along with some basic information (as
indicated) about the program itself.  A set of instructions and a
computer-printed roster (organized by field) are enclosed.  In
addition, you are given an opportunity to nominate other programs
at your institution that are not on the roster, but that you
believe have significant distinction and should be included in
our evaluation.  Any program you nominate must belong in one of
the 31 fields covered by the study.

The information supplied by your institution will be used for
two purposes.  First, a sample of the faculty members identified
with each program will be selected to evaluate research-doctorate
programs in their fields at other universities.  The selection
will be made in such a way as to ensure that all institutional
programs and faculty ranks are adequately represented in each
field category.  Secondly, a list of names of faculty and some of
the program information you supply will be provided to evaluators
selected from other institutions.  Thus, it is important that you
provide accurate and up-to-date information.  You may wish to ask
department chairmen or other appropriate persons at your institution to assist in providing the information requested.  If you do
so, we ask that your office coordinate the effort by collecting
the information on each program and sending a single package to
us in the envelope provided.
    We hope that you will be able to complete this request by
December 15.  Should you have any questions regarding our
request, please call (collect) Porter Coggeshall, the study
director, at (202)389-6552.  Thank you for your help in this
effort.

                    Sincerely,

                    Lyle V. Jones
                    Co-Chairman


                    Gardner. Lindzey
                    Co-Chairman

INSTRUCTIONS

## General Instructions

- Provided on the first page of the accompanying roster is a list of the 31 program fields to be covered in this study. Those program fields for which you are requested to furnish information have been designated with an asterisk (*).

- For every designated field there is a separate set of roster pages. Please provide all of the information requested on these pages.

- If your institution offers more than one research-doctorate program in a designated field, we ask that you copy the roster pages furnished for that field category and provide a separate set of information for each program. For example, if your university offers one doctoral program in statistics and another in biostatistics, these should be listed separately. For this purpose, programs offered by different departments (or other administative units) that are advertised as distinct programs in your catalogues would be listed separately. Do not consider different specialty areas within a department to be separate programs.

- If your institution currently does not offer a research-doctorate program in an asterisked field or if, in your judgment, a doctoral program offered fails to fit the designated field category, please so indicate on the roster pages provided for that field.

## List of Faculty Members (as of December 1, 1980)

- On each program roster please provide the names of faculty members who participate significantly in doctoral education.

- Included should be individuals who (a) are members of the regular academic faculty (typically holding the rank of assistant, associate, or full professor) and (b) regularly teach doctoral students and/or serve on doctoral committees.

- Members of the faculty who are currently on leave of absence but meet the above criteria should be included.

- Visiting faculty members should not be included.

- Emeritus or adjunct faculty members (or faculty with other comparable ranks) should also be excluded unless they currently participate significantly in doctoral education.

- Members of the faculty who participate significantly in doctoral education in more than one program should be listed on the roster for each program in which they participate.

- In many instances the list of faculty for a program may be identical to an institutional list of graduate faculty.

- Faculty names should be provided in the form in which they are most likely to be recognized by colleagues in the field. We prefer that, within each academic rank, you list faculty alphabetically by last name.

### Nomination of Faculty to Serve as Program Evaluators
### (Column 3 of Faculty Roster)

- Please check the names of at least two faculty members in each academic rank within each program who would be available and, in your opinion, well-qualified to evaluate research-doctorate programs in their field.

- A sample of evaluators will be selected from the list of faculty you provide for each program. In selecting evaluators preference will be given to those whose names you have checked. If no names are checked, a random sample will be selected from the faculty list.

### Faculty Who Do Not Hold Ph.D. Degrees From U.S. Universities
### (Column 4 of Faculty Roster)

- In order to help us match the faculty names you provide with records in the Doctorate Records File (maintained by the National Research Council), we ask that you identify those faculty members who do not hold a Ph.D. or equivalent research-doctorate from a university in the United States.

- This information will be used only for the purposes of collating records and will not be released to those who are selected to evaluate your institution's programs. Nor will this information affect in any way the selection of program evaluators from your institution's faculty.

### Nomination of Additional Programs

- We recognize the possibility that we may have omitted one or more research-doctorate programs at your institution that belong to (non-asterisked) fields listed on the first page of the roster and that you believe should be included in this study.

- The last two pages of the accompanying roster are provided for the nomination of an additional program. You are asked to provide the names of faculty and other information about each program you nominate. Should you decide to nominate more than one program, it will be necessary to make additional copies of these two pages of the roster.

- Please restrict your nominations to programs in your institution that you consider to be of uncommon distinction and that have awarded no fewer than two doctorates during the past two years.

- Only programs which fall under one of the 31 field categories listed on the first page of the accompanying roster will be considered for inclusion in the study.

PLEASE RETURN COMPLETED ROSTER IN
THE ENCLOSED ENVELOPE TO:

COMMITTEE ON AN ASSESSMENT OF
  QUALITY-RELATED CHARACTERISTICS
  OF RESEARCH-DOCTORATE PROGRAMS
NATIONAL RESEARCH COUNCIL, JH-711
2101 CONSTITUTION AVENUE, N.W.
WASHINGTON, D.C.  20418

FIELDS INCLUDED IN THE STUDY

ARTS AND HUMANITIES
* ART HISTORY
* CLASSICS
* ENGLISH LANGUAGE AND LITERATURE
* FRENCH LANGUAGE AND LITERATURE
* GERMAN LANGUAGE AND LITERATURE
  LINGUISTICS
  MUSIC
* PHILOSOPHY
* SPANISH AND PORTUGUESE LANGUAGE AND LITERATURE

BIOLOGICAL SCIENCES
* BIOCHEMISTRY
  BOTANY (INCLUDING PLANT PHYSIOLOGY, PLANT PATHOLOGY, MYCOLOGY)
* CELLULAR BIOLOGY/MOLECULAR BIOLOGY
* MICROBIOLOGY (INCLUDING IMMUNOLOGY, BACTERIOLOGY, PARASITOLOGY, VIROLOGY)
* PHYSIOLOGY (ANIMAL, HUMAN)
  ZOOLOGY

ENGINEERING
* CHEMICAL ENGINEERING
* CIVIL ENGINEERING
* ELECTRICAL ENGINEERING
* MECHANICAL ENGINEERING

PHYSICAL SCIENCES
* CHEMISTRY
* COMPUTER SCIENCES
* GEOSCIENCES (INCLUDING GEOLOGY, GEOCHEMISTRY, GEOPHYSICS, GENL EARTH SCI)
* MATHEMATICS
* PHYSICS (EXCLUDING ASTRONOMY, ASTROPHYSICS)
  STATISTICS (INCLUDING BIOSTATISTICS)

SOCIAL AND BEHAVIORAL SCIENCES
* ANTHROPOLOGY
* ECONOMICS
* HISTORY
* POLITICAL SCIENCE
* PSYCHOLOGY
* SOCIOLOGY

* DESIGNATES FIELDS FOR WHICH YOU ARE REQUESTED TO PROVIDE INFORMATION
  ON RESEARCH-DOCTORATE PROGRAMS IN YOUR INSTITUTION. (SEE INSTRUCTION
  SHEET REGARDING NOMINATION OF ADDITIONAL PROGRAMS TO BE INCLUDED IN
  THE STUDY) .

```
***********************************************
***                                - PART A  ***
***********************************************
```

PLEASE ANSWER EACH OF THE FOLLOWING QUESTIONS ABOUT THE RESEARCH-DOCTORATE PROGRAM

     IN _____

(1)   WHAT IS THE NAME OF THE DEPARTMENT (OR EQUIVALENT ACADEMIC UNIT) IN WHICH
      THIS RESEARCH-DOCTORATE PROGRAM IS OFFERED ?

          ........................................

(2)   HOW MANY PH.D.'S (OR EQUIVALENT RESEARCH-DOCTORATES) HAVE BEEN AWARDED IN THE
      PROGRAM IN EACH OF THE LAST FIVE ACADEMIC YEARS ?

          1975-76  ........

          1976-77  ........

          1977-78  ........

          1978-79  ........

          1979-80  ........

(3)   APPROXIMATELY HOW MANY FULL-TIME AND PART-TIME GRADUATE STUDENTS ENROLLED
      IN THE PROGRAM AT THE PRESENT TIME (FALL 1980) INTEND TO EARN DOCTORATES ?

          FULL-TIME STUDENTS  ........

          PART-TIME STUDENTS  ........

                  TOTAL  ........

(4)   IN APPROXIMATELY WHAT YEAR WAS THIS RESEARCH-DOCTORATE PROGRAM INITIATED ?
      (IF PROGRAM WAS DISCONTINUED AND SUBSEQUENTLY REINSTATED, PLEASE GIVE YEAR
      IT WAS REINSTATED) .

          ........

```
*********************************************************
***  _                                      - PART B  ***
*********************************************************
```

|  | (1) LIST BELOW ALL FACULTY WHO PARTICIPATE SIGNIFICANTLY IN DOCTORAL EDUCATION IN THIS PROGRAM (SEE INSTRUCTIONS SHEET). PLEASE PRINT OR TYPE NAMES IN FOLLOWING FORMAT:<br><br>EXAMPLE: MARY A. JONES<br>A. B. SMITH, JR. | (2) INDICATE THE ACADEMIC RANK OF EACH FACULTY MEMBER (PROF., ASSOC. PROF., ASST. PROF., ETC.). | (3) CHECK BELOW AT LEAST 2 FACULTY IN EACH RANK AVAILABLE AND WELL-QUALIFIED TO EVALUATE OTHER PROGRAMS (SEE INSTRUCTIONS SHEET). | (4) CHECK BELOW ANY FACULTY WHO DO NOT HOLD A PH.D. OR OTHER RESEARCH-DOCTORATE FROM A UNIVERSITY IN THE U.S. (SEE INSTRUCTIONS SHEET). |
|---|---|---|---|---|
| 01 | | .. | ( ) | ( ) |
| 02 | | .. | ( ) | ( ) |
| 03 | | .. | ( ) | ( ) |
| 04 | | .. | ( ) | ( ) |
| 05 | | .. | ( ) | ( ) |
| 06 | | .. | ( ) | ( ) |
| 07 | | .. | ( ) | ( ) |
| 08 | | .. | ( ) | ( ) |
| 09 | | .. | ( ) | ( ) |
| 10 | | .. | ( ) | ( ) |
| 11 | | .. | ( ) | ( ) |
| 12 | | .. | ( ) | ( ) |
| 13 | | .. | ( ) | ( ) |
| 14 | | .. | ( ) | ( ) |
| 15 | | .. | ( ) | ( ) |
| 16 | | .. | ( ) | ( ) |
| 17 | | .. | ( ) | ( ) |
| 18 | | .. | ( ) | ( ) |
| 19 | | .. | ( ) | ( ) |
| 20 | | .. | ( ) | ( ) |

( PART B CONTINUED ON NEXT PAGE )

APPENDIX B

SURVEY OF EARNED DOCTORATES

(Conducted by the National Research Council under the sponsorship
of the National Science Foundation, the Department of Education,
the National Institutes of Health, and the National Endowment for
the Humanities.)

This annual survey of new recipients of Ph.D. or equivalent
research doctorates in all fields of learning contains information
describing their demographic characteristics, educational background,
graduate training, and postgraduation plans.  The source file includes
nearly complete data from all 1958-81 doctorate recipients and partial
information for all 1920-57 doctoral graduates.

## SURVEY OF EARNED DOCTORATES

NSF Form 558 1977
OMB No. 99-R0290
Approval Expires June 30, 1979

This form is to be returned to the GRADUATE DEAN, for forwarding to ...................... Board on Human-Resource Data and Analyses
Commission on Human Resources
National Research Council

*Please print or type.*

2101 Constitution Avenue, Washington, D. C. 20418

A.  Name in full: ...................................................................................................................... (9-30)
(Last Name)                              (First Name)                        (Middle Name)

Cross Reference: Maiden name or former name legally changed ................................................... (31)

B.  Permanent address through which you could always be reached: (Care of, if applicable) ...............................
............................................................................................................................................
(Number)                              (Street)                              (City)

............................................................................................................................................
(State)                         (Zip Code)                    (Or Country if not U.S.)

C.  U.S. Social Security Number: __ __ __ – __ __ – __ __ __ __                                    (32-40)

D.  Date of birth: ..............................    Place of birth: ...............................................................
(41-45)     (Month)   (Day)   (Year)    (46-47)           (State)             (Or Country if not U.S.)

E.  Sex:        1 ☐ Male        2 ☐ Female                                                      (48)

F.  Marital status:    1 ☐ Married    2 ☐ Not married (including widowed, divorced)              (49)

G.  Citizenship:    0 ☐ U.S. native        2 ☐ Non U.S., Immigrant (Permanent Resident)
1 ☐ U.S. naturalized    3 ☐ Non-U.S., Non-Immigrant (Temporary Resident)    (50)
If Non-U.S., indicate country of present citizenship ................................................    (51-52)

H.  Racial or ethnic group: (Check all that apply.) *A person having origins in —*
0 ☐ American Indian or Alaskan Native .....any of the original peoples of North America, and who maintain cultural identification
through tribal affiliation or community recognition.
1 ☐ Asian or Pacific Islander ................any of the original peoples of the Far East, Southeast Asia, the Indian Subcontinent, or
the Pacific Islands. This area includes, for example, China, Japan, Korea, the Philippine
Islands, and Samoa.
2 ☐ Black, not of Hispanic Origin ..........any of the black racial groups of Africa.
3 ☐ White, not of Hispanic Origin ..........any of the original peoples of Europe, North Africa, or the Middle East.
4 ☐ Hispanic ...........................Mexican, Puerto Rican, Central or South American, or other Spanish culture or origins,
regardless of race.                                                    (53-55)

I.  Number of dependents: Do not include yourself. (Dependent = someone receiving at least one half of his or her support from you) ........(56)

J.  U.S. veteran status:    0 ☐ Veteran        1 ☐ On active duty        2 ☐ Non-veteran or not applicable    (57)

**EDUCATION**

K.  High school last attended: ........................................................................................ (58-59)
(School Name)              (City)              (State)
Year of graduation from high school: .................                                          (60-61)

L.  List in the table below all collegiate and graduate institutions you have attended including 2-year colleges. List chronologically, and include your doctoral institution as the last entry.

| Institution Name | Location | Years Attended | | Major Field | | Minor Field | Degree (if any) | | |
|---|---|---|---|---|---|---|---|---|---|
| | | From | To | Use Specialties List | | Number | Title of Degree | Granted | |
| | | | | Name | Number | | | Mo. | Yr. |
| | | | | | | | | | |
| | | | | | | | | | |
| | | | | | | | | | |
| | | | | | | | | | |
| | | | | | | | | | |
| | | | | | | | | | |

M.  Enter below the title of your doctoral dissertation and the most appropriate classification number and field. If a project report or a musical
or literary composition (not a dissertation) is a degree requirement, please check box. ☐    (44)

Title ........................................    Classify using Specialties List

........................................    Number        Name of field

........................................    ..........................................

N.  Name the department (or interdisciplinary committee, center, institute, etc.) and school or college of the university

which supervised your doctoral program: ........................................................................
(Department/Institute/Committee/Program)              (School)

O.  Name of your dissertation adviser: ...............................................................
(Last Name)              (First Name)    (Middle Initial)

*continued on next page*

## SURVEY OF EARNED DOCTORATES, Cont.

P. Please enter a "1" beside your primary source of support during graduate study. Enter a "2" beside your secondary source of support during graduate study. Check all other sources from which support was received.

| | | | |
|---|---|---|---|
| 58 ___ NSF Fellowship | 66 ___ GI Bill | 72 ___ Research Assistantship | 76 ___ Spouse's earnings |
| 59 ___ NSF Traineeship | 67 ___ Other Federal support | 73 ___ Educational fund of | 77 ___ Family contribu- |
| 60 ___ NIH Fellowship | (specify) ................. | industrial or | tions |
| 61 ___ NIH Traineeship | 68 ___ Woodrow Wilson Fellowship | business firm | 78 ___ Loans (NDSL |
| 62 ___ NDEA Fellowship | 69 ___ Other U.S. national fellowship | 74 ___ Other institutional | direct) |
| 63 ___ Other HEW | | funds (specify) | 79 ___ Other loans |
| 64 ___ AEC/ERDA | (specify) ........... ...... | ...................... | 80 ___ Other (specify) |
| Fellowship | 70 ___ University Fellowship | 75 ___ Own earnings | ..................... |
| 65 ___ NASA Traineeship | 71 ___ Teaching Assistantship | | |

Q. Please check the space which most fully describes your status during the year immediately preceding the doctorate.

0 ☐ Held fellowship
1 ☐ Held assistantship
2 ☐ Held own research grant
3 ☐ Not employed
4 ☐ Part-time employed

Full-time
Employed in:
(Other than
0, 1, 2)

5 ☐ College or university, teaching
6 ☐ College or university, non-teaching
7 ☐ Elem. or sec. school, teaching
8 ☐ Elem. or sec. school, non-teaching
9 ☐ Industry or business
(11) ☐ Other (specify) ...................................

(12) ☐ Any other (specify) .............................. **(9)**

R. How many years (full-time equivalent basis) of professional work experience did you have prior to the doctorate? (include assistantships as professional experience) ...................................................................................... (10-11)

## POSTGRADUATION PLANS

S. How well defined are your postgraduation plans?
0 ☐ Have signed contract or made definite commitment
1 ☐ Am negotiating with a specific organization, or more than one
2 ☐ Am seeking appointment but have no specific prospects

3 ☐ Other (specify) ............................... **(12)**

T. What are your immediate postgraduation plans?
0 ☐ Postdoctoral fellowship?
1 ☐ Postdoctoral research associateship?
2 ☐ Traineeship?
3 ☐ Other study (specify) ......................
4 ☐ Employment (other than 0, 1, 2, 3)
5 ☐ Military service?
6 ☐ Other (specify)....................**(13)**

} Go to Item "U"
} Go to Item "V"

U. If you plan to be on a postdoctoral fellowship, associateship, traineeship or other study

What will be the field of your postdoctoral study?
Classify using Specialties List.
Number                           Field

............    ........................(14-16)
What will be the primary source of support?
0 ☐ U.S. Government
1 ☐ College or university
2 ☐ Private foundation
3 ☐ Nonprofit, other than private foundation
4 ☐ Other (specify)
.............................................. (17)
6 ☐ Unknown
Go to Item "W"

V. If you plan to be employed, enter military service, or other —
What will be the type of employer?

0 ☐ 4-year college or university other than medical school
1 ☐ Medical school
2 ☐ Jr. or community college
3 ☐ Elem. or sec. school
4 ☐ Foreign government
5 ☐ U.S. Federal government
6 ☐ U.S. state government
7 ☐ U.S. local government
8 ☐ Nonprofit organization
9 ☐ Industry or business
(11) ☐ Self-employed
(12) ☐ Other (specify) ........................... **(18)**

Indicate *primary* work activity with "1" in appropriate box; *secondary* work activity (if any) with "2" in appropriate box.
0 ☐ Research and development
1 ☐ Teaching
2 ☐ Administration
3 ☐ Professional services to individuals
5 ☐ Other (specify)........................... **(19-20)**

In what field will you be working?
Please enter number from Specialties List .......... **(21-23)**

Go to Item "W"

W. What is the name and address of the organization with which you will be associated?

.................................................................................
(Name of Organization)

.................................................................................
(Street)            (City, State)     (Or Country if not U.S.)     **(24-29)**

## BACKGROUND INFORMATION

X. Please indicate, by circling the highest grade attained, the education of

| your father: | none | 1 2 3 4 5 6 7 8 | 9 10 11 12 | 1 2 3 4 | MA, MD PhD | Postdoctoral | (30) |
|---|---|---|---|---|---|---|---|
| | | Elementary school | High school | College | Graduate | | |
| your mother | none | 1 2 3 4 5 6 7 8 | 9 10 11 12 | 1 2 3 4 | MA, MD PhD | Postdoctoral | (31) |
| | 0 | 1          2          3 | 4          5 | 6          7 | 8          9 | (11) | |

Signature ................................................... Date completed ..................... 
(32-34)

# APPENDIX C

## LETTER TO EVALUATORS

COMMITTEE ON AN ASSESSMENT OF QUALITY-RELATED CHARACTERISTICS
OF RESEARCH-DOCTORATE PROGRAMS IN THE UNITED STATES

*Established by the Conference Board of Associated Research Councils*

*Office of the Staff Director  /  National Research Council  /  2101 Constitution Avenue, N.W.  /  Washington, D.C. 20418*

April 14, 1981

Dear

    As you may already know, our committee has undertaken an assessment
of research-doctorate programs in U.S. universities.  The study is examining approximately 2,650 programs in 31 fields in the arts and humanities,
biological sciences, engineering, physical and mathematical sciences, and
social sciences.  A study prospectus is provided on the reverse of this
page.  You have been selected from a faculty list furnished by your institution to evaluate programs offering research-doctorates in the field of
Chemistry.

    On the first page of the attached form is a list of the 145 programs
that are being evaluated in this field.  These programs produce more than 90
percent of the doctorate recipients in the field.  In order to keep the task
manageable, you are being asked to consider a randomly selected subset of
50 of these programs. These are designated with an asterisk in the list on
the next page and are presented in random sequence on the evaluation sheets
that follow. Please read the accompanying instructions carefully before
attempting your evaluations.

    We ask that you complete the attached survey form and return it in the
enclosed envelope within the next three weeks.  The evaluations you and
your colleagues render will constitute an important component of this study.
Your prompt attention to this request will be very much appreciated by our
committee.

                    Sincerely,

                    *[signature: Gardner Lindzey]*

                    Gardner Lindzey

                    *[signature: Lyle V. Jones]*

                    Lyle Jones
                    For the Study Committee

Enclosures

COMMITTEE MEMBERS

Lyle V. Jones, Co-Chairman
Gardner Lindzey, Co-Chairman
Paul A. Albrecht

Marcus Alexis
Robert M. Bock
Philip E. Converse
James H. M. Henderson
Ernest S. Kuh

Winfred P. Lehmann
Saunders Mac Lane
Nancy S. Milburn
Lincoln E. Moses
James C. Olson

Kumar Patel
Michael J. Pelczar, Jr.
Jerome B. Schneewind
Duane C. Spriestersbach
Harriet A. Zuckerman

**RESEARCH-DOCTORATE PROGRAMS IN THE FIELD OF CHEMISTRY**
(\* DESIGNATES THE PROGRAMS WHICH YOU ARE ASKED TO EVALUATE ON THE FOLLOWING PAGES.)

INSTITUTION - DEPARTMENT/ACADEMIC UNIT

\* UNIVERSITY OF AKRON - CHEMISTRY
\* UNIVERSITY OF AKRON - POLYMER SCIENCE
  UNIVERSITY OF ALABAMA, TUSCALOOSA - CHEMISTRY
  AMERICAN UNIVERSITY - CHEMISTRY
\* ARIZONA STATE UNIVERSITY, TEMPE - CHEMISTRY
  UNIVERSITY OF ARIZONA, TUCSON - CHEMISTRY
\* UNIVERSITY OF ARKANSAS, FAYETTEVILLE - CHEMISTRY
\* ATLANTA UNIVERSITY - CHEMISTRY
  AUBURN UNIVERSITY - CHEMISTRY
  BAYLOR UNIVERSITY, WACO - CHEMISTRY
  BOSTON COLLEGE - CHEMISTRY
  BOSTON UNIVERSITY - CHEMISTRY
  BRANDEIS UNIVERSITY - CHEMISTRY
  BRIGHAM YOUNG UNIVERSITY - CHEMISTRY
  BROWN UNIVERSITY - CHEMISTRY
  BRYN MAWR COLLEGE - CHEMISTRY
\* CALIFORNIA INSTITUTE OF TECHNOLOGY - CHEMISTRY AND CHEMICAL ENGINEERING
\* UNIVERSITY OF CALIFORNIA, BERKELEY - CHEMISTRY
  UNIVERSITY OF CALIFORNIA, DAVIS - CHEMISTRY
  UNIVERSITY OF CALIFORNIA, IRVINE - CHEMISTRY
\* UNIVERSITY OF CALIFORNIA, LOS ANGELES - CHEMISTRY
  UNIVERSITY OF CALIFORNIA, RIVERSIDE - CHEMISTRY
  UNIVERSITY OF CALIFORNIA, SAN DIEGO - CHEMISTRY
  UNIVERSITY OF CALIFORNIA, SANTA BARBARA - CHEMISTRY
  UNIVERSITY OF CALIFORNIA, SANTA CRUZ - CHEMISTRY
  CARNEGIE-MELLON UNIVERSITY - CHEMISTRY
\* CASE WESTERN RESERVE UNIVERSITY - CHEMISTRY
  CATHOLIC UNIVERSITY OF AMERICA - CHEMISTRY
  UNIVERSITY OF CHICAGO - CHEMISTRY
  UNIVERSITY OF CINCINNATI - CHEMISTRY
  CUNY, THE GRADUATE SCHOOL - CHEMISTRY
  CLARK UNIVERSITY - CHEMISTRY
\* CLARKSON COLLEGE OF TECHNOLOGY - CHEMISTRY
  CLEMSON UNIVERSITY - CHEMISTRY AND GEOLOGY
  COLORADO STATE UNIVERSITY, FT COLLINS - CHEMISTRY
\* UNIVERSITY OF COLORADO, BOULDER - CHEMISTRY
\* COLUMBIA UNIV-GRAD SCHOOL OF ARTS & SCI - CHEMISTRY
\* UNIVERSITY OF CONNECTICUT, STORRS - CHEMISTRY
  CORNELL UNIVERSITY, ITHACA - CHEMISTRY
  UNIVERSITY OF DELAWARE, NEWARK - CHEMISTRY
\* UNIVERSITY OF DENVER - CHEMISTRY
  DREXEL UNIVERSITY - CHEMISTRY
\* DUKE UNIVERSITY - CHEMISTRY
  EMORY UNIVERSITY - CHEMISTRY
  GEORGETOWN UNIVERSITY - CHEMISTRY
  GEORGIA INSTITUTE OF TECHNOLOGY - CHEMISTRY
  UNIVERSITY OF GEORGIA, ATHENS - CHEMISTRY
\* HARVARD UNIVERSITY - CHEMISTRY/CHEMICAL PHYSICS
\* UNIVERSITY OF HAWAII - CHEMISTRY
  UNIVERSITY OF HOUSTON - CHEMISTRY
  HOWARD UNIVERSITY - CHEMISTRY
  UNIVERSITY OF IDAHO, MOSCOW - CHEMISTRY
  ILLINOIS INSTITUTE OF TECHNOLOGY - CHEMISTRY
\* UNIV OF ILLINOIS AT URBANA-CHAMPAIGN - CHEMISTRY
  UNIVERSITY OF ILLINOIS, CHICAGO CIRCLE - CHEMISTRY
\* INDIANA UNIVERSITY, BLOOMINGTON - CHEMISTRY
  INST OF PAPER CHEMISTRY (APPLETON, WI) - CHEMISTRY
  IOWA STATE UNIVERSITY, AMES - CHEMISTRY
\* UNIVERSITY OF IOWA, IOWA CITY - CHEMISTRY
  JOHNS HOPKINS UNIVERSITY - CHEMISTRY
  KANSAS STATE UNIVERSITY, MANHATTAN - CHEMISTRY
\* UNIVERSITY OF KANSAS - CHEMISTRY
  UNIVERSITY OF KANSAS - PHARMACEUTICAL CHEMISTRY
  KENT STATE UNIVERSITY - CHEMISTRY
  UNIVERSITY OF KENTUCKY - CHEMISTRY
  LOUISIANA STATE UNIVERSITY, BATON ROUGE - CHEMISTRY
  UNIVERSITY OF NEW ORLEANS - CHEMISTRY
  UNIVERSITY OF LOUISVILLE - CHEMISTRY
  LOYOLA UNIVERSITY OF CHICAGO - CHEMISTRY
  UNIVERSITY OF MARYLAND, COLLEGE PARK - CHEMISTRY
\* MASSACHUSETTS INSTITUTE OF TECHNOLOGY - CHEMISTRY
\* UNIVERSITY OF MASSACHUSETTS, AMHERST - CHEMISTRY
\* UNIVERSITY OF MIAMI (FLORIDA) - CHEMISTRY

```
  MICHIGAN STATE UNIVERSITY, EAST LANSING - CHEMISTRY
  UNIVERSITY OF MICHIGAN, ANN ARBOR - CHEMISTRY
  UNIVERSITY OF MINNESOTA - CHEMISTRY
* UNIVERSITY OF MISSOURI, COLUMBIA - CHEMISTRY
  UNIVERSITY OF MISSOURI, KANSAS CITY - CHEMISTRY
  UNIVERSITY OF MISSOURI, ROLLA - CHEMISTRY
* MONTANA STATE UNIVERSITY, BOZEMAN - CHEMISTRY
  UNIVERSITY OF NEBRASKA, LINCOLN - CHEMISTRY
* UNIVERSITY OF NEW HAMPSHIRE - CHEMISTRY
  UNIVERSITY OF NEW MEXICO, ALBUQUERQUE - CHEMISTRY
  NEW YORK UNIVERSITY - CHEMISTRY
  UNIVERSITY OF NORTH CAROLINA, CHAPEL HILL - CHEMISTRY
* NORTH CAROLINA STATE UNIVERSITY, RALEIGH - CHEMISTRY
* NORTH DAKOTA STATE UNIVERSITY, FARGO - CHEMISTRY/POLYMERS COATINGS
  UNIVERSITY OF NORTH DAKOTA, GRAND FORKS - CHEMISTRY
  NORTH TEXAS STATE UNIVERSITY, DENTON - CHEMISTRY
  NORTHEASTERN UNIVERSITY - CHEMISTRY
  NORTHERN ILLINOIS UNIVERSITY, DE KALB - CHEMISTRY
  NORTHWESTERN UNIVERSITY - CHEMISTRY
* UNIVERSITY OF NOTRE DAME - CHEMISTRY
* OHIO STATE UNIVERSITY - CHEMISTRY
  OHIO UNIVERSITY - CHEMISTRY
* OKLAHOMA STATE UNIVERSITY, STILLWATER - CHEMISTRY
  UNIVERSITY OF OKLAHOMA - CHEMISTRY
  UNIVERSITY OF OREGON, EUGENE - CHEMISTRY
  OREGON STATE UNIVERSITY, COVALLIS - CHEMISTRY
* PENNSYLVANIA STATE UNIVERSITY - CHEMISTRY
  UNIVERSITY OF PENNSYLVANIA - CHEMISTRY
  UNIVERSITY OF PITTSBURGH - CHEMISTRY
* POLYTECHNIC INSTITUTE OF NEW YORK - CHEMISTRY
  PRINCETON UNIVERSITY - CHEMISTRY
  PURDUE UNIVERSITY, WEST LAFAYETTE - CHEMISTRY
  RENSSELAER POLYTECHNIC INSTITUTE - CHEMISTRY
  UNIVERSITY OF RHODE ISLAND - CHEMISTRY
  RICE UNIVERSITY - CHEMISTRY
  UNIVERSITY OF ROCHESTER - CHEMISTRY
* RUTGERS UNIVERSITY, NEW BRUNSWICK - CHEMISTRY
  RUTGERS UNIVERSITY, NEWARK - CHEMISTRY
  UNIVERSITY OF SOUTH CAROLINA, COLUMBIA - CHEMISTRY
  UNIVERSITY OF SOUTHERN CALIFORNIA - CHEMISTRY
* SOUTHERN ILLINOIS UNIVERSITY, CARBONDALE - CHEMISTRY AND BIOCHEMISTRY
* UNIV OF SOUTHERN MISSISSIPPI, HATTIESBURG - CHEMISTRY
  STANFORD UNIVERSITY - CHEMISTRY
  UNIVERSITY OF FLORIDA, GAINESVILLE - CHEMISTRY
* FLORIDA STATE UNIVERSITY, TALLAHASSEE - CHEMISTRY
  UNIVERSITY OF SOUTH FLORIDA, TAMPA - CHEMISTRY
  SUNY AT BINGHAMTON - CHEMISTRY
* SUNY AT BUFFALO - CHEMISTRY
  SUNY AT STONY BROOK - CHEMISTRY
* SYRACUSE UNIVERSITY - CHEMISTRY
* SUNY, COL OF ENVIR SCI & FORESTRY (SYRACUSE) - CHEMISTRY
* TEMPLE UNIVERSITY - CHEMISTRY
* UNIVERSITY OF TENNESSEE, KNOXVILLE - CHEMISTRY
  TEXAS A&M UNIVERSITY - CHEMISTRY
  TEXAS TECH UNIVERSITY, LUBBOCK - CHEMISTRY
  UNIVERSITY OF TEXAS, AUSTIN - CHEMISTRY
* TULANE UNIVERSITY - CHEMISTRY
  UNIVERSITY OF UTAH, SALT LAKE CITY - CHEMISTRY
  UTAH STATE UNIVERSITY, LOGAN - CHEMISTRY AND BIOCHEMISTRY
* VANDERBILT UNIVERSITY - CHEMISTRY
  UNIVERSITY OF VERMONT - CHEMISTRY
* VIRGINIA POLYTECHNIC INSTITUTE AND STATE UNIV - CHEMISTRY
* UNIVERSITY OF VIRGINIA - CHEMISTRY
* WASHINGTON STATE UNIVERSITY, PULLMAN - CHEMISTRY
  WASHINGTON UNIVERSITY (ST LOUIS) - CHEMISTRY
  UNIVERSITY OF WASHINGTON, SEATTLE - CHEMISTRY
  WAYNE STATE UNIVERSITY - CHEMISTRY
* WESTERN MICHIGAN UNIVERSITY - CHEMISTRY
  UNIVERSITY OF WISCONSIN, MADISON - CHEMISTRY
  UNIVERSITY OF WISCONSIN, MILWAUKEE - CHEMISTRY
  UNIVERSITY OF WYOMING - CHEMISTRY
* YALE UNIVERSITY - CHEMISTRY
```

## INSTRUCTIONS

At the top of the next page please provide the information requested on the highest degree you hold and your current field of specialization. You may be assured that all information you furnish on the survey form is to be used for purposes of statistical description only and that the confidentiality of your responses will be protected.

On the pages that follow you are asked to judge 50 programs (presented in random sequence) that offer the research-doctorate. Each program is to be evaluated in terms of: (1) scholarly quality of program faculty; (2) effectiveness of program in educating research scholars/scientists; and (3) change in program quality in the last five years (see below). Although the assessment is limited to these factors, our committee recognizes that other factors are relevant to the quality of doctoral programs, and that graduate programs serve important purposes in addition to that of educating doctoral candidates.

A list of the faculty members signficantly involved in each program, the name of the academic unit in which the program is offered, and the number of doctorates awarded in that program during the last five years have been printed on the survey form (whenever available). Although this information has been furnished to us by the institution and is believed to be accurate, it has not been verified by our study committee and may have a few omissions, misspellings, or other errors.

Before marking your responses on the survey form, you may find it helpful to look over the full set of programs you are being asked to evaluate. In making your judgments about each program, please keep in mind the following instructions:

(1) <u>Scholarly Quality of Program Faculty.</u> Check the box next to the term that most closely corresponds to your judgment of the quality of faculty in the research-doctorate program described. Consider only the scholarly competence and achievements of the faculty. It is suggested that no more than five programs be designated "distinguished."

(2) <u>Effectiveness of Program in Educating Research Scholars/Scientists.</u> Check the box next to the term that most closely corresponds to your judgment of the doctoral program's effectiveness in educating research scholars/scientists. Consider the accessibility of the faculty, the curricula, the instructional and research facilities, the quality of graduate students, the performance of the graduates, and other factors that contribute to the effectiveness of the research-doctorate program.

(3) <u>Change in Program Quality in Last Five Years.</u> Check the box next to the term that most closely corresponds to your estimate of the change that has taken place in the research-doctorate program in the last five years. Consider both the scholarly quality of the program faculty and the effectiveness of the program in educating research scholars/scientists. Compare the quality of the program today with its quality five years ago--<u>not</u> the change in the program's relative standing among other programs in the field.

In assessing each of these factors, mark the category "Don't know well enough to evaluate" if you are unfamiliar with that aspect of the program. It is quite possible that for some programs you may be knowledgeable about the scholarly quality of the faculty, but not about the effectiveness of the program or change in program quality.

For each of the programs identified, you are also asked to indicate the extent to which you are familiar with the work of members of the program faculty. For example, if you recognize only a very small fraction of the faculty, you should mark the category "Little or no familiarity."

Please be certain that you have provided a set of responses for each of the programs identified on the following pages. The fully completed survey form should be returned in the enclosed envelope to:

Committee on an Assessment of Quality-Related
Characteristics of Research-Doctorate Programs
National Research Council, JH-638
2101 Constitution Avenue, N.W.
Washington, D.C. 20418

Our committee will be most appreciative of your thoughtful assessment of these research-doctorate programs. We welcome any comments you may wish to append to the completed survey form.

PLEASE PROVIDE THE FOLLOWING INFORMATION:                    FORM NO. SAMP-66

HIGHEST DEGREE YOU HOLD:  ( ) PH.D.    ( ) OTHER (PLEASE SPECIFY): _____

YEAR OF HIGHEST DEGREE: _____

INSTITUTION OF HIGHEST DEGREE: _____

YOUR CURRENT FIELD OF SPECIALIZATION (CHECK ONLY ONE):

        A. ( ) ANALYTICAL CHEMISTRY

        B. ( ) BIOCHEMISTRY

        C. ( ) INORGANIC CHEMISTRY

        D. ( ) ORGANIC CHEMISTRY

        E. ( ) PHARMACEUTICAL CHEMISTRY

        F. ( ) PHYSICAL CHEMISTRY

        G. ( ) POLYMER CHEMISTRY

        H. ( ) THEORETICAL CHEMISTRY

        I. ( ) CHEMISTRY, GENERAL

        J. ( ) OTHER (PLEASE SPECIFY):

        _____

---

INSTITUTION:              UNIVERSITY OF ARKANSAS, FAYETTEVILLE        FORM NO. SAMP-01
DEPARTMENT/ACADEMIC UNIT:                      CHEMISTRY
TOTAL DOCTORATES AWARDED 1976-80:  32

PROFESSORS:  Robbin C. ANDERSON, George D. BLYHOLDER, A. Wallace CORDES, Arthur J. FRY,
  James F. HINTON, Lester C. HOWICK, Dale A. JOHNSON, P. K. KURODA, Walter L. MEYER,
  Francis S. MILLETT, Lothar SCHAFER, Samuel SIEGEL, Leslie B. SIMS, John A. THOMA

ASSOCIATE PROFESSORS:  Collis R. GEREN

ASSISTANT PROFESSORS:  Neil T. ALLISON, Danny J. DAVIS, Bill DURHAM, Robert B. GREEN,
  Roger E. KOEPPE, David W. PAUL, Norbert J. PIENTA

SCHOLARLY QUALITY OF PROGRAM FACULTY

1. ( )  DISTINGUISHED
2. ( )  STRONG
3. ( )  GOOD
4. ( )  ADEQUATE
5. ( )  MARGINAL
6. ( )  NOT SUFFICIENT FOR DOCTORAL EDUCATION

0. ( )  DON'T KNOW WELL ENOUGH TO EVALUATE

FAMILIARITY WITH WORK OF PROGRAM FACULTY

1. ( )  CONSIDERABLE FAMILIARITY
2. ( )  SOME FAMILIARITY
3. ( )  LITTLE OR NO FAMILIARITY

EFFECTIVENESS OF PROGRAM IN EDUCATING RESEARCH
SCHOLARS/SCIENTISTS

1. ( )  EXTREMELY EFFECTIVE
2. ( )  REASONABLY EFFECTIVE
3. ( )  MINIMALLY EFFECTIVE
4. ( )  NOT EFFECTIVE

0. ( )  DON'T KNOW WELL ENOUGH TO EVALUATE

CHANGE IN PROGRAM QUALITY IN LAST FIVE YEARS

1. ( )  BETTER THAN FIVE YEARS AGO
2. ( )  LITTLE OR NO CHANGE IN LAST FIVE YEAR
3. ( )  POORER THAN FIVE YEARS AGO

0. ( )  DON'T KNOW WELL ENOUGH TO EVALUATE

INSTITUTION: SUNY, COL OF ENVIR SCI & FORESTRY (SYRACUSE)   FORM NO. SAMP-02
DEPARTMENT/ACADEMIC UNIT:                 CHEMISTRY
TOTAL DOCTORATES AWARDED 1976-80:  21

PROFESSORS:  Robert T. LALONDE, John A. MEYER, Anatole SARKO, Conrad SCHUERCH,
  Robert M. SILVERSTEIN, Johannes SMID, Kenneth J. SMITH Jr, Stuart W. TANENBAUM, Tore E. TIMELL

ASSOCIATE PROFESSORS:  Paul M. CALUWE, Wilbur M. CAMPBELL, Michael FLASHNER, Gideon LEVIN

ASSISTANT PROFESSORS:  David L. JOHNSON

SCHOLARLY QUALITY OF PROGRAM FACULTY

1. ( )  DISTINGUISHED
2. ( )  STRONG
3. ( )  GOOD
4. ( )  ADEQUATE
5. ( )  MARGINAL
6. ( )  NOT SUFFICIENT FOR DOCTORAL EDUCATION

0. ( )  DON'T KNOW WELL ENOUGH TO EVALUATE

FAMILIARITY WITH WORK OF PROGRAM FACULTY

1. ( )  CONSIDERABLE FAMILIARITY
2. ( )  SOME FAMILIARITY
3. ( )  LITTLE OR NO FAMILIARITY

EFFECTIVENESS OF PROGRAM IN EDUCATING RESEARCH
SCHOLARS/SCIENTISTS

1. ( )  EXTREMELY EFFECTIVE
2. ( )  REASONABLY EFFECTIVE
3. ( )  MINIMALLY EFFECTIVE
4. ( )  NOT EFFECTIVE

0. ( )  DON'T KNOW WELL ENOUGH TO EVALUATE

CHANGE IN PROGRAM QUALITY IN LAST FIVE YEARS

1. ( )  BETTER THAN FIVE YEARS AGO
2. ( )  LITTLE OR NO CHANGE IN LAST FIVE YEAR
3. ( )  POORER THAN FIVE YEARS AGO

0. ( )  DON'T KNOW WELL ENOUGH TO EVALUATE

---

INSTITUTION: VIRGINIA POLYTECHNIC INSTITUTE AND STATE UNIV   FORM NO. SAMP-03
DEPARTMENT/ACADEMIC UNIT:                 CHEMISTRY
TOTAL DOCTORATES AWARDED 1976-80:  55

PROFESSORS:  H. J. ACHE, L. K. BRICE Jr, A. F. CLIFFORD, R. F. DESSY, J. G. DILLARD, J. D. GRAYBEAL
  M. HUDLICKY, D. G. KINGSTON, J. G. MASON, J. E. MCGRATH, H. M. MCNAIR, M. A. OGLIARUSO,
  J. C. SCHUG, L. T. TAYLOR, J. P. WIGHTMAN, J. F. WOLFE

ASSOCIATE PROFESSORS:  H. M. BELL, H. C. DORN, P. E. FIELD, G. SANZONE, H. D. SMITH, T. C. WARD

ASSISTANT PROFESSORS:  B. R. BARTSCHMID, H. O. FINKLEA, B. E. HANSON, P. J. HARRIS, R. A. HOLTON,
  J. W. VIERS

OTHER STAFF:  D. G. LARSEN, F. M. VANDAMME

SCHOLARLY QUALITY OF PROGRAM FACULTY

1. ( )  DISTINGUISHED
2. ( )  STRONG
3. ( )  GOOD
4. ( )  ADEQUATE
5. ( )  MARGINAL
6. ( )  NOT SUFFICIENT FOR DOCTORAL EDUCATION

0. ( )  DON'T KNOW WELL ENOUGH TO EVALUATE

FAMILIARITY WITH WORK OF PROGRAM FACULTY

1. ( )  CONSIDERABLE FAMILIARITY
2. ( )  SOME FAMILIARITY
3. ( )  LITTLE OR NO FAMILIARITY

EFFECTIVENESS OF PROGRAM IN EDUCATING RESEARCH
SCHOLARS/SCIENTISTS

1. ( )  EXTREMELY EFFECTIVE
2. ( )  REASONABLY EFFECTIVE
3. ( )  MINIMALLY EFFECTIVE
4. ( )  NOT EFFECTIVE

0. ( )  DON'T KNOW WELL ENOUGH TO EVALUATE

CHANGE IN PROGRAM QUALITY IN LAST FIVE YEARS

1. ( )  BETTER THAN FIVE YEARS AGO
2. ( )  LITTLE OR NO CHANGE IN LAST FIVE YEAR
3. ( )  POORER THAN FIVE YEARS AGO

0. ( )  DON'T KNOW WELL ENOUGH TO EVALUATE

APPENDIX D

THE ARL LIBRARY INDEX

(SOURCE: Mandel, Carol A., and Mary P. Johnson, ARL Statistics 1979-80, Association of Research Libraries, Washington, D.C., 1980, pp. 23-24.)

The data tables at the beginning of the ARL Statistics display figures reported by ARL member libraries in 22 categories that, with the exception of the measures of interlibrary loan activity, describe the size of ARL libraries in terms of holdings, expenditures, and personnel. The rank order tables provide an overview of the ranges and medians for 14 of these categories, or variables, among ARL academic libraries as well as quantitatively comparing each library with other ARL member institutions. However, none of the 22 variables provides a summary measure of a library's relative size within ARL or characterizes the ARL libraries as a whole.

The ARL Library Index has been derived as a means of providing this summary characterization, permitting quantitative comparisons of ARL academic libraries, singly and as a group, with other academic libraries. Through the use of statistical techniques known as factor analysis, it can be determined that 15 of the variables reported to ARL are more closely correlated with each other than with other categories. Within this group of 15 variables, some are subsets or combinations of materials. When the subsets and combinations are eliminated, 10 variables emerge as characteristic of ARL libary size. These are: volumes held, volumes added (gross), microform units held, current serials received, expenditures for library materials, expenditures for binding, total salary and wage expenditures, other operating expenditures, number of professional staff, and number of nonprofessional staff.

These 10 categories delineate an underlying dimension, or factor, of library size. By means of principal component analysis, a technique that is a variant of factor analysis, it is possible to calculate the correlations of each of the variables with this hypothetical factor of library size. From this analysis a weight for each variable can be determined based on how closely that variable is correlated with the overall dimension of library size defined by all 10 categories. A high correlation indicates that much of the variation in ARL library size is accounted for by the variable in question, implying a characteristic in which ARL libraries are relatively alike. The component score coefficients, or weights, for

213

the 1979-80 ARL academic library data are as follows:

| | |
|---|---|
| Volumes held | .12108 |
| Volumes added (gross) | .11940 |
| Microforms held | .07509 |
| Current serials received | .12253 |
| Expenditures for library materials | .12553 |
| Expenditures for binding | .11266 |
| Expenditures for salaries and wages | .12581 |
| Other operating expenditures | .10592 |
| Number of professional staff | .12347 |
| Number of nonprofessional staff | .11297 |

From these weights an individual library can compute an index score that will indicate its relative position among ARL libraries with respect to the overall factor of library size. The data for each of the 10 variables are converted to standard normal form and multiplied by the appropriate weight. The resulting scores are expressed in terms of the number of standard deviations above or below the mean index score for ARL academic libraries. Thus, the formula* for calculating a library's 1979-80 index score is as follows:

```
 .12108 (log of volumes held - 6.2916)/.2172
+.11940 (log of volumes added gross - 4.8412)/.2025
+.07509 (log of microforms - 6.0950)/.1763
+.12253 (log of current serials - 4.3432)/.2341
+.12553 (log of expenditures for materials - 6.2333)/.1636
+.11266 (log of expenditures for binding - 5.0480)/.2475
+.12581 (log of total salaries - 6.4675)/.2103
+.10592 (log of operating expenditures - 5.6773)/.2635
+.12347 (log of professional staff - 1.8281)/.1968
+.11297 (log of nonprofessional staff - 2.1512)/.2046
```

The index scores for the 99 academic libraries that were members of ARL during 1979-80 are shown on the following page. It is important to emphasize that these scores are only a summary descrip- tion of library size, distributing ARL libraries along a normal curve, based on 10 quantitative measures that are positively correlated with one another in ARL libraries. The scores are in no way a qualitative assessment of the collections, services, or operations of these libraries.

---

*For calculation on a hand calculator, the formula can be mathemati- cally simplied to: (.55746 x log of volumes held) + (.58963 x log of volumes added gross) + (.42592 x log of microforms) + (.52341 x log of current serials) + (.76730 x log of expenditures for materials) + (.45519 x log of expenditures for binding) + (.59824 x log of total salries) + (.40197 x log of operating expenditures) + (.62739 x log of professional staff) + (.55215 x log of nonprofessional staff) - 26.79765.

# APPENDIX E

## FACULTY RESEARCH SUPPORT

The names of National Science Foundation (NSF) research grant awardees were obtained from a file maintained by the NSF Division of Information Systems. The file provided to the committee covered all research grant awards made in FY1978, FY1979, and FY1980 and included the names of the principal investigator and co-principal investigators for each award. Also available from this file was information concerning the field of science/engineering of the research grant and the institution with which the investigator was affiliated. This information was used in identifying which research grant recipients were on the program faculty lists provided by institutional coordinators.

The names of National Institutes of Health (NIH) and Alcohol, Drug Abuse, and Mental Health Administration (ADAMHA) research grant recipients (principal investigators only) were obtained from the NIH Information for Management Planning, Analysis, and Coordination System. This system contains a detailed record of all applications and awards in the various training and research support programs of these agencies. For the purposes of this study, information analogous to that available from the NSF file was extended for FY1978-80 research grant awardees and their records were matched with the program faculty lists. Measure 13 constitutes the fraction of program faculty members who had received one or more research grant awards from NSF (including both principal investigators and co-principal investigators), NIH, or ADAMHA during the FY1978-80 period.

## R&D EXPENDITURES

Total university expenditures for R&D activities are available from the NSF Survey of Scientific and Engineering Expenditures at Universities and Colleges. A copy of the survey form appears on the following pages.

215

NSF FORM 411 (Dec. 1979)

FORM APPROVED
OMB No. 99-R0279

**NATIONAL SCIENCE FOUNDATION**
Washington, D.C. 20550

## SURVEY OF SCIENTIFIC AND ENGINEERING
## EXPENDITURES AT UNIVERSITIES AND COLLEGES, FY 1979

(Current and Capital Expenditures for Research,
Development, and Instruction in the Sciences and Engineering)

Organizations are requested to complete and return this form to:

**NATIONAL SCIENCE FOUNDATION**
1800 G Street, N.W.
Washington, D.C. 20550
Attn: UNISG

This form should be returned by February 1, 1980. Your cooperation in returning the survey questionnaire promptly is very important.

Financial data are requested for your institution's 1979 fiscal year.

This information is solicited under the authority of the National Science Foundation Act of 1950, as amended. All information you provide will be used for statistical purposes only. Your response is entirely voluntary and your failure to provide some or all of the information will in no way adversely affect your institution.

All financial data requested on this form should be reported in thousands of dollars; for example, an expenditure of $25,342 should be rounded to the nearest thousand dollars and reported as $25.

Where exact data are not available, estimates are acceptable. Your estimates will be better than ours.

Please correct if name or address has changed

(Includes aggregate data from 567 universities and colleges but excludes 19 university-administered FFRDC's)

Include data for branches and all organizational units of your institution, such as medical schools and agricultural experiment stations. Also include hospitals or clinics owned, operated, or controlled by universities, and integrated operationally with the clinical programs of your medical schools. Exclude data for federally funded research and development centers (FFRDC's). A separate questionnaire is included in this package if your institution administers an FFRDC. **If you have any questions please contact Jim Hoehn (202-634-4674).**

Please enter the beginning and ending dates of your institution's fiscal year for which you are reporting on this form:

_____ through _____

Please note in space below:
    (1)    Any suggestions to improve the design of the survey questionnaire, (2) any suggestions to improve the instructions, or (3) any comments on significant change in R&D in your institution.

(Attach additional sheets, if necessary.)

| PLEASE TYPE OR PRINT<br>NAME OF PERSON SUBMITTING THIS FORM | TITLE | AREA CODE | EXCH | NO. | EXT |
|---|---|---|---|---|---|
| | | | | | |

| NAME OF PERSON WHO PREPARED THIS<br>SUBMISSION (If different from above) | TITLE | AREA CODE | EXCH | NO. | EXT |
|---|---|---|---|---|---|
| | | | | | |

| Please check and correct if necessary the name and address of your institution shown on the mailing label. | DATE |
|---|---|

## ITEM 1. CURRENT EXPENDITURES FOR SEPARATELY BUDGETED RESEARCH AND DEVELOPMENT (R&D) IN THE SCIENCES AND ENGINEERING, BY SOURCE OF FUNDS AND BASIC RESEARCH, FY 1979 (Include indirect costs)

### ITEMS 1. & 2. INSTRUCTIONS

Separately budgeted research and development (R&D) includes all funds expended for activities specifically organized to produce research outcomes and commissioned by an agency either external to the institution or separately budgeted by an organizational unit within the institution. **Include** equipment purchased under research project awards as part of "current funds." Research funds subcontracted to outside organizations should also be included. **Exclude** training grants, public service grants, demonstration projects, etc.

Under a. **Federal Government.** Report grants and contracts for R&D by all agencies of the Federal Government including indirect costs from these sources.

Under b. **State and local governments.** Include funds for R&D from State, county, municipal, or other local governments and their agencies. Include here State funds which support R&D at agricultural experiment stations.

Under c. **Industry.** Include all grants and contracts for R&D from profitmaking organizations, whether engaged in production, distribution, research, service, or other activities. Do not include grants and contracts from nonprofit foundations financed by industry, which should be reported under **All other sources.**

Under d. **Institutional funds.** Report funds which your institution spent for R&D activities including indirect costs from the following sources: (1) General-purpose State or local government appropriations; (2) general-purpose grants from industry, foundations, or other outside sources; (3) tuition and fees; (4) endowment income. In addition, estimate your institution's contribution to unreimbursed indirect costs incurred in association with R&D projects financed by outside organizations, and mandatory cost sharing on Federal and other grants. To estimate unreimbursed indirect costs, many institutions use a university-wide negotiated indirect cost rate multiplied by the base (e.g., direct salaries and wages, etc.) minus actual indirect cost recoveries. If your institution now separately budgets what was previously classified as departmental research, these data should be included in line d.

Under e. **All other sources.** Include foundations and voluntary health agencies grants for R&D, as well as all other sources not elsewhere classified. Funds from foundations which are affiliated with or grant solely to your institution should be included under d. Institutional funds. Funds for R&D received from a health agency that is a unit of a State or local government should be reported under State and local governments. Also include gifts from individuals that are restricted by the donor to research.

Please exclude from your response any R&D expenditures in the fields of education, law, humanities, music, the arts, physical education, library science, and all other nonscience fields.

| Source of funds | | (1) Total R&D expenditures (Dollars in thousands) | (2) Basic research (Percent of column 1) |
|---|---|---|---|
| a. Federal Government | 1110 | $ 3,431,538 | 73.4 % |
| *b. State and local governments | 1125 | 467,311 | |
| c. Industry | 1150 | 193,794 | **Basic research** is directed toward an increase of knowledge; it is research where the primary aim of the **investigator** is a fuller knowledge or understanding of the subject under study rather than a practical application thereof. |
| d. Institutional funds | 1160 | 716,241 | |
| (1) Separately budgeted | 1161 | 357,926 | |
| (2) Underrecovery of indirect costs and cost sharing | 1162 | 358,315 | |
| *e. All other sources | 1175 | 373,845 | |
| f. TOTAL (sum of a through e) | 1100 | $ 5,182,729 | 68.5 % |

*Combined data cell (See instructions for b and e).

Total R&D expenditures reported in line 1100 column (1) and line 1400 column (1) should be the same.
Federally financed R&D expenditures reported in line 1100 column (1) and line 1400 column (2) should be the same.

| ITEM 2. TOTAL AND FEDERALLY FINANCED EXPENDITURES FOR SEPARATELY BUDGETED RESEARCH AND DEVELOPMENT, BY FIELD OF SCIENCE, FY 1979 (Include indirect costs and equipment). | | | | |
|---|---|---|---|---|
| Field of science | Illustrative disciplines | | (Dollars in thousands) | |
| | | | (1) Total | (2) Federal |
| a. ENGINEERING (TOTAL) | Aeronautical, agricultural, chemical, civil, electrical, industrial, mechanical, metallurgical, mining, nuclear, petroleum, bio- and biomedical, energy, textile, architecture | 1410 | $ 715,454 | $ 474,866 |
| b. PHYSICAL SCIENCES (TOTAL) | | 1420 | 559,566 | 448,992 |
| (1) Astronomy | Astrophysics, optical and radio, x-ray, gamma-ray, neutrino | 1421 | 39,026 | 26,862 |
| (2) Chemistry | Inorganic, organo-metallic, organic, physical, analytical, pharmaceutical, polymer science (exclude biochemistry) | 1422 | 204,062 | 154,031 |
| (3) Physics | Acoustics, atomic and molecular, condensed matter, elementary particles, nuclear structure, optics, plasma | 1423 | 275,680 | 236,872 |
| (4) Other | Used for multidisciplinary projects within physical sciences and for disciplines not requested separately | 1424 | 40,798 | 31,227 |
| c. ENVIRONMENTAL SCIENCES (TOTAL) | ATMOSPHERIC SCIENCES: Aeronomy, solar weather modification, meteorology, extra-terrestrial atmospheres GEOLOGICAL SCIENCES: Engineering geophysics, geology, geodesy, geomagnetism, hydrology, geochemistry, paleomagnetism, paleontology, physical geography, cartography, seismology, soil sciences OCEANOGRAPHY: Chemical, geological, physical, marine geophysics, marine biology, biological oceanography | 1430 | 429,129 | 307,493 |
| d. MATHEMATICAL AND COMPUTER SCIENCES (TOTAL) | | 1440 | 145,087 | 94,534 |
| (1) Mathematics | Algebra, analysis, applied mathematics, foundations and logic, geometry, numerical analysis, statistics, topology | 1441 | 65,637 | 49,043 |
| (2) Computer sciences | Design, development, and application of computer capabilities to data storage and manipulation, information science | 1442 | 79,450 | 45,491 |
| e. LIFE SCIENCES (TOTAL) | | 1450 | 2,814,824 | 1,810,729 |
| (1) Biological sciences | Anatomy, biochemistry, biophysics, biogeography, ecology, embryology, entomology, genetics, immunology, microbiology, nutrition, parasitology, pathology, pharmacology, physical anthropology, physiology, botany, zoology | 1451 | 949,993 | 690,805 |
| (2) Agricultural | Agricultural chemistry, agronomy, animal science, conservation, dairy science, plant science, range science, wildlife | 1452 | 565,697 | 168,849 |
| (3) Medical | Anesthesiology, cardiology, endocrinology, gastroenterology, hematology, neurology, obstetrics, opthalmology, preventive medicine and community health, psychiatry, radiology, surgery, veterinary medicine, dentistry, pharmacy | 1453 | 1,214,442 | 890,612 |
| (4) Other | Used for multidisciplinary projects within life sciences | 1454 | 84,692 | 60,463 |
| f. PSYCHOLOGY (TOTAL) | Animal behavior, clinical, educational, experimental, human development and personality, social | 1460 | 99,732 | 72,256 |
| g. SOCIAL SCIENCES (TOTAL) | | 1470 | 290,057 | 153,674 |
| (1) Economics | Econometrics, international, industrial, labor, agricultural, public finance and fiscal policy | 1471 | 85,415 | 40,641 |
| (2) Political science | Regional studies, comparative government, international relations, legal systems, political theory, public administration | 1472 | 39,029 | 18,452 |
| (3) Sociology | Comparative and historical, complex organizations, culture and social structure, demography, group interactions, social problems and welfare, theory | 1473 | 72,669 | 46,739 |
| (4) Other | History of science, cultural anthropology, linguistics, socio-economic geography | 1474 | 92,944 | 47,842 |
| h. OTHER SCIENCES, n.e.c. (TOTAL)* | To be used when the multidisciplinary and interdisciplinary aspects make the classification under one primary field impossible | 1480 | 128,880 | 68,994 |
| i. TOTAL (SUM of a through h) Check to insure that column totals are identical with data reported in item 1. | | 1400 | 5,182,729 | 3,431,538 |

*PLEASE EXCLUDE FROM YOUR RESPONSE ANY R&D EXPENDITURES IN THE FIELDS OF EDUCATION, LAW, HUMANITIES, MUSIC, THE ARTS, PHYSICAL EDUCATION, LIBRARY SCIENCE, AND ALL OTHER NONSCIENCE FIELDS.

ITEM 3. CAPITAL EXPENDITURES FOR SCIENTIFIC AND ENGINEERING FACILITIES AND
EQUIPMENT FOR RESEARCH, DEVELOPMENT, AND INSTRUCTION,
BY FIELD OF SCIENCE AND SOURCE OF FUNDS, FY 1979

ITEM 3. INSTRUCTIONS

Report funds for facilities which were in process or completed during FY 1979. Expenditures for administration buildings, steam plants, residence halls, and other such facilities should be excluded unless utilized principally for research, development, or instruction in engineering or in the sciences. Land costs should be **excluded**. Exclude small equipment items in your current fund account costing approximately $300 or less per unit or as recommended by the Joint Accounting Group (JAG) or as determined by your institutional policy; these are to be reported under items 1 and 2.

Facilities and equipment expenditures **include** the following: (a) Fixed equipment such as built-in equipment and furnishings; (b) movable scientific equipment such as oscilloscopes and pulse-height analyzers; (c) movable furnishings such as desk; (d) architect's fees, site work, extension of utilities, and the building costs of service functions such as integral cafeterias and bookstores of a facility; (e) facilities constructed to house separate components such as medical schools and teaching hospitals; and (f) special separate facilities used to house scientific apparatus such as accelerators, oceanographic vessels, and computers.

| Field of science | | Total (1) | Federal (2) | All other sources (3) |
|---|---|---|---|---|
| | | (Dollars in thousands) | | |
| a. Engineering | 1710 | $ 95,399 | $ 22,060 | $ 73,339 |
| b. Physical sciences | 1720 | 64,551 | 32,439 | 32,112 |
| c. Environmental sciences | 1730 | 25,293 | 8,970 | 16,323 |
| d. Mathematical and computer sciences | 1740 | 27,465 | 3,049 | 24,416 |
| e. Life sciences | 1750 | 456,477 | 92,567 | 363,910 |
| f. Psychology | 1760 | 7,803 | 1,767 | 6,036 |
| g. Social sciences | 1770 | 20,932 | 2,069 | 18,863 |
| h. Other sciences, n.e.c. | 1780 | 31,984 | 5,054 | 26,930 |
| i. **Total (sum of a through h)** | 1700 | $ 729,904 | $ 167,975 | $ 561,929 |

APPENDIX F

DATA ON PUBLICATION RECORDS

Data for these measures were provided by a subcontractor, Computer Horizons, Inc. A detailed description of the derivation of these measures and examples of their use is given in:

Francis Narin, <u>Evaluative Bibliometrics: The Use of Publications and Citations Analysis in the Evaluation of Scientific Activity</u>, Report to the National Science Foundation, March 1976.

The following pages have been excerpted from Chapters VI and VII of this report and describe operational considerations in compiling the publication records included here (measure 15) and the methodology used in determining the "influence" of published articles (measure 16).

VI.    OPERATIONAL CONSIDERATIONS

A.    Basics of Publication and Citation Analysis

The first section of this chapter discusses the major stages of publication and citation analysis techniques  in evaluative bibliometrics.  Later sections of the chapter consider publication and citation count parameters in further detail, including discussions of data bases, of field-dependent characteristics of the literature, and of some cautions and hazards in performing citation analyses for individual scientists.

The basic stages which must be kept in mind when doing a publication or citation analysis are briefly summarized in Figure 6-1.

1.    Type of Publication

For a publication analysis the fundamental decision is which type of publication to count.  A basic count will include all regular scientific articles.  However, notes are often counted since some engineering and other journals often contain notes with significant technical content.  Reviews may be included. Letters-to-the-editor must also be considered as a possible category for inclusion, since some important journals are sometimes classified as letter journals.  For example, publications in Physical Review Letters were classified as letters by the Science Citation Index prior to 1970, although they are now classified as articles.

For most counts in the central core of the scientific literature, articles, notes and reviews are used as a measure of scientific output.  When dealing with engineering fields, where many papers are presented at meetings accompanied by reprints and published proceedings, meeting presentations must also be considered.  In some applied fields, i.e., agriculture, aerospace and nuclear engineering, where government support has been particularly comprehensive, the report literature may also be important.  Unfortunately, reports generally contain few references, and citations to them are limited so they are not amenable to the normal citation analyses.

Books, of course, are a major type of publication, especially in the social sciences where they are often used instead of a series of journal articles.  In bibliometrics a weighting of n articles equal to one book is frequently used; no uniformly acceptable value of n is available.  A few of the papers discussed in Chapter V contain such measures.

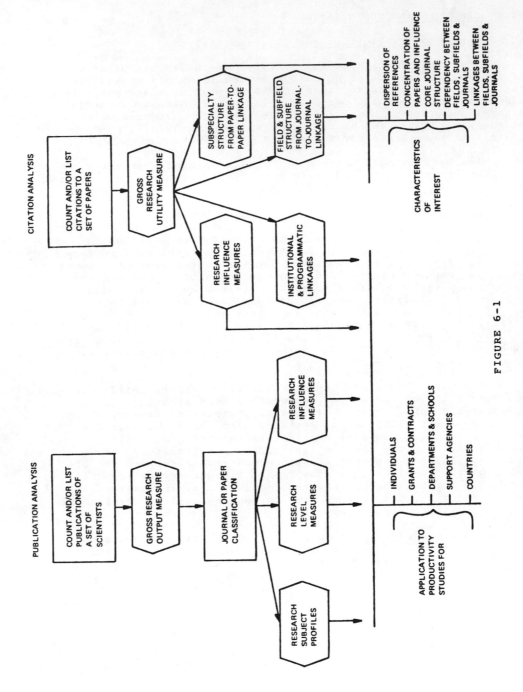

FIGURE 6-1

STAGES OF PUBLICATION AND CITATION ANALYSIS

## 2.    Time Spans

A second important decision in making a publication count is to select the time span of interest.  In the analysis of the publications of an institution a fixed time span, usually one year or more, is most appropriate.  In comparing publication histories of groups of scientists, their professional ages (normally defined as years since attaining the PhD degree) must be comparable so that the build-up of publications at the beginning of a career or the decline at the end will not complicate the results.  A typical scientist's first publication appears soon after his dissertation; if he continued working as a scientist, his publications may continue for thirty or more years.

The accurate control of the time span of a count is not as trivial as it might seem.  Normally, the publication count is made from secondary sources (abstracting or indexing services) rather than from scanning the publications individually.  Since most abstracting and indexing sources have been expanding their coverage over time, any publication count covering more than a few years must give careful consideration to changes in coverage.  Furthermore, the timeliness of the secondary sources varies widely, with sources dependent on outside abstractors lagging months or even years behind.  Since these abstracting lags may depend upon language, field and country of origin, they are a particular problem in international publication counts.

The Science Citation Index is one of the most current secondary sources, with some 80% to 90% of a given year's publications in the SCI for that year.

Of course, no abstracting or indexing service can be perfect, since some journals are actually published months after their listed publication dates.  Nevertheless, variations in timeliness are large from one service to another.

## 3.    Comprehensiveness of Source Coverage

An important consideration in making a publication count is the comprehensiveness of the source coverage.  Most abstracting and indexing sources cover some journals completely, cover other journals selectively, and omit some journals in their field of interest.  The Science Citation Index is an exception in that it indexes each and every important entry from any journal it covers.  This is one of the major advantages in using the SCI as a data base.  Chemical Abstracts and Biological Abstracts have a group of journals which they abstract completely, coupled with a much larger set of journals from which they abstract selectively, based upon the appropriateness of the article to the subject coverage.  In some cases the abstractor or indexer may make a quality judgment, based on his estimate of the importance or the quality of the article or upon his knowledge of whether similar information has appeared elsewhere; Excerpta Medica is a comprehensive abstracting service for which articles are included only if they meet the indexers' quality criteria.

Some data on the extent of coverage of the major secondary sources is presented in Section D of this chapter.

4.    Multiple Authorships and Affiliations

Attributing credits for multiple authorships and affiliations is a significant problem in publication and citation analysis. In some scientific papers the authors are listed alphabetically; in others the first author is the primary author; still others use different conventions. These conventions have been been discussed by Crane[1] and by other social scientists.[2] There does not seem to be any reasonable way to deal with the attribution problem, except to attribute a fraction of a publication to each of the authors. For example, an article which has three authors would have one-third of an article attributed to each author. The amount of multiple authorship unfortunately differs from country to country and from field to field. Several studies have investigated the problem, but no comprehensive data exists.[3]

Multiple authorship takes on particular importance when counting an individual's publications since membership on a large research team may lead to a single scientist being a co-author of ten or more publications per year. This number of publications is far in excess of the normal publication rate of one to two articles per year per scientist.

Multiple authorship problems arise less often in institutional publication counts since there are seldom more than one or two institutions involved in one publication.

A particularly vexing aspect of multiple authorship is the first author citation problem: almost all citations are to the first author in a multi-authored publication. As a result, a researcher who is second author of five papers may receive no

---

[1] Diana Crane, "Social Structure in a Group of Scientists: A Test of the 'Invisible College' Hypothesis," American Sociological Review 34 (June 1969):335-352.

[2] James E. McCauly, "Multiple Authorship," Science 141 (August 1963):579.

Beverly L. Clark, "Multiple Authorship Trends in Scientific Papers," Science 143 (February 1964):822-824.

[3] Harriet Zuckerman, "Nobel Laureates in Science: Patterns of Productivity, Collaboration, and Authorship," American Sociolgoical Review 32 (June 1967):391-403.

citations under his own name, even though the papers he co-author-
ed may be highly cited.  Because of this, a citation count for a
person must account for the citations which appear under the
names of the first authors of publications for which the author
of interest was a secondary author.  This can lead to a substantial amount of tedious additional work, since a list of first
authors must be generated for all of the subjects' multi-authored papers.  Citations to each of these first authors must then
be found, the citations of interest noted, and these citations
fractionally attributed to the original author.  Since multiple
years of the Citation Index are often involved, the amount of
clerical work searching from volume to volume and from author
to author, and citation to citation can be quite large.

A note of caution about the handling of multiple authorship in the Corporate Index of the Science Citation Index: SCI
lists a publication giving all the corporate affiliations, but
always with the first author's name.  Thus a publication by
Jones and Smith where Jones is at Harvard and Smith is at Yale
would be listed in the Corporate Index under Harvard with the
name Jones and also under Yale with the name Jones.  To find
the organization with which the various authors are affiliated,
the original article must be obtained.

Although the publisher of the Science Citation Index, the
Institute for Scientific Information, tries to maintain a consistent policy in attributing institutional affiliations, when
authors have multiple affiliations the number of possible var-
iants is large.  In the SCI data base on magnetic tape, sufficient information is included to assign a publication with authors from a number of different institutions in a reasonably
fair way to those institutions; however, in the printed Corporate Index, one has to refer to the Source Index to find the
actual number of authors, or to the paper itself to find the
affiliations of each of the authors.

5.    Completeness of Available Data

Another consideration in a publication analysis is the
completeness of data available in the secondary source, since
looking up hundreds or thousands of publications individually
is tedious and expensive.  One difficulty here is that most of
the abstracting and indexing sources are designed for retrieval
and not for analysis.  As a result, some of the parameters which
are of greatest analytical importance, such as the affiliation
of the author and his source of financial support, are often
omitted.  Furthermore, some of the abstracting sources are
cross-indexed in complex ways, so that a publication may only
be partially described at any one point, and reference must be
made to a companion  volume to find even such essential data
as the author's name.  While intellectually trivial, these

searches can be exceedingly time consuming when analyzing large numbers of publications.

The specific data which are consistently available in the secondary sources are the basic bibliographic information: i.e., authors' name, journal or report title, volume, page, etc. This information is the basic data used for retrieval, and since the abstracting and indexing services are retrieval oriented, this bibliographic information is always included.

Data which are less consistently available in the secondary source are the authors' affiliation and the authors' rank or title. Both of these are of interest in analysis. For example, the ranking of universities based on publication in a given subject area is often of interest. This ranking can be tabulated only from a secondary source which gives the authors' university affiliation.

6.    Support Acknowledgements

The source of the authors' financial support is seldom given in any secondary source, although it is now being added to the MEDLARS data base. Since this financial data can be used to define the fraction of a subject literature which is being supported by a particular corporate body such as a governmental agency, the data are of substantial evaluative interest.

The amount of acknowledgement of agency support in the scientific literature has changed over time. In a Computer Horizons study completed in 1973 the amount of agency support acknowledgement was tabulated in twenty major journals from five different fields.[4]    Table 6-1 summarizes those support acknowledgements for 1969 and 1972.

In 1969, only 67% of the articles in 20 major journals acknowledged financial support. By 1972, the percentage of articles acknowledging financial support had risen to approximately 85%. The table shows that the sources of support differ from one field to another and also shows that the fields of interest to these sources differ as well. For example, the National Science Foundation is the major source of acknowledged support in mathematics, while the National Institutes of Health clearly dominate the support of biology. Chemistry is the field with the largest amount of non-government (private sector) support in the U.S.

Note also that the 20 journals used were major journals in their fields; as less prestigious journals are examined, the amount of support acknowledgement generally decreases.

---

[4] Computer Horizons, Inc., Evaluation of Research in the Physical Sciences Based on Publications and Citations, Washington, D.C., National Science Foundation, Contract No. NSF-C627, November, 1973.

TABLE 6-1

AGENCY SUPPORT ACKNOWLEDGEMENTS IN 20 LEADING JOURNALS
FROM 5 MAJOR FIELDS - 1969 and 1972

| Agency Acknowledged | Mathematics 1969 | 1972 | Physics 1969 | 1972 | Chemistry 1969 | 1972 | Biochemistry 1969 | 1972 | Biology 1969 | 1972 | All Fields 1969 | 1972 |
|---|---|---|---|---|---|---|---|---|---|---|---|---|
| NSF | 18% | 37% | 14% | 19% | 18% | 21% | 8% | 8% | 8% | 8% | 13% | 16% |
| NIH | 2 | 1 | 1 | 1 | 11 | 10 | 37 | 39 | 23 | 32 | 13 | 16 |
| AEC | 1 | 1 | 21 | 15 | 10 | 8 | 3 | 2 | 3 | 2 | 11 | 8 |
| DOD | 15 | 7 | 19 | 15 | 10 | 10 | 1 | 1 | 2 | 3 | 10 | 9 |
| NASA | 1 | 1 | 7 | 9 | 2 | 2 | 1 | 1 | 1 | 2 | 3 | 4 |
| Other U.S. Government | 1 | 2 | 1 | 2 | 2 | 2 | 1 | 1 | 1 | 3 | 1 | 2 |
| Other U.S. | 3 | 10 | 3 | 14 | 8 | 21 | 10 | 10 | 9 | 13 | 7 | 14 |
| Foreign | 5 | 4 | 5 | 15 | 7 | 8 | 16 | 25 | 10 | 24 | 8 | 16 |
| Unacknowledged | 55 | 37 | 31 | 11 | 32 | 18 | 25 | 13 | 42 | 14 | 33 | 15 |

In an attempt to account for the 15% of unacknowledged papers, a questionnaire was sent to all U.S. authors in the 1972 sample who did not acknowledge agency support. Almost 70% of the authors who had not listed sources of support responded to the questionnaire. Of the authors who responded, over two-thirds were supported by their institutions as part of their regular duties; approximately 20% of the respondents cited specific governmental agencies as sources of support, even though they had not acknowledged these in the article itself. Twelve percent of the respondents listed no agency or institutional support; research done as fulfillment of graduate studies was included in this category.

Overall, the 1972 tabulation and survey showed that 88% of the research reported in these prestigious journals was externally supported, and that 97% of the externally supported work was acknowledged as such.

### 7.  Subject Classification

Having constructed a basic list of publications, the next step in analysis is normally to subject classify the publications. Either the journals or the papers themselves may be classified. When a large number of papers is to be analyzed, classification of the papers by the field of the journal can be very convenient. Such a classification implies, of course, a degree of homogeneity of publication which is normally adequate when analyzing hundreds of papers. Such a classification may not be sufficient for the analysis of the scientific publications of one or a few individuals.

Subject classification schemes differ from one abstracting and indexing service to another. Therefore, a comparison of a collection of papers based on the classification schemes of more than one abstracting and indexing service is almost hopeless. A classification of papers at the journal level has been used in the influence methodology discussed in Chapters VII through X.

### 8.  Citation Counts

Citation counts are a tool in evaluative bibliometrics second in importance only to the counting and classification of publications. Citation counts may be used directly as a measure of the utilization or influence of a single publication or of all the publications of an individual, a grant, contract, department, university, funding agency or country. Citation counts may be used to link individuals, institutions, and programs, since they show how one publication relates to another publication.

In addition to these evaluative uses, citations also have important bibliometric uses, since the references from one paper to another define the structure of the scientific literature. Chapter III discusses how this type of analysis may be carried out at a detailed, micro-level to define closely related papers through bibliographic coupling and co-citation. That chapter also describes how citation analysis may be used at a macro-level to link fields and subfields through journal-to-journal mapping. The bibliometric characteristics of the literature also provide a numeric base against which evaluative parameters may be normalized.

Some of the characteristics of the literature which are revealed by citation analysis are noted on Figure 6-1. These characteristics include:

The dispersion of references: a measure of scientific "hardness", since in fields that are structured and have a central core of accepted knowledge, literature references tend to be quite concentrated.

The concentration of papers and influence: another measure of centrality in a field, dependent upon whether or not a field has a core journal structure.

The hierarchic dependency relationships between field, subfield and journals, including the comparison of numbers of references from field A to field B, compared with number of references from field B to field A: this comparison provides a major justification for the pursuit of basic research as a foundation of knowledge utilized by more applied areas.

The linkages between fields, subfields and journals: a measure of the flow of information, and of the importance of one sector of the scientific mosaic to another.

VII.  THE INFLUENCE METHODOLOGY

A.    Introduction

In this chapter an influence methodology will be described which allows advanced publication and citation techniques to be applied to institutional aggregates of publications, such as those of departments, schools, programs, support agencies and countries, without performing an individual citation count.  In essence, the influence procedure ascribes a weighted average set of properties to a collection of papers, such as the papers in a journal, rather than determining the citation rate for the papers on an individual basis.

The influence methodology is completely general, and can be applied to journals, subfields, fields, institutions or countries.

There are three separate aspects of the influence methodology which are particularly pertinent to journals.  These are

1.    A subject classification for each journal

2.    A research type (level) classification for the biomedical journals, and

3.    Citation influence measures for each journal.

It is the third of these, the citation influence measures, which add a quality or utilization aspect to the analysis.  The influence methodology assumes that, although citations to papers vary within a given journal, aggregates of publications can be characterized by the influence measures of the journals in which they appear.  Chapter IX discusses this assumption in some detail.

Older measures of influence all suffer from some defect which limits their use as evaluative measures.

The total number of publications of an individual, school or country is a measure of total activity only; no inferences concerning importance may be drawn.

The total number of citations to a set of publications, while incorporating a measure of peer group recognition, depends on the size of the set involved and has no meaning on an absolute scale.

The journal "impact factor" introduced by Garfield is a size-independent measure, since it is defined as the ratio of the number of citations the journal receives to the number of publications in a specified earlier time period.[1]   This

---

[1]Eugene Garfield, "Citation Analysis As a Tool in Journal Evaluation," Science 178 (November 3, 1972):471.

measure, like the total number of citations, has no meaning on an absolute scale. In addition the impact factor suffers from three more significant limitations. Although the size of the journal, as reflected in the number of publications, is corrected for, the average length of individual papers appearing in the journal is not. Thus, journals which publish longer papers, namely review journals, tend to have higher impact factors. In fact the nine highest impact factors obtained by Garfield were for review journals. This measure can therefore not be used to establish a "pecking order" for journal prestige.

The second limitation is that the citations are unweighted, all citations being counted with equal weight, regardless of the citing journal. It seems more reasonable to give higher weight to a citation from a prestigious journal than to a citation from a peripheral one. The idea of counting a reference from a more prestigious journal more heavily has also been suggested by Kochen.[2]

A third limitation is that there is no normalization for the different referencing characteristics of different segments of the literature: a citation received by a biochemistry journal, in a field noted for its large numbers of references and short citation times, may be quite different in value from a citation in astronomy, where the overall citation density is much lower and the citation time lag much longer.

In this section three related influence measures are developed, each of which measures one aspect of a journal's influence, with explicit recognition of the size factor. These measures are:

(1) The influence weight of the journal: a size-independent measure of the weighted number of citations a journal receives from other journals, normalized by the number of references the journal gives to other journals.

(2) The influence per publication for the journals: the weighted number of citations each article, note or review in a journal receives from other journals.

(3) The total influence of the journal: the influence per publication times the total number of publications.

[2] M. Kochen, Principles of Information Retrieval, (New York: John Wiley & Sons, Inc. 1974), 83.

B.     Development of the Weighting Scheme

1.     The Citation Matrix

A citation matrix may be used to describe the interactions among members of a set of publishing entities.  These entities may, for example, be journals, institutions, individuals, fields of research, geographical subdivisions or levels of research methodology.  The formalism to be developed is completely general in that it may be applied to any such set.  To emphasize this generality, a member of a set will be referred to as a unit rather than as a specific type of unit such as a journal.

The citation matrix is the fundamental entity which contains the information describing the flow of influence among units.

The matrix has the form

$$C = \begin{pmatrix} c_{11} & c_{12} & \ldots & c_{1n} \\ c_{21} & c_{22} & \ldots & c_{2n} \\ \cdot & \cdot & & \cdot \\ \cdot & \cdot & & \cdot \\ \cdot & \cdot & & \cdot \\ c_{n1} & c_{n2} & \ldots & c_{nn} \end{pmatrix}$$

A distinction is made between the use of the terms "reference" and "citation" depending on whether the issuing or receiving unit is being discussed.  Thus, a term $c_{ij}$ in the citation matrix indicates both the number of references unit i gives to unit j and the number of citations unit j receives from unit i.

The time frame of a citation matrix must be clearly understood in order that a measure derived from it be given its proper interpretation.  Suppose that the citation data are based on references issued in 1973.  The citations received may be to papers in any year up through 1973.  In general, the papers issuing the references will not be the same as those receiving the citations.  Thus, any conclusions drawn from such a matrix assume an on-going, relatively constant nature for each of the units.  For instance, if the units of study are journals, it is assumed that they have not changed in size relative to each other and represent a constant subject area.  Journals in rapidly changing fields and new journals would therefore have to be treated with caution.

A citation matrix for a specific time lag may also be formulated.  This would link publications in one time period with publications in some specified earlier time period.

## 2.  Influence Weights

For each unit in the set a measure of the influence of that unit will be extracted from the citation matrix.  Because total influence is clearly a size-dependent quantity, it is essential to distinguish between a size-independent measure of influence, to be called the influence weight, and the size-dependent total influence.

To make the idea of a size-independent measure more precise, the following property of such a measure may be specified: if a journal were randomly subdivided into smaller entities, each entity would have the same measure as the parent journal.

The citation matrix may be thought of as an "input-output" matrix with the medium of exchange being the citation. Each unit gives out references and receives citations; it is above average if it has a "positive citation balance", i.e., receives more than it gives out.  This reasoning provides a first order approximation to the weight of each unit, which is

$$W_i^{(1)} = \frac{\text{total number of citations to the ith unit from other units}}{\text{total number of references from the ith unit to other units}}.$$

This is the starting point for the iterative procedure for the calculation of the influence weights to be described below.

The denominator of this expression is the row sum

$$S_i = \sum_{j=1}^{n} c_{ij}$$

corresponding to the ith unit of the citation matrix; it may be thought of as the "target size" which this unit presents to the referencing world.

The influence weight, $W_i$, of the ith unit is defined as

$$W_i = \sum_{k=1}^{n} \frac{W_k \, c_{ki}}{S_i}$$

In the sum, the number of cites to the ith unit from the kth unit is weighted by the weight of kth (referencing) unit. The number of cites is also divided by the target  size $S_i$  of the unit i being cited.  The n equations, one for each unit, provide a self consistent "bootstrap" set of relations in which each unit plays a role in determining the weight of every other unit. The following summarizes the derivation of those weights.

The equations defining the weights,

$$W_i = \sum_{k=1}^{n} \frac{W_k \, C_{ki}}{S_i}, \qquad i = 1, \ldots, n \qquad (1)$$

are a special case of a more general system of equations which may be written in the form

$$\left\{ \sum_{k=1}^{n} W_k \, \gamma_{ki} \right\} - \lambda W_i = 0, \qquad i = 1, \ldots, n \qquad (2)$$

Here $\gamma_{ki} = \dfrac{C_{ki}}{S_i}$ and Equation 1 is shown to be

a special case of Equation 2 corresponding to $\lambda = 1$.  As will be explained shortly the system of equations given in (1) will not, in general, possess a non-zero solution; only for certain values of $\lambda$ called the <u>eigenvalues</u> of the system, will there be non-zero solutions.

With the choice of target size $S_i$, the value $\lambda = 1$ is in fact an eigenvalue so that Equation 1 itself does possess a solution.

Using the notation $\gamma^T$ for the transpose of $\gamma$,

$$\gamma^T_{ik} = \gamma_{ki} \; ; \text{ introducing the Kronecker delta symbol}$$

defined by
$$\delta_{ik} = \begin{cases} 1 & i = k \\ 0 & i \neq k \end{cases}$$

the equation can then be written

$$\sum_{k=1}^{n} \left( \gamma_{ik}^{T} - \lambda \delta_{ik} \right) W_k = 0 . \qquad (3)$$

This is a system of n homogeneous equations for the weights. In order that a solution for such a system exists, the determinant of the coefficients must vanish. This gives an nth order equation for the eigenvalues

$$\begin{vmatrix} \gamma_{11} - \lambda & \gamma_{21} \ldots . & \gamma_{n1} \\ \gamma_{12} & \gamma_{22} - \lambda \ldots & \gamma_{n2} \\ \cdot & \cdot & \cdot \\ \cdot & \cdot & \cdot \\ \cdot & \cdot & \cdot \\ \gamma_{1n} & \gamma_{2n} \ldots . & \gamma_{nn} - \lambda \end{vmatrix} = 0 \qquad (4)$$

called the characteristic equation.

Only for values of $\lambda$ which satisfy this equation, does a non-zero solution for the W's exist. Moreover, Equation 3 does not determine the values of the $W_k$ themselves, but at best determines their ratios. Equivalently the eigenvalue equation may be thought of as a vector equation for the vector unknown $\underline{W} = \left\{ W_1, \ldots, W_n \right\}$

$$\underline{\underline{\gamma}}^{T} \cdot \underline{W} = \lambda \underline{W} \qquad (5)$$

from which it is clear that only the direction of $\underline{W}$ is determined.

The normalization or scale factor is then fixed by the condition that the size-weighted average of the weights is 1, or

$$\frac{\displaystyle\sum_1^n s_k \, w_k}{\displaystyle\sum_1^n s_k} = 1 \qquad\qquad (6)$$

This normalization assures that the weight values have an absolute as well as a relative meaning, with the value  1  representing an average value.

Each root of the characteristic equation determines a solution vector or eigenvector of the equation, but the weight vector being sought is the eigenvector corresponding to the largest eigenvalue.  This can be seen from the consideration of an alternative procedure for solving the system of equations, a procedure which also leads to the algorithm of choice.

Consider an iterative process starting with equal weights for all units.  The values $w_i^{(0)} = 1$ can be thought of as zeroth order approximations to the weights.  The first order weights are then

$$w_i^{(1)} = \frac{\displaystyle\sum_{k=1}^n c_{ki}}{s_i}$$

This ratio (total cites to a unit divided by the target size of the unit) is the simplest size-corrected citation measure and, in fact, corresponds to the impact measure used by Garfield. These values are then substituted into the right hand side of Equation 1 to obtain the next order of approximation.  In general, the mth order approximation is

$$W_i^{(m)} = \sum_{k=1}^{n} \frac{W_k^{(m-1)} c_{ki}}{S_i} = \sum_{k=1}^{n} W_k^{(m-1)} \times \gamma_{ki} = \sum_{j=1}^{n} \left(\gamma\right)_{ji}^{m}$$

The exact weights are therefore

$$W_i = W_i^{(\infty)} = \sum_{j=1}^{n} \left(\lim_{m\to\infty} \gamma^m\right)_{ji}$$

This provides the most convenient numerical procedure for finding the weights, the whole iteration procedure being reduced to successive squarings of the $\gamma$ matrix.

This procedure is closely related to the standard method for finding the dominant eigenvalue of a matrix. Since $\lambda = 1$ is the largest eigenvalue, repeated squarings are all that is needed. If the largest eigenvalue had a value other than 1, the normalization condition, Equation 6, would have to be reimposed with each squaring. Convergence to three decimal places usually occurs with six squarings, corresponding to raising $\gamma$ to the 64th power.

APPENDIX G

CONFERENCE ON THE ASSESSMENT OF
QUALITY OF GRADUATE EDUCATION PROGRAMS

September 27-29, 1976
Woods Hole, Massachusetts

## Participants

| | |
|---|---|
| Robert A. ALBERTY | Dean of Science, Massachusetts Institute of Technology |
| Charles ANDERSEN | Coordinator, Education Statistics, American Council on Education |
| Richard C. ATKINSON | Acting Director, National Science Foundation |
| R. H. BING | Chairman, Department of Mathematics, University of Texas at Austin |
| David W. BRENEMAN | Senior Fellow, The Brookings Institution |
| John E. CANTLON | Vice-President for Research and Graduate Studies, Michigan State University |
| Henry E. COBB | Professor, Department of History, Southern University |
| Monroe D. DONSKER | Professor, Courant Institute of Mathematical Sciences, New York University |
| David E. DREW | Senior Scientist, Rand Corporation |
| E. Alden DUNHAM | Program Officer, Carnegie Corporation of New York |
| David A. GOSLIN | Executive Director, Assembly of Behavioral and Social Sciences, National Research Council |
| Hanna H. GRAY | Provost, Yale University |
| Norman HACKERMAN | President, Rice University |
| Philip HANDLER | President, National Academy of Sciences |
| David D. HENRY | President Emeritus, University of Illinois |
| Roger W. HEYNS | President, American Council on Education |
| Lyle V. JONES | Vice Chancellor and Dean, Graduate School, University of North Carolina at Chapel Hill |
| Charles V. KIDD | Executive Secretary, Association of American Universities |
| Winfred P. LEHMANN | Professor, Department of Linguistics, University of Texas at Austin |
| Charles T. LESTER | Vice-President of Arts and Sciences, Emory University |

| | |
|---|---|
| Gardner LINDZEY | Director, Center for Advanced Study in the Behavioral Sciences (<u>Chairman</u>) |
| Raymond P. MARIELLA | Dean of the Graduate School, Loyola University |
| Cora B. MARRETT | Center for Advanced Study in the Behavioral Sciences |
| Peter S. McKINNEY | Acting Dean, Graduate School of Arts and Sciences, Harvard University |
| Doris H. MERRITT | Dean, Research and Sponsored Programs, Indiana University/Purdue University |
| John Perry MILLER | Corporation Officer for Institutional Development, The Campaign for Yale |
| Lincoln E. MOSES | Professor, Department of Family, Community and Preventive Medicine, Stanford University Medical Center |
| Frederick W. MOTE | Professor, Department of East Asian Studies, Princeton University |
| Thomas A. NOBLE | Executive Associate, American Council of Learned Societies |
| J. Boyd PAGE | President, The Council of Graduate Schools in the United States |
| C. K. N. PATEL | Director, Physical Research Laboratory, Bell Laboratories |
| Michael J. PELCZAR, Jr. | Vice-President for Graduate Studies and Research, University of Maryland, College Park |
| Frank PRESS | Chairman, Department of Earth and Planetary Sciences, Massachusetts Institute of Technology |
| John J. PRUIS | President, Ball State University |
| Lorene L. ROGERS | President, University of Texas at Austin |
| John SAWYER | President, The Andrew W. Mellon Foundation |
| Robert L. SPROULL | President, University of Rochester |
| Eliot STELLAR | Provost, University of Pennsylvania |
| Alfred S. SUSSMAN | Dean, Horace H. Rackham School of Graduate Studies, University of Michigan |
| Donald C. SWAIN | Academic Vice-President, University of California System |
| Mack E. THOMPSON | Executive Director, American Historical Association |
| Charles V. WILLIE | Professor of Education and Urban Studies, The Graduate School of Education, Harvard University |
| H. Edwin YOUNG | Chancellor, University of Wisconsin, Madison |
| Harriet A. ZUCKERMAN | Associate Professor, Department of Sociology, Columbia University |

SUMMARY

September 27-29, 1976, Woods Hole, Massachusetts

Report of the Conference

A substantial majority of the Conference believes that the earlier assessments of graduate education have received wide and important use:  by students and their advisors, by the institutions of higher education as aids to planning and the allocation of educational functions, as a check of unwarranted claims of excellence, and in social science research.

The recommendations which follow attempt to distill the main points of consensus within the conference.  This report does not in any sense adequately represent the rich diversity of points of view revealed during the Conference nor the deep and real differences in belief among the participants.

Recommendations

1.  A new assessment of graduate programs is needed, and we believe that the Conference Board is an appropriate sponsor.  While we do not propose to specify the details of this assessment, we are prepared to suggest the following guidelines.

2.  The assessment should include a modified replication of the Roose-Andersen study, with the addition of some fields and the subdivision of others.

3.  It is important to provide additional indices relevant to program assessment such as some of those cited by Breneman, Drew, and Page.  The Conference directs specific attention to the CGS/ETS Study currently nearing completion and urges that the results of that study be carefully examined and used to the fullest possible extent.

4.  The initial assessment study should be one of surveying the quality of scholarship and research and the effectiveness of Ph.D. programs in the fields selected for inclusion.

    a.  It is intended that the study be carried forward on a continuing basis to provide valuable longitudinal data.  This should be implemented along the lines suggested by Moses, involving annual assessment of subsets of programs.

    b.  Every eligible institution should be given the choice of whether to be included in the study.

    c.  Each program is to be characterized by a set of scores, one for each selected index.  The presentation of scores for all reported indices should be accompanied by a discussion of their substantive meaning. In addition, appropriate measures of uncertainty should accompany all tables of results.

5. We propose a simultaneous study exploring ways of reviewing goals of graduate education other than research and scholarship. This would involve review of other doctoral programs and selected master's programs.

# APPENDIX H

## PLANNING COMMITTEE FOR THE
## STUDY OF THE QUALITY OF RESEARCH-DOCTORATE PROGRAMS

### September 1978

Robert M. Bock
Dean of the Graduate School
University of Wisconsin
  at Madison

Philip E. Converse
Institute for Social Research
University of Michigan

Richard A. Goldsby
Department of Genetics
Stanford University

Hugh Holman
Department of English
University of North Carolina
  at Chapel Hill

Lyle V. Jones
Vice Chancellor and Dean,
  Graduate School
University of North Carolina
  at Chapel Hill

Gardner Lindzey, <u>Co-Chairman</u>
Director
Center for Advanced Study in the
  Behavioral Sciences
Stanford, California

Sterling McMurrin
Dean of the Graduate School
University of Utah

Lincoln E. Moses
Administrator
Energy Information
  Administration
Washington, D.C.

George Pake
Xerox Corporation
Palo Alto, California

C. K. N. Patel
Director, Physical Research
Bell Laboratories

Cornelius Pings
Dean of the Graduate School
California Institute of
  Technology

Gordon Ray
President
The John Simon Guggenheim
  Memorial Foundation

Harriet A. Zuckerman
  <u>Co-Chairman</u>
Department of Sociology
Columbia University

APPENDIX I

REGION AND STATE CODES
FOR THE UNITED STATES AND POSSESSIONS
(and U.S. Government)

REGION 1 - NEW ENGLAND

11  Maine
12  New Hampshire
13  Vermont
14  Massachusetts
15  Rhode Island
16  Connecticut

REGION 2 - MIDDLE ATLANTIC

21  New York
22  New Jersey
23  Pennsylvania

REGION 3 - EAST NORTH CENTRAL

31  Ohio
32  Indiana
33  Illinois
34  Michigan
35  Wisconsin

REGION 4 - WEST NORTH CENTRAL

41  Minnesota
42  Iowa
43  Missouri
44  North Dakota
45  South Dakota
46  Nebraska
47  Kansas

REGION 5 - SOUTH ATLANTIC

51  Delaware
52  Maryland
53  District of Columbia
54  Virginia
55  West Virginia
56  North Carolina
57  South Carolina
58  Georgia
59  Florida

REGION 6 - EAST SOUTH CENTRAL

61  Kentucky
62  Tennessee
63  Alabama
64  Mississippi

REGION 7 - WEST SOUTH CENTRAL

71  Arkansas
72  Louisiana
73  Oklahoma
74  Texas

REGION 8 - MOUNTAIN

81  Montana
82  Idaho
83  Wyoming
84  Colorado
85  New Mexico
86  Arizona
87  Utah
88  Nevada

REGION 9 - PACIFIC

90  Guam
91  Washington
92  Oregon
93  California
94  Alaska
95  Hawaii
96  Virgin Islands
97  Panama Canal Zone
98  Puerto Rico